MICROBIOLOGY:
A Laboratory Manual

MICROBIOLOGY:
A Laboratory Manual

James G. Cappuccino • **Natalie Sherman**
Rockland Community College, State University of New York

 THE BENJAMIN/CUMMINGS PUBLISHING COMPANY, INC.

Menlo Park, California • Reading, Massachusetts • Don Mills, Ontario
Wokingham, U.K. • Amsterdam • Sydney • Singapore
Tokyo • Madrid • Bogotá • Santiago • San Juan

Copyright © 1987 by The Benjamin/Cummings Publishing Company, Inc.

Library of Congress Cataloging-in-Publication Data

Cappuccino, James G.
Microbiology, a laboratory manual.

   Includes index.
1. Microbiology—Laboratory manuals.   I. Sherman,
Natalie.   II. Title.
QR63.C34   1986   576′.07′8   86-14126
ISBN 0-201-11636-7

ABCDEFGHIJ-AL-89876

The Benjamin/Cummings Publishing Company, Inc.
2727 Sand Hill Road
Menlo Park, California 94025

PREFACE

Microbiology, by its very nature, is a dynamic science that is constantly evolving as more information is added to the continuum of knowledge. In view of the rapidity in which microbiological techniques are being modified and refined, the authors have endeavored to provide a blend of the staid traditional methodologies with the newer and more contemporary procedures in the second edition of *Microbiology: A Laboratory Manual*. In an attempt to meet the pedagogical needs of all students studying microbiology, the changes introduced in this edition include:

1. Expansion of the medical microbiology and immunology sections to include current, rapid testing methods for the isolation and identification of staphylococci, streptococci, and the etiological agents of sexually transmitted diseases, namely Herpes simplex and the chlamydiae. Also incorporated into the manual are new experiments pertaining to the preparation of blood cultures for the diagnosis of septicemia and the use of drug combinations to illustrate their synergistic effects.
2. Expansion of the introductory and principle sections of some experiments to provide greater clarity to facilitate student comprehension.
3. Consolidation of related experimetal procedures to produce a more workable sequence.
4. Addition of computer-assisted programs to the experiment concerned with the identification of unknown bacterial cultures for use in conjunction with the traditional protocol. IBM PC, Apple, or TRS-80 disk will be available to adopters upon request, from the publisher.
5. Incorporation of colored photographs of selected biochemical reactions to facilitate student interpretation of test observations.

The design of the manual has remained essentially the same as the earlier edition. Comprehensive introductory material is given at the beginning of each major area of study, and specific explanations and detailed directions precede each experimental exercise. We feel that this textual approach will augment, enhance, and reinforce course lectures, thereby enabling students more readily to comprehend the concepts and purposes of each experiment. This will be a further asset to students in institutions in which the laboratory and lecture sections are not taught concurrently. Finally, we feel that this book should reduce the time required for explanations at the beginning of each laboratory session and thus make more time available for students to spend on the experiments.

The experiments were critically selected and tested to instruct students effectively in the basic principles and techniques within a variety of microbiological areas. The protocols vary in content and complexity, providing the instructor with flexibility to mold the laboratory syllabus to the particular needs of the students within the confines and scope of the course. Furthermore, it provides a wide spectrum of exercises suitable for students in elementary and advanced general microbiology courses as well as those in allied health programs. Lastly, the design of each procedure was carefully planned so that the supplies, equipment, and instrumentation commonly found in undergraduate institutions would suffice for their successful execution.

The manual, consisting of 77 exercises, is arranged into 15 major sections covering the following topics essential to a general microbiology course:

Section I, on **microscopy,** introduces the students to the use and care of the microscope for the study of microorganisms.

Section II, on **bacterial staining,** instructs the students in procedures for visualization and differentiation of microorganisms and cell structures.

Section III, on **microbiological cultivation techniques,** stresses the various procedures used for the isolation and enumeration of microorganisms.

Section IV, on **cultivation of microorganisms,** acquaints students with the nutritional and physical requirements for microbial growth.

Section V, on **biochemical activities,** introduces the varied cellular enzymatic activities that may be used for differentiation and identification of specific groups of microorganisms.

Sections VI, VII, and VIII introduce the areas of **protozoology, mycology,** and **virology.**

Sections IX, **control of microbial growth,** acquaints students with the antimicrobial activities of various physical and chemical agents.

Sections X and XI are concerned with the sanitary aspects of **water** and **food,** as well as the fermentative role of microorganisms in the production of some beverages and food products.

Section XII, on **microbiology of soil,** discusses the role of soil microorganisms in the nitrogen cycle and antibiotic production.

Section XIII, on **bacterial genetics,** presents selected experiments to illustrate genetic principles using bacterial systems.

Sections XIV and XV, on **medical microbiology** and **immunology,** acquaint students with both the conventional and the more recent, rapid, clinical screening methodologies used for the isolation and identification of pathogenic microorganisms.

The format of each exercise is intended to facilitate presentation of the material by the instructor and to maximize the learning experience for the student. To this end, each experiment is designed as follows:

Purpose: Defines the specific principles and/or techniques to be mastered.

Principle: An in-depth discussion of the microbiological concept or technique and the specific experimental procedure.

Materials: To facilitate the preparation of all laboratory sessions, each procedure contains a list of the following:

Cultures: The selected test organisms chosen to demonstrate effectively the experimental principle or technique under study, as well as their ease of cultivation and maintenance in stock culture. A complete listing of the experimental cultures and prepared slides is presented in Appendix 4.

Media: The specific media and their quantities per designated student group. Appendix 1 lists the composition and method of preparation of all the media used in this manual.

Reagents: Biological stains and test reagents. The chemical composition and preparation of the reagents is presented an Appendixes 2 and 3.

Equipment: Supplies and instrumentation. The suggested equipment was selected to minimize expense.

Procedure: Explicit instructions are augmented by diagrams to aid students in the execution and interpretation of the experiments.

Observations and results: Tear-out sheets are located at the end of each exercise to facilitate interpretation of data by students and collection of the report sheets by the instructor.

Review questions: Questions are located on the report sheet to aid the instructor in determining the student's ability to understand the experimetal concepts and techniques.

Finally, it is our hope that this manual will serve as a vehicle for the development of manipulative skills and techniques essential for understanding the integrated complexity of the biochemical structure and function of the single cell. This will enable students to extend these principles to the better understanding of the more complex, higher forms of life. Ultimately, it is our hope that some might become sufficiently stimulated and excited by the happenings in the microbial world to pursue further the study of life at the molecular level, or to use and apply these laboratory skills in the vocational fields of applied microbiology and allied health.

Suffern, New York   J.G.C.
                                                         N.S.

ACKNOWLEDGMENTS

We wish to express our sincere gratitude to the reviewers of this manuscript for their constructive criticisms and invaluable suggestions: Susan McMahon, State University of New York at Geneseo; Frank L. Binder, Marshall University, Huntington, West Virginia; Richard C. Tilton, University of Connecticut Health Center, Farmington, Connecticut; Randall Kotell, Mesa College, San Diego, California; and Charline Mims, San Antonio College, San Antonio, Texas. We would also like to thank Susan Morrison, College of Charleston, and James Millam, University of Florida, who were kind enough to respond to our request for suggestions and criticisms during the preparation of this revision.

Special thanks are extended to our microbiology laboratory technician, Joan Grace, for her assistance in prerunning these experiments to ensure their success when performed by the students. We are also indebted to Michael McKeon and Betty Sinowitz, Rockland Community College, for their unselfish assistance in the preparation of the computer-assisted programs for the identification of unknown bacterial cultures. Finally, we wish to extend our appreciation to our families for their patience and support during the initial writing and revision of this manual.

LABORATORY SAFETY: GENERAL RULES AND REGULATIONS

1. Upon entering the laboratory, coats, books, and other paraphernalia should be place in specified locations and never on bench tops.
2. At the beginning and termination of each laboratory session bench tops are to be wiped with a disinfectant solution provided by the instructor.
3. Smoking, eating, and drinking in the laboratory are absolutely prohibited.
4. Students should wear a lab coat or apron while working in the laboratory to protect clothing from contamination or accidental discoloration by staining solutions.
5. Long hair should be placed inside a paper cap or tied back to minimize fire hazard or contamination of experiments.
6. Removal of media, equipment, and especially bacterial cultures from the laboratory is absolutely prohibited.
7. All cultures should be handled as being potentially pathogenic, and the following precautions should be observed at all times:
   a. Cultures must always be carried in a test tube rack when the student is moving around the laboratory.
   b. Cultures must be kept in a test tube rack on the bench tops when not in use.
   c. Broth cultures must never be pipetted by mouth.
   d. Spilled cultures should be covered with paper towels and then saturated with disinfectant solution. Following 15 minutes of reaction time, the towels should be removed and disposed of in the manner indicated by the instructor.
8. Use of glassware markers:
   a. Non–water-soluble markers must be used (i.e., sharpies).
   b. If wax pencils are used, care must be taken not to grind broken tips into benchtops and the laboratory floor.
9. On completion of the laboratory session all cultures and materials to be discarded should be placed in the disposal area designated by the instructor.
10. Hands must be washed with detergent prior to leaving the laboratory.

CONTENTS

VI THE PROTOZOA 185

VII THE FUNGI 197

VIII THE VIRUSES 213

IX PHYSICAL AND CHEMICAL AGENTS FOR THE CONTROL OF MICROBIAL GROWTH 231

X MICROBIOLOGY OF FOOD 265

XI MICROBIOLOGY OF WATER 283

MICROBIOLOGY:
A Laboratory Manual

I

Microscopy

PURPOSES

1. To acquaint students with the evolutionary diversity of microscopic instruments.

2. To familiarize students with the components, use, and care of the compound brightfield microscope.

3. To instruct students in the correct use of the microscope for observation and measurement of microorganisms.

INTRODUCTION

Microbiology, the branch of science that has so vastly extended and expanded our knowledge of the living world, owes its existence to Antonj van Leeuwenhoek. In 1673, with the aid of a crude microscope consisting of a biconcave lens enclosed in two metal plates, Leeuwenhoek introduced the world to the existence of microbial forms of life. Over the years, microscopes have evolved from the simple, single-lens instrument of Leeuwenhoek, with a magnification of 300, to the present-day electron microscopes capable of magnifications greater than 250,000.

They are designated as either light microscopes or electron microscopes. The former use visible light or ultraviolet rays to illuminate specimens. They include the brightfield, darkfield, phase-contrast, and fluorescent instruments. Fluorescent microscopes use ultraviolet radiations whose wavelengths are shorter than those of visible light and not directly perceptible to the human eye. Electron microscopes use electron beams instead of light rays, and magnets instead of lenses to observe submicroscopic particles.

ESSENTIAL FEATURES
OF VARIOUS MICROSCOPES

BRIGHTFIELD MICROSCOPE This instrument contains two main lens systems: The ocular lens in the eyepiece and the objective lens located in the nosepiece. The specimen is illuminated by a beam of tungsten light focused on it by a substage lens designated as a condenser. A major limitation of this system is the absence of contrast between the specimen and the surrounding medium, which makes it difficult to observe living cells. Therefore nonviable, stained preparations are used instead.

1

DARKFIELD MICROSCOPE This is similar to the ordinary light microscope; however, the condenser system is modified so that the specimen is not illuminated directly. The condenser directs the light obliquely so that the light is deflected or scattered from the specimen, which then appears bright against a dark background. Living specimens may be observed more readily with darkfield than with brightfield microscopy.

PHASE-CONTRAST MICROSCOPE Observation of microorganisms in an unstained state is possible with this microscope. Its optics include special phase-contrast objectives and a condenser that make visible cellular components that differ only slightly in their refractive indexes. As light is transmitted through a specimen with a refractive index different than that of the surrounding medium, a portion of this light is refracted (bent) due to slight variations in density and thickness of the cellular components. The special optics convert the difference between transmitted light and refracted rays, resulting in a significant variation in the intensity of light and thereby produce a discernible image of the structure under study. The image appears dark against a light background.

FLUORESCENT MICROSCOPE This microscope is used to visualize specimens that are chemically tagged with a fluorescent dye. The source of illumination is an ultraviolet (UV) light obtained from a high-pressure mercury lamp or hydrogen quartz lamp. The ocular lens is fitted with a filter that permits the longer ultraviolet wavelengths to pass, while the shorter wavelengths are blocked or eliminated. Ultraviolet radiations are absorbed by the fluorescent label and the energy is re-emitted in the form of a different wavelength in the visible light range. The fluorescent dyes absorb at wavelengths between 230 and 350 nm and emit orange, yellow, or greenish light. This microscope is used primarily for the detection of antigen-antibody reactions. Antibodies conjugated with a fluorescent dye are excited in the presence of ultraviolet light and the fluorescent portion of the dye becomes visible against a black background.

ELECTRON MICROSCOPE The instrument provides a revolutionary method of microscopy, with magnifications up to one million that permit visualization of submicroscopic cellular particles as well as viral agents. The lens and illumination system are similar to those of the light microscope except that in the electron microscope the specimen is illuminated by a beam of electrons rather than light, and the focusing is carried out by electromagnets instead of a set of optics. These components are sealed in a tube in which a complete vacuum is established. Transmission electron microscopes require thinly prepared, fixed, and dehydrated specimens for the electron beam to pass freely through them. As the electrons pass through the specimen, images are formed by directing the electrons onto x-ray film, thus internal cellular structures are made visible. Scanning electron microscopes are used for visualizing surface characteristics rather than intracellular structures. A narrow beam of electrons scans back and forth producing a three-dimensional image as the electrons are reflected off the specimen's surface.

It is apparent that the microscopist has a variety of optical instruments with which to perform routine laboratory procedures as well as sophisticated research. The compound brightfield microscope might justly be termed the workhorse, however, as it is commonly found in all biological laboratories. Although students should be familiar with the basic principles of microscopy, it is generally not possible for the average undergraduate to be exposed to this diverse array of complex and expensive equipment. Therefore only the compound brightfield microscope will be discussed in depth and used to examine specimens.

Microscopic Examination of Stained Cell Preparations

PURPOSES

To familiarize the students with:

1. The theoretical principles of brightfield microscopy.

2. The component parts of the compound microscope.

3. The use and care of the compound microscope.

4. The practical use of the compound microscope for visualization of cellular morphology from stained slide preparations.

PRINCIPLE

Microbiology is a science that studies living organisms that are too small to be seen with the naked eye. Needless to say, such a study must involve the use of a good compound microscope. There are many types and variations, however, they all fundamentally consist of a two-lens system, a variable but controllable light source, and mechanical adjustable parts for determining focal length between the lenses and specimen (Figure 1.1 on page 4).

Components of the Microscope

STAGE A fixed platform with an opening in the center allows for the passage of light from an illuminating source below to the lens system above the stage. This platform provides a surface for the placement of a slide with its specimen over the central opening. In addition to the fixed stage, most microscopes have a **mechanical stage** that can be moved vertically or horizontally by means of adjustment knobs. Less sophisticated microscopes have clips on the fixed stage and the slide must be positioned over the central opening manually.

ILLUMINATION The light source is positioned in the base of the instrument. Some microscopes are equipped with a built-in light source to provide direct illumina-

tion. Those that are not are provided with a mirror, one side flat and the other concave. An external light source such as a lamp is placed in front of the mirror to direct the light upward into the lens system. The flat side of the mirror is used for artificial light, and the concave side for sunlight.

ABBE CONDENSER This is found directly under the stage and contains two sets of lenses that collect and concentrate light passing upward from the light source into the lens systems. The condenser is equipped with an **iris diaphragm,** which is a shutter controlled by a lever and used to regulate the amount of light entering the lens system.

BODY TUBE Above the stage and attached to the arm of the microscope is the body tube. This structure houses the lens system that magnifies the specimen. The upper end of the tube contains the **ocular** or **eyepiece** lens. The lower portion consists of a movable **nosepiece** containing the **objective lenses.** Rotation of the nosepiece positions objectives above the stage opening. The body tube may be raised or lowered with the aid of **coarse-adjustment and fine-adjustment knobs** that are located above or below the stage, depending on the type and make of instrument.

Theoretical Principles of Microscopy

To use the microscope with efficiency and minimal frustration, students should understand the basic principles of microscopy: magnification, resolution, numerical aperture, illumination, and focusing.

MAGNIFICATION Enlargement or magnification of a specimen is the function of a two-lens system; the **ocular lens** is found in the eyepiece and the **objective lens** is situated in a revolving nosepiece. These lenses are separated by the **body tube.** The objective lens is nearer the specimen and magnifies it, producing the **real image** that is projected up into the focal plane and then magnified by the ocular lens to produce the final image.

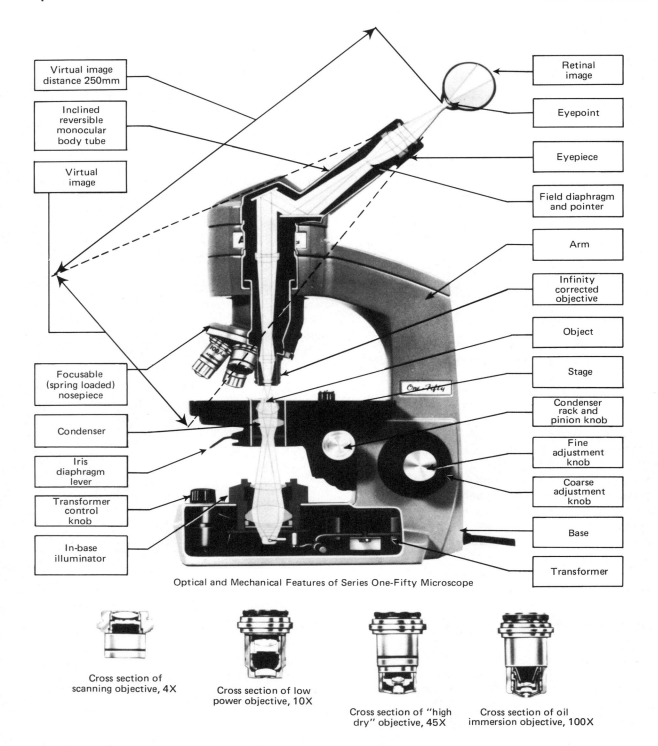

Virtual image distance 250mm

Inclined reversible monocular body tube

Virtual image

Focusable (spring loaded) nosepiece

Condenser

Iris diaphragm lever

Transformer control knob

In-base illuminator

Retinal image

Eyepoint

Eyepiece

Field diaphragm and pointer

Arm

Infinity corrected objective

Object

Stage

Condenser rack and pinion knob

Fine adjustment knob

Coarse adjustment knob

Base

Transformer

Optical and Mechanical Features of Series One-Fifty Microscope

Cross section of scanning objective, 4X

Cross section of low power objective, 10X

Cross section of "high dry" objective, 45X

Cross section of oil immersion objective, 100X

FIGURE 1.1. Optical and mechanical features of the microscope. (Courtesy of Reichert Scientific Instruments.)

TABLE 1.1. Overall linear magnification

Magnification		Total Magnification
Objective lenses	Ocular lens	Objective multiplied by ocular
Scanning 4 ×	10 ×	40 ×
Low-power 10 ×	10 ×	100 ×
High-power 45 ×	10 ×	450 ×
Oil immersion 97 ×	10 ×	970 ×

The most commonly used microscopes are equipped with a revolving nosepiece containing four objective lenses possessing different degrees of magnification. When these are combined with the magnification of the ocular lens, the total or overall linear magnification of the specimen is obtained. This is shown in Table 1.1.

RESOLVING POWER OR RESOLUTION Although magnification is important, students must be aware that unlimited enlargement is not possible by merely increasing the magnifying power of the lenses or by the use of additional lenses, because lenses are limited by a property called **resolving power.** By definition, resolving power is the ability of a lens to show two adjacent objects as discrete entities. When a lens cannot discriminate, that is, when the two objects appear as one, it has lost resolution. Increased magnification will not rectify the loss, and will, in fact blur the object. The resolving power of a lens is dependent on the wavelength of light used and a characteristic of the lens called the numerical aperture. This is expressed by the following formula:

$$\text{Resolving power} = \frac{\text{Wavelength of light}}{\text{Numerical aperture}}$$

With this formula it can be shown that the shorter the wavelength, the greater the resolving power of the lens. Thus short wavelengths of the electromagnetic spectrum are better suited than longer wavelengths in terms of the numerical aperture.

As with magnification, resolving power is also not without limits. The student might rationalize that merely decreasing the wavelength will automatically increase the resolving power of a lens. Such is not the case, because the visible portion of the electromagnetic spectrum is very narrow and borders on the very short wavelengths found in the ultraviolet portion of the spectrum. Resolving power, therefore, may only be increased by increasing the numerical aperture. The **numerical aperture** is defined as a function of the diameter of the objective lens in relation to its focal length. It is doubled by use of the substage condenser,

which illuminates the object with rays of light that pass through the specimen obliquely as well as directly. Thus the larger the numerical aperture the greater the resolving power, which may be expressed mathematically, as follows:

$$\text{Resolving power} = \frac{\text{Wavelength of light}}{2 \text{ (numerical aperture)}}$$

The relationship between wavelength and numerical aperture is only valid for increased resolving power when light rays are parallel. Therefore the resolving power is dependent on another factor, the **refractive index.** This is the bending power of light passing through air from the glass slide to the objective lens. The refractive index of air is lower than that of glass, and as light rays pass from the glass slide into the air they are bent or refracted so that they do not pass into the objective lens. This would cause a loss of light, which would reduce the numerical aperture and diminish the resolving power of the objective lens. Loss of refracted light can be compensated for by interposing mineral oil, which has the same refractive index as glass, between the slide and the objective lens. In this way decreased light refraction occurs and more light rays enter directly into the objective lens, producing a vivid image with high resolution (Figure 1.2 on page 6).

ILLUMINATION Effective illumination is required for efficient magnification and resolving power. Since the intensity of daylight is an uncontrolled variable, artificial light from a tungsten lamp is the most commonly used light source in microscopy. The light is passed through the condenser located beneath the stage. The condenser contains two lenses that are necessary to produce a maximum numerical aperture. The height of the condenser can be adjusted with the **condenser knob.** The condenser should always be kept close to the stage, especially when using the oil-immersion objective.

Between the light source and the condenser is the iris diaphragm that can be opened and closed by means of a lever, thereby regulating the amount of light en-

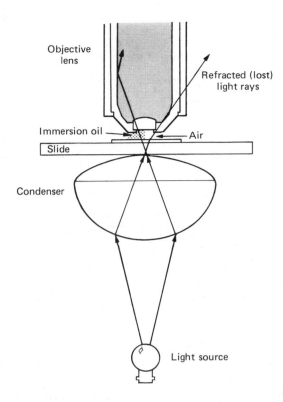

FIGURE 1.2. Refracted index in air and mineral oil

tering the condenser. Excessive illumination may actually obscure the specimen because of lack of contrast. The amount of light entering the microscope differs with each objective lens used. A rule of thumb is that **as the magnification of the lens increases, the distance between the objective lens and slide, called working distance, decreases and the angle of aperture of the objective increases** (Figure 1.3).

Use and Care of the Microscope

Students will be responsible for the proper care and use of microscopes. Since microscopes are expensive items of equipment the following regulations and procedures must be observed.

The instruments are housed in special cabinets and must be transported by students to the laboratory bench. The correct and only acceptable way to do this is to grip the microscope arm firmly with the right hand and the base with the left hand, and lift the instrument from the cabinet shelf. It is carried close to the body and gently placed on the laboratory bench. This will prevent collision with furniture or co-workers, and protect the instrument against damage.

Once the microscope is placed on the laboratory bench the following rules must be observed:

1. All unnecessary materials such as books, papers, purses, and hats should be removed from the laboratory bench.

2. The microscope's electric wire is uncoiled and plugged into an electrical outlet.

3. All lens systems are cleaned, as the smallest bit of dust, oil, lint, or eyelashes will decrease the efficiency of the microscope. The ocular, scanning, low-power, and high-power lenses may be cleaned by wiping several times with acceptable lens tissue. Paper toweling or cloth should never be used on a lens surface. If the oil-immersion lens is gummy or tacky, a moistened piece of lens paper saturated with xylol is used to wipe it clean. The xylol is immediately removed with a tissue moistened with 95 percent alcohol, and the lens is wiped dry with lens paper. **This procedure should only be used when necessary, as consistent use of xylol may loosen the lens.**

The following routine procedures must be followed to ensure correct and efficient use of the microscope.

1. Place the microscope slide with the specimen within the stage clips on the fixed stage. Move the slide to center the specimen over the opening in the stage directly over the light source.

2. Rotate the scanning lens or low-power lens into position. Lower the body tube with the coarse-adjustment knob to its lowest position. **Never lower the body tube while looking through the ocular lens.**

3. While looking into the ocular lens, with the coarse-adjustment knob slowly raise the stage until the specimen comes into focus. Using the fine-adjustment knob, bring the specimen into sharp focus.

4. Adjust the substage condenser and iris diaphragm to produce optimum illumination.

5. Most microscopes are **par focal,** which means that when one lens is in focus other lenses will also have the same focal length and can be rotated into position without further major adjustment. In practice, however, usually a half-turn of the fine-adjustment knob in either direction is necessary for sharp focus.

6. Once the specimen has been brought into sharp focus with a low-powered lens, preparation may

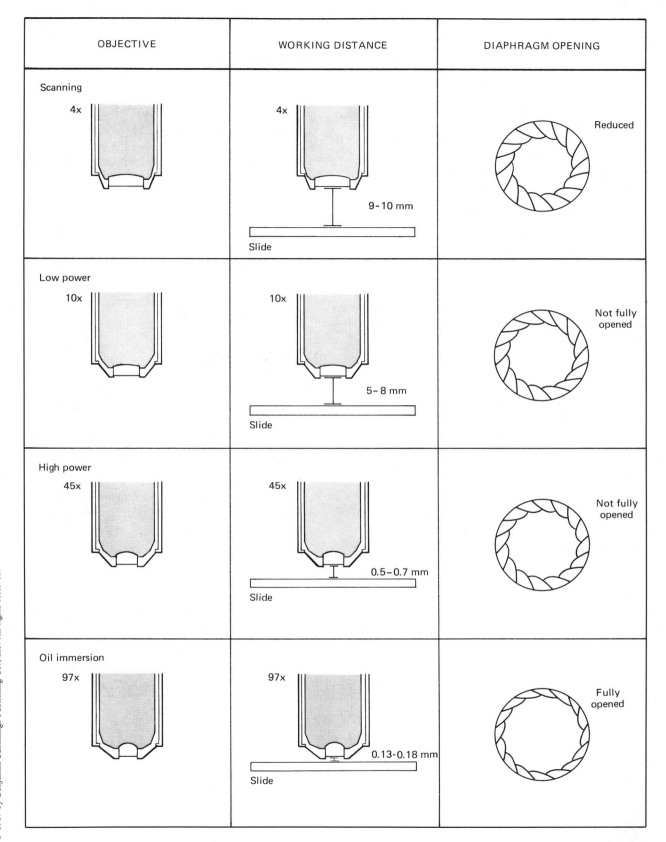

OBJECTIVE	WORKING DISTANCE	DIAPHRAGM OPENING
Scanning 4x	4x 9-10 mm Slide	Reduced
Low power 10x	10x 5-8 mm Slide	Not fully opened
High power 45x	45x 0.5-0.7 mm Slide	Not fully opened
Oil immersion 97x	97x 0.13-0.18 mm Slide	Fully opened

FIGURE 1.3. Relationship between working distance objective and diaphragm opening

be made for visualizing the specimen under oil immersion. A drop of oil is placed on the slide directly over the area to be viewed. The nosepiece is rotated until the oil-immersion objective locks into position. The slide is observed from the side as the objective is rotated slowly into position. This will ensure that the objective will be properly immersed in the oil. The fine-adjustment knob is readjusted to bring the image into sharp focus.

7. During microscopic examination of microbial organisms it is always necessary to observe several areas of the preparation. This is accomplished by scanning the slide without the application of additional immersion oil.

On the completion of the laboratory exercise the microscope is returned to its cabinet in its original condition. The following steps are recommended:

1. All lenses are cleaned with dry, clean lens paper. Xylol is used to remove oil from the stage and other microscope parts if necessary.

2. The low-power objective is placed in position and the body tube is lowered completely.

3. The mechanical stage is centered.

4. The electric wire is coiled around the body tube and the stage.

5. The microscope is carried to its position in its cabinet in the manner previously described.

MATERIALS

Slides

Commercially prepared slides of *Staphylococcus aureus, Bacillus subtilis, Aquaspirillum itersonii, Saccharomyces cerevisiae, Paramecium caudatum,* and a human blood smear.

Equipment

Compound microscope, lens paper, immersion oil, and xylol.

PROCEDURE

1. Review instructions for the use of the microscope, giving special attention for the use of the oil-immersion objective.

2. Examine the prepared slides, noting the size of the microorganisms under low-power and high-power objectives.

3. With the exception of the *Paramecium* slide, examine the prepared slides under the oil-immersion objective, noting the size and shape of the cells.

Observations and Results

EXPERIMENT **1**

In the chart provided:

1. Draw several cells from a typical microscopic field as viewed under each magnification.
2. Give total magnification for each objective.

	Low power	*High power*	*Oil immersion*
S. aureus Magnification	◯ _____	◯ _____	◯ _____
B. subtilis Magnification	◯ _____	◯ _____	◯ _____
A. itersonii Magnification	◯ _____	◯ _____	◯ _____
S. cerevisiae Magnification	◯ _____	◯ _____	◯ _____
P. caudatum Magnification	◯ _____	◯ _____	

Low power	High power	Oil immersion
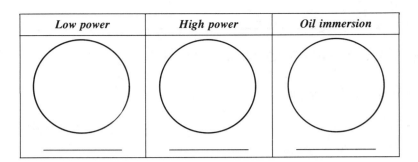		

Blood smear

Magnification _____   _____   _____

REVIEW QUESTIONS

1. Explain why the body tube of the microscope should not be lowered while looking through the ocular lens.

2. For what purpose would you adjust each of the following microscope components during a microscopy exercise?

   a. Iris diaphragm:

   b. Coarse adjustment:

   c. Fine adjustment:

   d. Condenser:

3. Discuss the procedure used to maintain the optical system of the microscope properly.

Microscopic Examination of Living Bacterial Preparations

PURPOSE

To perform the hanging-drop procedure for microscopic observation of living bacteria.

PRINCIPLE

Bacteria, because of their small size and a refractive index that closely approximates that of water, do not lend themselves readily to microscopic examination in a living, unstained state. Examination of living microorganisms is useful, however, to:

1. Observe cell activities such as motility and binary fission.

2. Observe the natural size and shape of the cells, as heat fixation and exposure to chemicals during staining cause some degree of distortion.

In this experiment, students will use a mixed culture of *Pseudomonas aeruginosa* (motile bacilli), *Bacillus cereus* (large bacilli), and *Staphylococcus aureus* (cocci), for the hanging-drop preparation (Fig. 2.1). The preparation will be observed microscopically for differences in the size and shape of the cells, as well as motility. It is essential to differentiate between actual motility and **Brownian movement,** a vibratory movement of the cells due to their bombardment by water molecules in the suspension.

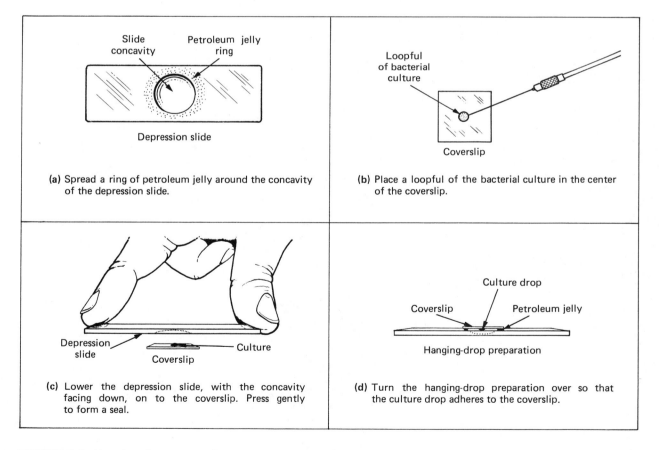

(a) Spread a ring of petroleum jelly around the concavity of the depression slide.

(b) Place a loopful of the bacterial culture in the center of the coverslip.

(c) Lower the depression slide, with the concavity facing down, on to the coverslip. Press gently to form a seal.

(d) Turn the hanging-drop preparation over so that the culture drop adheres to the coverslip.

FIGURE 2.1. Hanging-drop preparation

MATERIALS

Cultures

24-hour broth culture of *Pseudomonas aeruginosa, Bacillus cereus,* and *Staphylococcus aureus.*

Equipment

Bunsen burner, inoculating loop, depression slide, coverslip, microscope, and petroleum jelly.

PROCEDURE

1. Apply a ring of petroleum jelly around the concavity of the depression slide.

2. Using sterile technique, place a loopful of the mixed culture in the center of a clean coverslip.

3. Place the depression slide, with the concave surface facing down, over the coverslip so that the depression covers the drop of culture. Press the slide gently to form a seal between the slide and the coverslip.

4. Quickly turn the slide right side up so that the drop continues to adhere to the inner surface of the coverslip.

5. For microscopic examination, first focus in on the drop of culture under the low-power objective with reduced light. Place a drop of oil on the coverslip and use the oil-immersion objective for detailed observation.

Observations and Results

1. Examine the hanging-drop preparation as to the size, shape, and motility of the different bacteria present. Record your results in the chart.

Organisms	Size and shape	True motility or brownian movement
S. aureus		
P. aeruginosa		
B. cereus		

2. Draw a representative field:

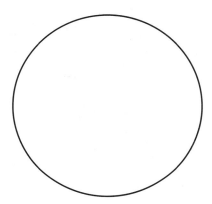

REVIEW QUESTIONS

1. Why are living, unstained bacterial preparations more difficult to observe microscopically than stained preparations?

2. For what purposes is it essential that living specimens be observed?

3. Distinguish between true motility and Brownian movement.

The Microscopic Measurement of Microorganisms

PURPOSES

1. To instruct the student in calibration of an ocular micrometer.

2. To perform an experimental procedure in the measurement of microorganisms.

PRINCIPLE

Determination of microbial size is not as simple as one might assume. Before an accurate measurement of cells can be made the diameter of the microscopic field must be established by means of optic devices, namely, an **ocular micrometer** and a **stage micrometer.** The ocular micrometer (Figure 3.1a), which is placed on a circular shelf inside the eyepiece, is a glass disc with graduations etched on its surface. The distance between these graduations will vary depending on the objective being used, which determines the size of the field. This distance is determined by using a stage micrometer (Figure 3.1b), which is a special glass slide with etched graduations that are 0.01 mm or 10 μm apart, since there are 1000 micrometers in 1 millimeter.

The calibration procedure for the ocular micrometer first requires that the graduations on both micrometers be superimposed on each other (Figure 3.1c), which is accomplished by rotating the ocular lens. A determination is then made of the number of ocular divisions per known distance on the stage micrometer. Finally, the calibration factor for one ocular division (O.D.) is calculated as follows:

1 division on ocular micrometer in mm =
$$\frac{\textit{Known distance between 2 lines on stage micrometer}}{\textit{Number of divisions on ocular micrometer}}$$

Example: If 13 ocular divisions coincide with two stage divisions (2 × 0.01 = known distance of 0.02), then,

$$\textit{1 ocular division} = \frac{0.02}{13}$$
$$= 0.00154 \text{ mm or } 1.54 \text{ } \mu\text{m}.$$

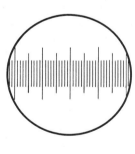

(a) Ocular micrometer: Graduations are arbitrary and must be calibrated for each objective on the microscope.

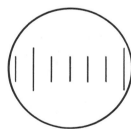

(b) Stage micrometer: Graduations are spaced 0.01 mm (10 μm) apart.

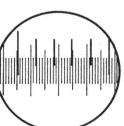

(c) Superimposed ocular micrometer (below) over stage micrometer (above): Thirteen graduations on the ocular micrometer coincide with two graduations on the stage micrometer.

(d) Measurement of microorganisms with the calibrated micrometer: Each cell occupies five divisions on the ocular micrometer.

FIGURE 3.1. Calibration and use of the ocular micrometer

Once the ocular micrometer is calibrated the size of a microorganism can easily be determined, first, by counting the number of spaces occupied by the organism (Figure 3.1d) and second, by multiplying this number by the calculated calibration factor for one ocular division.

Example: If an organism occupies five spaces on the ocular micrometer, then,

Length of organism = number of ocular divisions
                      occupied × calibration factor
                      for one ocular division

$$= 5 \times 1.54 \ \mu m = 7.70 \ \mu m.$$

In this experiment students will calibrate an ocular micrometer for the oil-immersion objective and determine the size of microorganisms such as bacteria, yeast, and protozoa.

MATERIALS

Slides

Prepared slides of yeast, a protozoan, bacterial cocci, and bacilli.

Equipment

Ocular micrometer, stage micrometer, microscope, immersion oil, xylene, and lens paper.

PROCEDURE

1. With the assistance of a laboratory instructor, carefully place the ocular micrometer into the eyepiece.

2. Place the stage micrometer on the microscope stage and center it over the illumination source.

3. With the stage micrometer in clear focus under the low-power objective, slowly rotate the eyepiece to superimpose the ocular micrometer graduations over those of the stage micrometer.

4. Add a drop of immersion oil to the stage micrometer, bring the oil-immersion objective into position, and focus, if necessary, with the fine adjustment only.

5. Move the mechanical stage so that a line on the stage micrometer coincides with a line on the ocular micrometer at one end. Find another line on the ocular micrometer that coincides with a line on the stage micrometer. Determine the distance on the stage micrometer (number of divisions × 0.01 mm) and the corresponding number of divisions on the ocular micrometer.

6. Determine the value of the calibration factor for the oil-immersion objective.

7. Remove the stage micrometer from the stage.

8. Determine the size of the cocci on the prepared slides under the oil-immersion objective by calculating the number of ocular divisions a single cell occupies.

9. Determine the size of the other microorganisms by observing the remaining prepared slides under oil immersion. Since these organisms are not round, both length and width measurements are required.

Observations and Results

1. Calibration of ocular micrometer for the oil-immersion objective:
   Distance on stage micrometer _____
   Divisions on ocular micrometer _____
   Calibration factor for one ocular division _____

2. Record your observations and measurements of the microorganisms studied in the chart.

| | Size of microorganisms in millimicrons (μm) | | | | | |
| | Width | | | Length | | |
Organism	*Number of ocular divisions*	*Calibration factor*	*Size*	*Number of ocular divisions*	*Calibration factor*	*Size*
Cocci						
Bacilli						
Yeast						
Protozoa						

REVIEW QUESTIONS

1. Can the same calibration factor be used to determine the size of a microorganism under all objectives? Explain.

2. If 1 stage micrometer division contains 12 ocular divisions, then the distance between 2 lines on the ocular micrometer is _____. Show your calculation.

3. If 1 μm is equal to 1/25,400 inch, convert the size of a bacterium measuring 3 μm × 1.5 μm to inches. _____ Show your calculations.

4. Would you expect the measurements for given organisms to be the same if a size determination is made from a stained and unstained preparation? _____ Explain.

Bacterial Staining

PURPOSES

This section is designed to instruct students in:

1. The chemical and theoretical bases of biological staining.

2. The manipulative techniques of smear preparation.

3. The performance of simple staining and negative staining procedures.

4. The performance of differential staining procedures such as the Gram, acid-fast, capsule, and spore stains.

INTRODUCTION

Visualization of microorganisms in the living state is most difficult, not only because they are minute but also because they are transparent and practically colorless when suspended in an aqueous medium. To study their properties and to differentiate microorganisms into specific groups for diagnostic purposes, biological stains and staining procedures in conjunction with light microscopy have become major tools in microbiology.

Chemically, a stain (dye) may be defined as an organic compound containing a benzene ring plus a chromophore and auxochrome group.

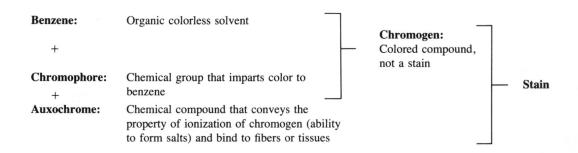

Benzene: Organic colorless solvent

   +

Chromophore: Chemical group that imparts color to benzene

   +

Auxochrome: Chemical compound that conveys the property of ionization of chromogen (ability to form salts) and bind to fibers or tissues

Chromogen: Colored compound, not a stain

Stain

The stain picric acid may be used to illustrate this definition:

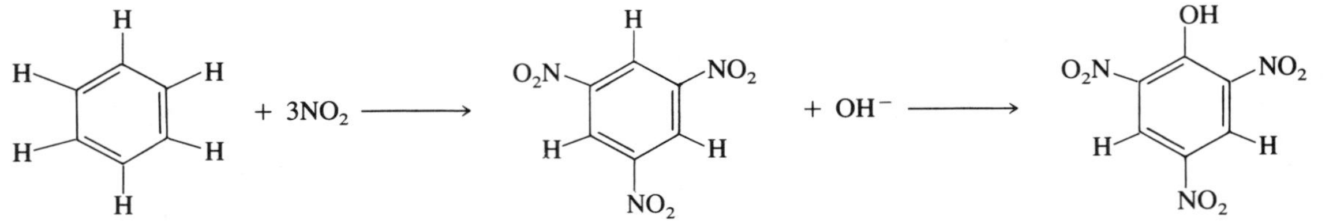

Benzene — colorless   **Nitrite** — chromophores   **Trinitrobenzene** chromogen (not a stain) yellow in color   Auxochrome   **Trinitrohydroxybenzene** (picric acid) yellow stain

The ability of a stain to bind to macromolecular cellular components such as proteins or nucleic acids depends on the electrical charge found on the chromogen portion as well as on the cellular component to be stained.

Acidic dyes are anionic, which means that on ionization of the stain the chromogen portion exhibits a negative charge and therefore has a strong affinity for the positive constituents of the cell. Proteins, positively charged cellular components, will readily bind to and accept the color of the negatively charged, anionic, chromogen of an acidic stain. Structurally, picric acid is an example of an acidic stain that produces an anionic chromogen as illustrated:

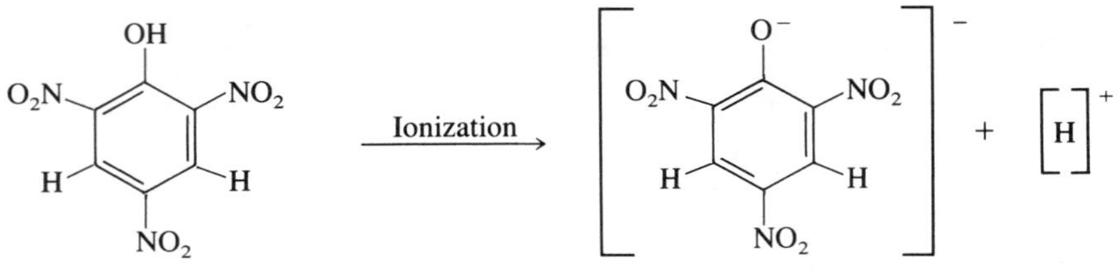

Picric acid   **Anionic chromogen**

Basic dyes are cationic, because on ionization the chromogen portion exhibits a positive charge and therefore has a strong affinity for the negative constituents of the cell. Nucleic acids, negatively charged, cellular components, will readily bind to and accept the color of the positively charged, cationic chromogen of a basic stain. Structurally, methylene blue is a basic stain that produces a cationic chromogen as illustrated:

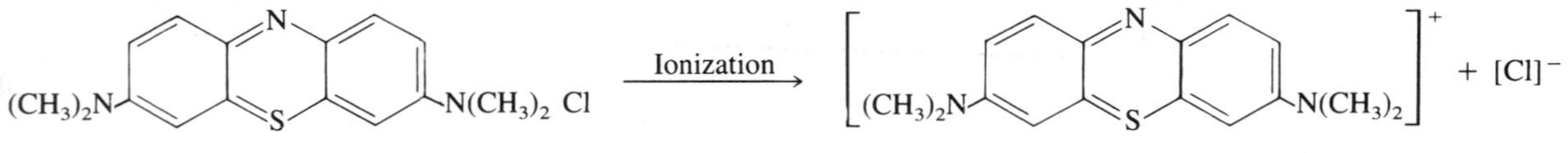

Methylene blue   **Cationic chromagen**

The following is a summary of acidic and basic stains:

Stains
- **Acidic** — Sodium, potassium, calcium, or ammonium salts of colored acids ionize to give a negatively charged chromogen
- **Basic** — The chloride or sulfate salts of colored bases ionize to give a positively charged chromogen

Numerous staining techniques are available for visualization, differentiation, and separation of bacteria in terms of morphological characteristics and cellular structures. A summary of commonly used procedures and their purposes is outlined as follows:

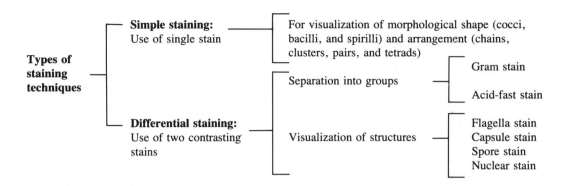

All microbiological staining procedures require preparation of smears prior to the execution of any of the specific staining techniques listed above. The technique, although not difficult, requires adequate care in its preparation. The following basic rules should be followed meticulously.

1. **PREPARATION OF GLASS SLIDES:** Clean slides are essential for preparation of microbial smears. Grease or oil from fingers on slides must be removed by washing slides with soap and water or scouring powders, followed by a water rinse and a rinse of 95 percent alcohol. Slides should be dried and placed on laboratory towels until ready for use.

2. **PREPARATION OF SMEAR:** Avoidance of thick, dense smears is absolutely essential. A good smear is one that, when dried, appears as a thin whitish layer or film. Those made from broth cultures or cultures from solid medium require variations in technique.

   a. **Broth cultures:** One or two loopsful of suspended cells should be directly applied to the glass slide with a sterile inoculating loop and spread evenly over an area about the size of a dime.

   b. **Cultures from solid medium:** Organisms cultured in a solid medium produce thick, dense surface growth and are not amenable to direct transfer to the glass slide. These cultures must be diluted by placing a loopful of water on the slide in which the cells will then be emulsified. Transfer of cells from the culture requires the use of a sterile inoculating needle. Only the tip of the needle should touch the culture to prevent the transfer of too many cells. Suspension is accomplished by spreading the cells in a circular motion in the drop of water with the needle tip. The finished smear should occupy an area about the size of a nickel and should appear as a semitransparent, confluent, whitish film. Before proceeding further, the smear must be allowed to dry completely. **Do not blow on the slide or wave it in the air.**

3. **HEAT FIXATION:** Unless fixed on the glass slide, the bacterial smear will wash away during the staining procedure. This is avoided by heat fixation, during which the bacterial proteins are coagulated and fixed to the glass surface. Heat fixation is performed by the rapid passage of the air-dried smear two to three times over the flame of the Bunsen burner.

From liquid media   From solid media

(a) Place one to two loopfuls of the cell suspension on the clean slide.

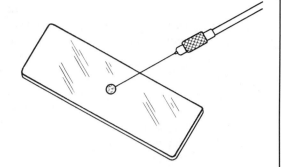

(a) Take one drop of water on the loop and place it on the center of the slide.

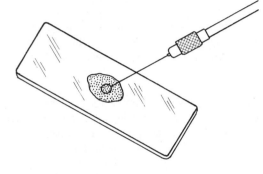

(b) With a circular movement of the loop, spread the suspension into a thin area approximately the size of a dime.

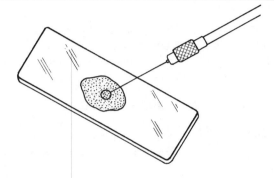

(b) Transfer a small amount of the bacterial inoculum from the slant culture into the drop of water. Spread both into a thin area approximately the size of a nickel.

Fixation

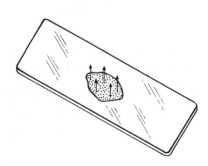

(c) Allow the smear to air dry.

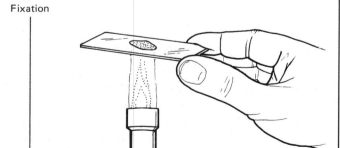

(d) While holding the slide at one end, quickly pass the smear over the flame of the Bunsen burner two to three times.

Bacterial smear preparation

Simple Staining

PURPOSE

To perform the simple staining procedure to compare morphological shapes and arrangements of bacterial cells.

PRINCIPLE

In **simple staining,** the bacterial smear is stained with a single reagent. Basic stains with a positively charged chromogen are preferred, since bacterial nucleic acids and certain cell-wall components carry a negative charge that strongly attracts and binds to the cationic chromogen. The purpose of simple staining is to elucidate the morphology and arrangement of bacterial cells. The most commonly used basic stains are methylene blue, crystal violet, and carbol fuchsin. It should be noted that exposure times differ for each of these stains; carbol fuchsin requires 15 to 30 seconds, crystal violet 2 to 60 seconds, and methylene blue 1 to 2 minutes.

MATERIALS

Cultures

24-hour nutrient agar slant cultures of *Escherichia coli,* *Bacillus cereus,* and *Staphylococcus aureus.*

Reagents

Methylene blue, crystal violet, and carbol fuchsin.

Equipment

Bunsen burner, inoculating loop, staining tray, microscope, lens paper, bibulous paper, and glass slides.

PROCEDURE

1. Prepare separate bacterial smears of the organisms following the procedure described. Note: All smears must be heat fixed prior to staining.

2. Place a slide on the staining tray and flood the smear with one of the indicated stains, using the appropriate exposure time.

3. Wash smear with tap water to remove excess stain. During this step the slide should be held parallel to the stream of water, thereby reducing the loss of organisms from the preparation.

4. Using bibulous paper, blot dry but do not wipe the slide.

5. This procedure is repeated with the remaining two organisms, using a different stain for each.

6. Examine all stained slides under oil immersion.

Observations and Results

In the space provided:

1. Draw a representative field for each organism.

2. Describe the morphology of the organisms with reference to their shape (bacilli, cocci, spirilli) and arrangement (chains, clusters, pairs).

	Methylene blue	*Crystal violet*	*Carbol fuchsin*
Drawing of a representative field	◯	◯	◯

Organism
Cell morphology
  Shape
    Arrangement
Cell color
	_____
	_____
	_____
	_____
		_____
		_____
		_____
		_____
			_____
			_____
			_____
			_____

REVIEW QUESTIONS

1. What is the purpose of heat fixation?

2. Are compounds that consist only of a chromophore considered to be stains? _____. Explain.

3. Why are basic dyes more effective for bacterial staining than acidic dyes?

4. Can simple staining techniques be used to identify more than the morphological characteristics of microorganisms? _____ Explain.

Negative Staining

PURPOSE

To perform a negative staining procedure (Fig. 5.1).

PRINCIPLE

Negative staining requires the use of an acidic stain such as eosin or nigrosin. The acidic stain, with its negatively charged chromogen, will not penetrate the cells because of the negative charge on the surface of bacteria. Therefore the unstained cells are easily discernible against the colored background.

The practical application of negative staining is twofold: First, since heat fixation is not required and the cells are not subjected to the distorting effects of chemicals, their natural size and shape can be seen; second, it is possible to observe bacteria that are difficult to stain, such as some spirilli.

MATERIALS

Cultures

24-hour agar slant mixed culture of *Micrococcus luteus, Bacillus cereus,* and *Aquaspirillum itersonii.*

Reagent

Nigrosin.

(a) Place a drop of nigrosin toward one end of the slide.

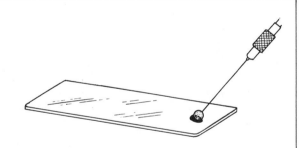

(b) Place a loopful of the inoculum into the drop of stain and mix with the loop.

(c) Place a slide against the drop of suspended organisms and allow the drop to spread along the edge of the applied slide.

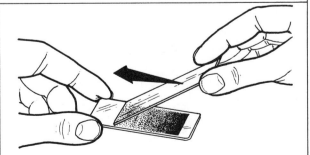

(d) Push the slide away from the previously spread drop of suspended organisms, forming a thin smear.

FIGURE 5.1. Negative staining procedure

Equipment

Bunsen burner, inoculating loop, staining tray, glass slides, lens paper, and microscope.

PROCEDURE

1. Place a small drop of nigrosin close to one end of a clean slide.

2. Using sterile technique, place a loopful of inoculum from the mixed culture into the drop of nigrosin and mix.

3. With the edge of a second slide held at a 30-degree angle and placed in front of the bacterial suspension, push the mixture to form a thin smear.

4. Air dry. **Do not heat fix the slide.**

5. Examine the slide under oil immersion.

Observations and Results EXPERIMENT **5**

1. Draw a representative field of your microscopic observation.

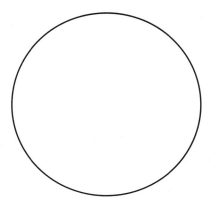

2. Describe the microscopic appearance of the different bacteria.

REVIEW QUESTIONS

1. Can methylene blue be used in place of the nigrosin for negative staining? _____
   Explain.

2. What are the practical advantages of negative staining?

3. Why do the bacteria remain unstained in the negative staining procedure?

Handwritten notes at top:
Stain or Die [usually colorless Saltant ex. Benzene } chromogene only implies color to the cell
chromophore (factor) +
Auxachrome = dealing ē fiber or tissue
chromophore = tri nitro benzene
bacteriological stain is made up of
$C_6 6_{12} O_6$ = Benzene

EXPERIMENT **6**

Gram Stain

PURPOSES

To familiarize students with:

1. The chemical and theoretical bases for differential staining procedures.

2. The chemical basis of the Gram stain.

3. Performance of the procedure for differentiating bacteria into two principal groups: gram-positive and gram-negative (Fig. 6.1).

PRINCIPLE

Differential staining requires the use of at least three chemical reagents that are applied sequentially to a heat-fixed smear. The first reagent is called the **primary stain.** Its function is to impart its color to all cells. In order to establish a color contrast, the second reagent used is the **decolorizing agent.** Based on the chemical composition of cellular components, the decolorizing agent may or may not remove the primary stain from the entire cell or only from certain cell structures. The final reagent, the **counterstain,** has a contrasting color to that of the primary stain. Following decolorization, if the primary stain is not washed out the counterstain cannot be absorbed and the cell or its components will retain the color of the primary stain. If the primary stain is removed, the decolorized cellular components will accept and assume the contrasting color of the counterstain. In this way cell types or their structures can be distinguished from each other on the basis of the stain that is retained.

The most important differential stain used in bacteriology is the **Gram stain,** named after Dr. Christian Gram. It divides bacterial cells into two major groups, gram-positive and gram-negative, which makes it an essential tool for classification and differentiation of microorganisms. The Gram stain uses four different reagents. These reagents and their mechanism of action follow.

Handwritten: Test I A
Primary Stain *is primary stain*

CRYSTAL VIOLET This violet stain is used first and stains all cells purple-blue.

Mordant *Handwritten: Test I A*

GRAM'S IODINE This reagent serves as a mordant, a substance that forms an insoluble complex by binding to the primary stain. The resultant crystal violet-iodine (CV-I) complex serves to intensify the color of the stain and all the cells will appear purple-black at this point. In gram-positive cells only, this CV-I complex binds to the magnesium-ribonucleic acid component of the cell wall, forming a magnesium-ribonucleic acid-crystal violet-iodine (Mg-RNA-CV-I) complex, which is more difficult to remove.

Decolorizing Agent  *Handwritten: Test I A*

ETHYL ALCOHOL, 95 PERCENT This reagent serves a dual function as a lipid solvent and as a protein-dehydrating agent. Its action is determined by the lipid concentration of the microbial cell walls. In gram-positive cells, the low lipid concentration is important to retention of the Mg-RNA-CV-I complex. Therefore, the small amount of lipid content is readily dissolved by the action of the alcohol, causing formation of minute cell wall pores. These are then closed by alcohol's dehydrating effect. As a consequence, the tightly bound primary stain is difficult to remove, and the cells remain blue. In gram-negative cells, the high lipid concentration found in outer layers of the cell wall is now instrumental in removal of the CV-I complex. Therefore, the high concentration of lipid is dissolved by the alcohol, creating large pores in the cell wall that do not close appreciably on dehydration of cell wall proteins. This facilitates release of the unbound CV-I complex, leaving these cells colorless or unstained.

Counterstain

SAFRANIN This is the final reagent to stain red those cells that have been previously decolorized. Since only gram-negative cells undergo decolorization, they may now absorb the counterstain. Gram-positive cells retain the blue color of the primary stain.

The preparation of adequately stained smears requires that students bear in mind the following precautions:

1. The most critical phase of the procedure is the decolorization step that is based on the ease with which the CV-I complex is released from the cell. It must be remembered that overdecolorization will result in loss of the primary stain, causing gram-positive organisms to appear gram-negative. Underdecolorization, however, will not completely remove the CV-I complex, causing gram-negative organisms to appear gram-positive. Strict adherence to all instructions will help remedy part of the difficulty, but individual experience and practice are the keys to correct decolorization.

2. It is imperative that slides be thoroughly washed under running tap water between applications of each reagent. This removes excess reagent and prepares the slide for application of the subsequent agent.

3. The best gram-stained preparations are made with fresh cultures, not older than 24 hours. As cultures age, especially in the case of gram-positive cells, the organisms tend to lose their ability to retain the primary stain and may appear to be **gram-variable;** that is, some cells will appear blue, while others will appear red.

MATERIALS

Cultures

24-hour nutrient agar slant cultures of *Escherichia coli, Staphylococcus aureus,* and *Bacillus cereus*.

Reagents

Crystal violet, Gram's iodine, 95 percent ethyl alcohol, and safranin.

Equipment

Bunsen burner, inoculating loop or needle, staining tray, glass slides, bibulous paper, lens paper, and microscope.

PROCEDURE

1. Clean four glass slides.

2. Using sterile technique, prepare a smear of each of the three organisms and on the remaining slide prepare a smear consisting of a mixture of *S. aureus* and *E. coli.* This is done by placing a drop of water on the slide and then transferring each organism separately to the drop of water on the slide with a sterile, cooled needle. Both organisms are then mixed and spread by means of a circular motion of the inoculating needle.

3. Allow smears to air dry, then heat fix in the usual manner.

4. Flood the smears with crystal violet and let stand for one minute.

5. Wash with tap water.

6. Flood smears with the Gram's iodine mordant and let stand for one minute.

7. Wash with tap water.

8. Decolorize with 95 percent ethyl alcohol. **Caution: Do not overdecolorize.** Add reagent drop by drop until crystal violet fails to wash from smear.

9. Wash with tap water.

10. Counterstain with safranin for 45 seconds.

11. Wash with tap water.

12. Blot dry with bibulous paper and examine under oil immersion.

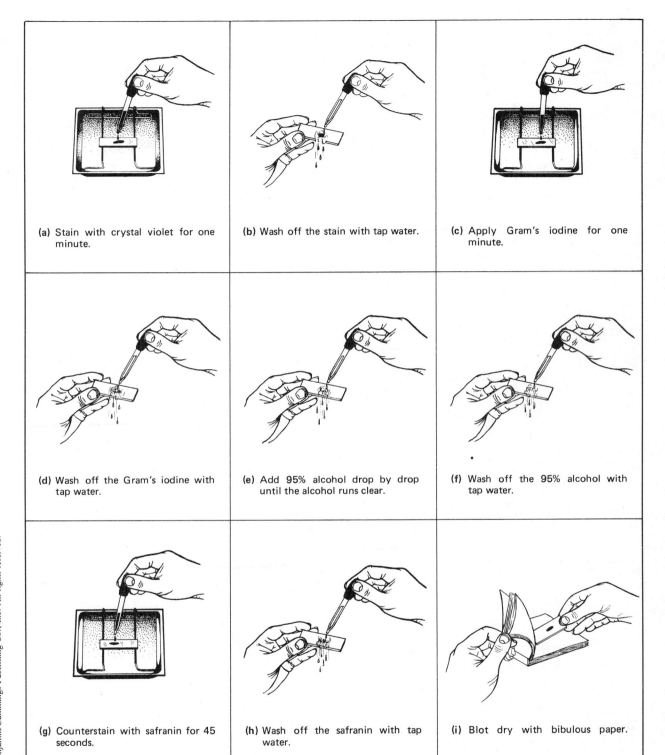

(a) Stain with crystal violet for one minute.

(b) Wash off the stain with tap water.

(c) Apply Gram's iodine for one minute.

(d) Wash off the Gram's iodine with tap water.

(e) Add 95% alcohol drop by drop until the alcohol runs clear.

(f) Wash off the 95% alcohol with tap water.

(g) Counterstain with safranin for 45 seconds.

(h) Wash off the safranin with tap water.

(i) Blot dry with bibulous paper.

FIGURE 6.1. Gram's staining procedure

Acid-Fast Stain (Ziel-Neelsen Method)

PURPOSES

To familiarize students with:

1. The chemical basis of the acid-fast stain.

2. Performance of the procedure for differentiation of bacteria into acid-fast and non-acid-fast groups (Fig. 7.1).

PRINCIPLE

While the majority of bacterial organisms are stainable by either simple or Gram-staining procedures, a few genera, particularly the members of the genus *Mycobacterium,* are resistant and can only be visualized by the **acid-fast** method. Since *M. tuberculosis* and *M. leprae* represent bacteria that are pathogenic to humans, the stain is of diagnostic value in identifying these organisms.

The characteristic difference between mycobacteria and other microorganisms is the presence of a thick waxy (lipoidal) wall that makes penetration by stains extremely difficult. Once the stain has penetrated, however, it cannot be readily removed even with the vigorous use of acid alcohol as a decolorizing agent. Because of this property, these organisms are called acid-fast, while all other microorganisms, which are easily decolorized by acid-alcohol, are non-acid-fast.

The acid-fast stain uses three different reagents.

Primary Stain

CARBOL FUCHSIN Unlike cells that are easily stained by ordinary aqueous stains, most species of mycobacteria are not sensitive to common dyes such as methylene blue and crystal violet. Carbol fuchsin, a phenolic stain that is soluble in the lipoidal materials that constitute the major portion of the mycobacterial cell wall, allows penetration and retention of this red stain. Penetration is further augmented by the application of heat, which drives the carbol fuchsin through the li-

poidal wall and into the cytoplasm. A modification of the Ziel-Neelsen method circumvents the use of heat by addition of a wetting agent (Turgitol) to this stain, which reduces surface tension between the cell wall of the mycobacteria and the stain. Following application of the primary stain, all cells will appear red.

Decolorizing Agent

ACID-ALCOHOL (3 percent HCl + 95 percent ethanol) Prior to decolorization the smear is cooled, which allows the waxy cell substances to harden. On application of acid-alcohol, acid-fast cells will be resistant to decolorization, since the primary stain is more soluble in the cellular waxes than in the decolorizing reagent. In this event, the primary stain is retained and the mycobacteria will stay red. This is not the case with non-acid-fast organisms that lack cellular waxes. The primary stain is more rapidly removed during decolorization, leaving these cells colorless or unstained.

Counterstain

METHYLENE BLUE This is used as the final reagent to stain previously decolorized cells. As only non-acid-fast cells undergo decolorization, they may now absorb the counterstain and take on its blue color, while acid-fast cells retain the red of the primary stain.

MATERIALS

Cultures

24-hour to 48-hour trypticase soy broth cultures of *Mycobacterium smegmatis* and *Staphylococcus aureus*.

Reagents

Carbol fuchsin, acid-alcohol, and methylene blue.

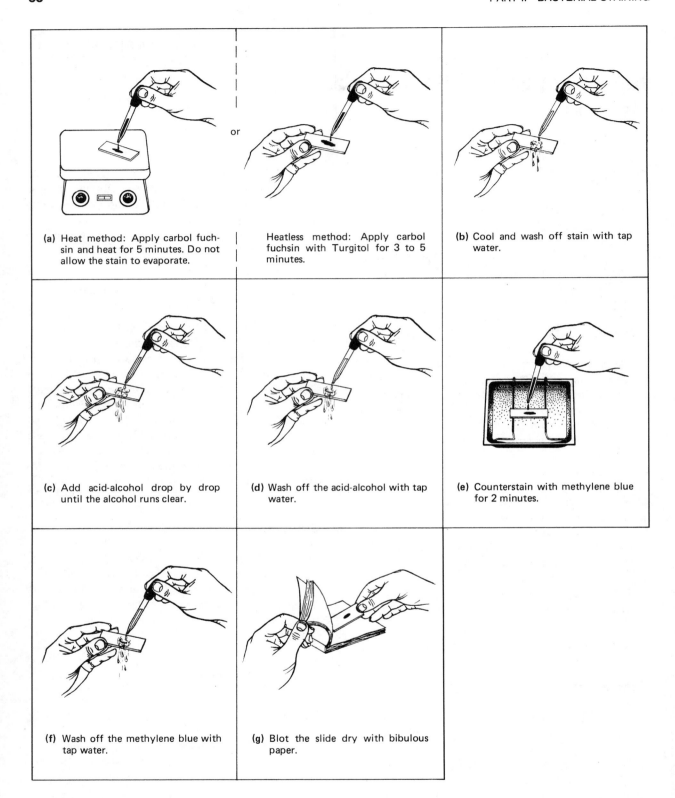

(a) Heat method: Apply carbol fuchsin and heat for 5 minutes. Do not allow the stain to evaporate.

or

Heatless method: Apply carbol fuchsin with Turgitol for 3 to 5 minutes.

(b) Cool and wash off stain with tap water.

(c) Add acid-alcohol drop by drop until the alcohol runs clear.

(d) Wash off the acid-alcohol with tap water.

(e) Counterstain with methylene blue for 2 minutes.

(f) Wash off the methylene blue with tap water.

(g) Blot the slide dry with bibulous paper.

FIGURE 7.1. Acid-fast staining procedure

Equipment

Bunsen burner, hot plate, inoculating loop, glass slides, bibulous paper, lens paper, staining tray, and microscope.

PROCEDURE

1. Clean three glass slides.

2. Using sterile technique, prepare a bacterial smear of each organism plus a third mixed smear of *M. smegmatis* and *S. aureus*.

3. Allow smears to air dry and then heat fix in the usual manner.

4. Flood smears with carbol fuchsin and place on a warm hot plate, allowing the preparation to steam for five minutes. **Caution: Do not allow stain to evaporate but replenish stain as needed; also, prevent stain from boiling by adjusting the hot plate to a proper temperature.** For heatless method, flood smear with carbol fuchsin containing Turgitol for three to five minutes.

5. Wash with tap water. Heated slides must be cooled prior to washing.

6. Decolorize with acid-alcohol, adding the reagent drop by drop until carbol fuchsin fails to wash from smear.

7. Counterstain with methylene blue for two minutes.

8. Wash smear with tap water.

9. Blot dry with bibulous paper and examine under oil immersion.

Spore Stain
(Schaeffer-Fulton Method)

PURPOSES

To familiarize students with:

1. The chemical basis of the spore stain.

2. Performance of the procedure for differentiation between bacterial spore and vegetative cell forms.

PRINCIPLE

Members of the anaerobic genera *Clostridium* and *Desulfomaculatum* and the aerobic genus *Bacillus* are examples of organisms that have the capacity to exist as either metabolically active **vegetative cells** or as highly resistant, metabolically inactive cell types called **spores.** When environmental conditions become unfavorable for continuing vegetative cellular activities, these cells have the capacity to undergo **sporogenesis** and give rise to a new intracellular structure called the **endospore,** which is surrounded by impervious layers called spore coats. As conditions continue to worsen, the endospore is released from the degenerating vegetative cell and becomes an independent cell called a **spore.** Due to the chemical composition of spore layers, the spore is resistant to the deleterious effects of excessive heat, freezing, radiation, desiccation, and chemical agents, as well as to the commonly employed microbiological stains. With the return of favorable environmental conditions, the free spore may now revert to a metabolically active and less resistant vegetative cell through **germination** (see Fig. 8.1 on p. 44). It should be emphasized that sporogenesis and germination are not means of reproduction, but merely mechanisms that ensure cell survival under all environmental conditions.

In practice, the spore stain uses two different reagents.

Primary Stain

MALACHITE GREEN Unlike most vegetative cell types that stain by common procedures, the spore, because of its impervious coats, will not accept the primary stain easily. To augment penetration, the application of heat is required. After the primary stain is applied and the smear is heated, both the vegetative cell and spore will appear green.

Decolorizing Agent

TAP WATER Once the spore accepts the malachite green it cannot be decolorized by tap water, which removes only the excess primary stain. The spore will remain green. On the other hand, the stain does not demonstrate a strong affinity for vegetative cell components, the water removes it, and these cells will be colorless.

Counterstain

SAFRANIN This contrasting red stain is used as the second reagent to color the decolorized vegetative cells, which will absorb the counterstain and appear red. The spores retain the green of the primary stain.

MATERIALS

Cultures

48-hour to 72-hour nutrient agar slant culture of *Bacillus cereus* and Thioglycollate culture of *Clostridium butyricum.*

Reagents

Malachite green and safranin.

Equipment

Bunsen burner, hot plate, staining tray, inoculating loop, glass slides, bibulous paper, lens paper, and microscope.

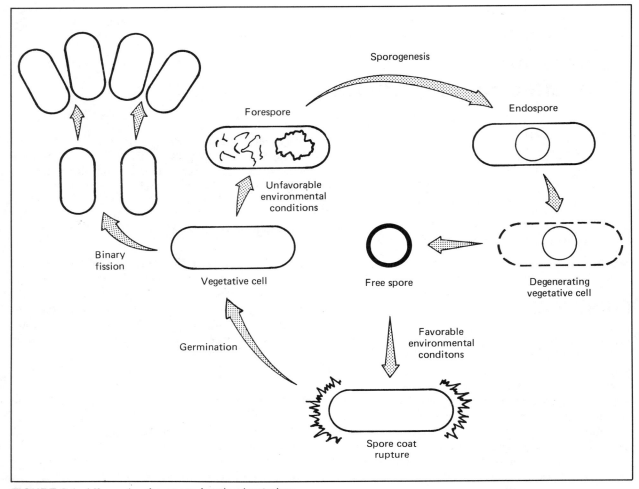

FIGURE 8.1. Life cycle of a spore-forming bacterium

PROCEDURE

1. Clean three glass slides.

2. Make individual smears in the usual manner using sterile technique.

3. Allow smear to air dry and heat fix in the usual manner.

4. Flood smears with malachite green and place on a warm hot plate, allowing the preparation to steam for two to three minutes. **Caution: Do not allow stain to evaporate; replenish stain as needed.** Prevent the stain from boiling by adjusting the hot plate to a proper temperature.

5. Remove slides from hot plate, cool, and wash under running tap water.

6. Counterstain with safranin for 30 seconds.

7. Wash with tap water.

8. Blot dry with bibulous paper and examine under oil immersion.

NAME_____

Observations and Results   EXPERIMENT **8**

Following the observation of all slides under oil immersion, record your results in the chart.

1. Make drawings of a representative microscopic field of each preparation.

2. Describe the location of the endospore within the vegetative cell as being central, subterminal, or terminal on each preparation.

3. Indicate the color of the spore and vegetative cell on each preparation.

	C. butyricum	*B. cereus*
Drawing of a representative field		
Color of spores: Color of vegetative cells: Location of endospore:		

REVIEW QUESTIONS

1. Why is heat necessary in spore staining?

2. Explain the function of tap water in spore staining.

3. Explain the properties that differentiate the spore from the vegetative cell.

Copyright © 1987 by Benjamin/Cummings Publishing Co., Inc. All rights reserved.

4. Distinguish between sporogenesis and germination.

5. Compare the appearance of the endospore following simple staining and the use of the spore stain.

Capsule Stain

PURPOSES

To familiarize students with:

1. The chemical basis of the capsule stain.

2. Performance of the procedure to distinguish capsular material from the bacterial cell.

PRINCIPLE

A **capsule** is a gelatinous outer layer secreted by the cell that surrounds and adheres to the cell wall. It is not common to all organisms. Cells that have a heavy capsule are known to be virulent and capable of producing disease, as the structure protects bacteria against the normal phagocytic activities of host cells. Chemically, the capsular material is a polysaccharide, a glycoprotein, or a polypeptide.

   Capsule staining is more difficult than other types of differential staining procedures, as the capsular materials are water-soluble and may be dislodged and removed with vigorous washing. Smears should not be heated, as the resultant cell shrinkage may create a clear zone around the organism that is an artifact that can be mistaken for the capsule.

   The capsule stain uses two reagents.

Primary Stain

CRYSTAL VIOLET A purple-blue stain is applied to a non–heat-fixed smear. At this point the cell and the capsular material will take on the dark color.

Decolorizing Agent and Counterstain

COPPER SULFATE (20 PERCENT) Unlike the cell proper, the capsule is non-ionic and therefore the primary stain, although it adheres, will not be absorbed. Since capsular substances are water-soluble and may be lost during washing, copper sulfate is used as a decolorizing agent. It removes the excess primary stain as well as color that was adsorbed to the capsule. At the same time, it acts as a counterstain as it is adsorbed to the decolorized capsular material. The capsule will now appear light blue in contrast to the deep purple of the cell.

MATERIALS

Cultures

48-hour-old skimmed milk cultures of *Alcaligenes viscolactis*, *Klebsiella pneumoniae*, and *Enterobacter aerogenes*.

Reagents

1 percent crystal violet and 20 percent copper sulfate ($CuSO_4 \cdot 5H_2O$).

Equipment

Bunsen burner, inoculating loop or needle, staining tray, bibulous paper, lens paper, glass slides, and microscope.

PROCEDURE

1. Clean three glass slides.

2. Using sterile technique, prepare a heavy smear of each organism.

3. Allow smears to air dry. **Caution: Do not heat fix.**

4. Flood smears with crystal violet and let stand for five to seven minutes.

5. Wash smears with 20 percent copper sulfate solution.

6. Gently blot dry and examine under oil immersion.

Observations and Results   EXPERIMENT **9**

Following the observation of all slides under oil immersion, record your results in the chart.

1. Make drawings of a representative microscopic field of each preparation.

2. Record the comparative size of the capsule, that is, small, moderate, or large.

3. Indicate the color of the capsule and the cell on each preparation.

	A. viscolactis	*K. pneumoniae*	*E. aerogenes*
Drawing of a representative field			
Capsule size: _____ Color of capsule: _____ Color of cell: _____	_____ _____ _____	_____ _____ _____	_____ _____ _____

REVIEW QUESTIONS

1. Explain the medical significance of a capsule.

2. Why must heat fixation be omitted during the preparation of a smear for capsular staining?

3. Explain the function of copper sulfate in this procedure.

4. Explain why capsule formation depends on environmental factors, nutritional status, and age
   of the culture.

Microbiological Equipment and Cultivation Techniques for the Isolation and Enumeration of Microorganisms

PURPOSES

This section is designed to instruct students in:

1. The types of laboratory equipment and culture media needed to develop and maintain pure cultures.

2. The concept of sterility and the procedures necessary for successful subculturing of microorganisms.

3. Streak-plate and spread-plate inoculations for separation of microorganisms in a mixed microbial population for subsequent pure culture isolation.

4. The serial dilution agar plating technique for enumeration of viable microorganisms.

5. Cultural and morphological characteristics of microorganisms grown in pure culture.

INTRODUCTION

Microorganisms are ubiquitous. They are found in soil, air, water, food, sewage, and on body surfaces. In short, every area of our environment is replete with them. The microbiologist separates these mixed populations into individual species for study. A culture containing a single unadulterated species of cells is called a **pure culture.** To isolate and study microorganisms in pure culture, the microbiologist requires basic laboratory apparatus and the application of specific techniques, as illustrated below.

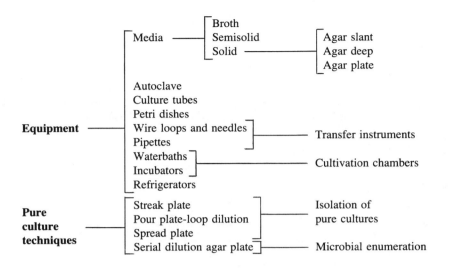

MEDIA

The survival and continued growth of microorganisms depends on an adequate supply of nutrients and a favorable growth environment. For the former, most must use soluble low-molecular-weight substances that are frequently derived from the enzymatic degradation of complex nutrients. A solution containing these nutrients is a **culture medium.** Basically, all culture media are liquid, semisolid, or solid. A liquid medium lacks a solidifying agent and is called a **broth medium.** A broth medium supplemented with a solidifying agent called **agar** results in a solid or semisolid medium. Agar is an extract of seaweed, a complex carbohydrate composed mainly of galactose, and is without nutritional value. A completely solid medium requires an agar concentration of about 1.5 percent to 1.8 percent. A concentration of less than 1 percent agar results in a **semisolid medium.** Agar serves as an excellent solidifying agent as it liquefies at 100 degrees C and solidifies at 40 degrees C. Because of these properties, organisms, especially pathogens, can be cultivated at temperatures of 37.5 degrees C or slightly higher without fear of the medium liquefying. Therefore, a solid medium has the advantage in that it presents a hardened surface on which microorganisms can be grown using specialized techniques for the isolation of discrete colonies. Each individual colony is a cluster of cells that originates from the multiplication of a single cell and represents the growth of a single species of microorganisms. Such a defined and well-isolated colony can serve as the source for the establishment of a pure culture. Also, while in the liquefied state, solid media can be placed in test tubes, which are then allowed to cool and harden in a slanted position producing **agar slants.** These are useful for maintaining pure cultures for sub-culturing purposes. Similar tubes that, following preparation, are not slanted but allowed to harden in the upright position are designated as **agar deep tubes.** Agar deep tubes are used primarily for the study of the gaseous requirements of microorganisms. However, they may be liquefied in a boiling water bath and poured into Petri dishes, producing **agar plates,** which provide large surface areas for the isolation and study of microorganisms. The various forms of solid media are illustrated on page 53.

In addition to nutritional needs, the environmental factors must also be regulated, including proper pH, temperature, gaseous requirements, and osmotic pressure. A more detailed explanation of these are presented in Part IV, the section dealing with cultivation of microorganisms; this description is merely for introductory purposes and students should bear in mind that numerous types of media are available.

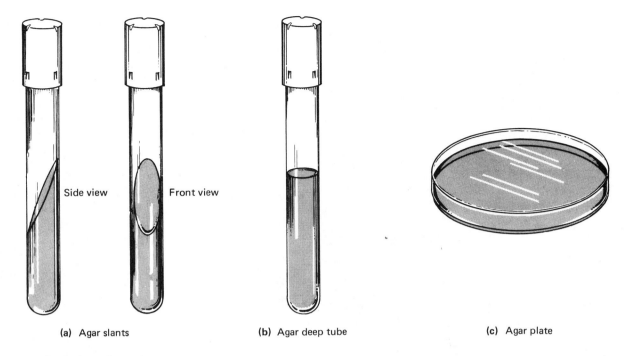

(a) Agar slants   (b) Agar deep tube   (c) Agar plate

Forms of Solid (Agar) Media

STERILIZATION

Sterility is the hallmark for successful work in the microbiology laboratory. To achieve this, it is mandatory that sterile equipment and sterile techniques are used. **Sterilization** is the process of rendering a medium or material free of all forms of life. Although a more detailed discussion is presented in Part IX, the section dealing with control of microorganisms, the following is a brief outline of the routine techniques used in the microbiology laboratory.

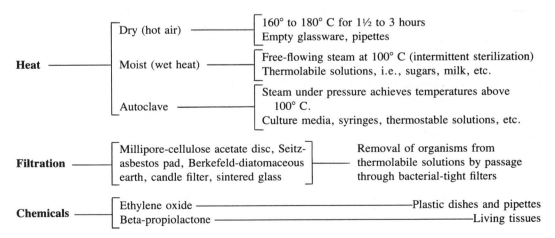

CULTURE TUBES AND PETRI DISHES

Glass **test tubes** and glass or plastic **Petri dishes** are used to cultivate microorganisms. A suitable nutrient medium in the form of broth or agar may be added to the tubes, while only a solid medium is used in Petri dishes. A sterile environment is maintained in culture tubes by various

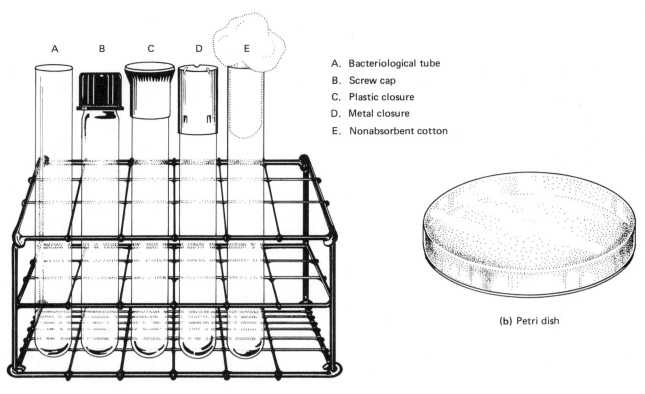

A. Bacteriological tube
B. Screw cap
C. Plastic closure
D. Metal closure
E. Nonabsorbent cotton

(b) Petri dish

(a) Test tube rack with tubes showing various closures

Culture Vessels

types of closures. Historically, the first type, a cotton plug, was developed by Schroeder and von Dusch in the nineteenth century. Today most laboratories use sleevelike caps made of metals, such as stainless steel, or heat-resistant plastics. The advantage of these closures over the cotton plug is that they are labor-saving and, most of all, slip on and off the test tubes easily.

Petri dishes provide a larger surface area for growth and cultivation. They consist of a bottom dish portion that contains the medium and a larger top portion that serves as a loose cover. Petri dishes are manufactured in various sizes to meet different experimental requirements. For routine purposes, dishes approximately 15 cm in diameter are used. The sterile agar medium is dispensed to previously sterilized dishes from molten agar deep tubes containing 15 to 20 ml of medium, or from a molten sterile medium prepared in bulk and contained in 250-ml to 500-ml flasks. When cooled to 40 degrees C, the medium will solidify. Students must remember that **after inoculation, Petri dishes are incubated in an inverted position,** top down. This is to prevent condensation that forms on the cover during solidification from dropping down on to the surface of the hardened agar.

TRANSFER INSTRUMENTS

Microorganisms must be transferred from one vessel to another or from stock cultures to various media for maintenance and study. Such a transfer is called **subculturing** and must be carried out under sterile conditions to prevent possible contamination.

Wire loops and needles are made from inert metals such as nichrome or platinum and are inserted into metal shafts that serve as handles. They are extremely durable instruments and are easily sterilized by incineration in the blue (hottest) portion of the Bunsen burner flame.

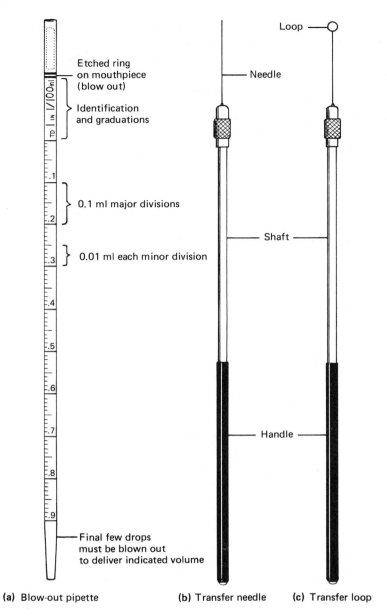

Etched ring
on mouthpiece
(blow out)

Identification
and graduations

0.1 ml major divisions

0.01 ml each minor division

Final few drops
must be blown out
to deliver indicated volume

Loop

Needle

Shaft

Handle

(a) Blow-out pipette   **(b)** Transfer needle   **(c)** Transfer loop

Transfer Instruments

A **pipette** is another instrument used for sterile transfers. Pipettes are similar in function to straws, that is, they draw up liquids. They are made of glass or plastic drawn out to a tip at one end and with a mouthpiece forming the other end. They are calibrated to deliver different volumes depending on requirements. Pipettes may be sterilized in bulk inside canisters or they may be wrapped individually in brown paper and sterilized in an autoclave or dry-heat oven.

CULTIVATION CHAMBERS

The specific temperature requirements for growth are discussed in detail in Part IV. However, a prime requirement for the cultivation of microorganisms is that they be grown at their optimum temperature. An **incubator** is used to maintain optimum temperature during the necessary growth period. It resembles an oven and is thermostatically controlled so that temperatures can be varied depending on the requirements of specific microorganisms. Most incubators use dry heat. Moisture is supplied by placing a beaker of water in the incubator during the growth period. A moist environment retards dehydration of the medium and thereby avoids spurious experimental results.

A thermostatically controlled, **shaking waterbath** is another piece of apparatus used to cultivate microorganisms. Its advantage is that it provides a rapid and uniform transfer of heat to the culture vessel, and its agitation provides increased aeration, resulting in acceleration of growth. The single disadvantage of this instrument is that it can be used only for cultivation of organisms in a broth medium.

REFRIGERATOR

A refrigerator is used for a wide variety of purposes such as maintenance and storage of stock cultures between subculturing periods, storage of sterile media to prevent dehydration, and to serve as a repository for thermolabile solutions, antibiotics, serums, and biochemical reagents.

Culture Transfer Techniques

PURPOSE

To instruct students in the technique of aseptic removal and transfer of microorganisms for subculturing.

PRINCIPLE

Microorganisms are transferred from one medium to another by **subculturing.** This technique is of basic importance and is used routinely in preparing and maintaining stock cultures, as well as in microbiological test procedures. At the bottom of this page is a schema of the types of transfers that are commonly performed.

Microorganisms are always present in the air and on laboratory surfaces, benching, and equipment. They can serve as a source of external contamination and thus interfere with experimental results unless proper techniques are used during subculturing. The following are essential steps that must be remembered for aseptic transfer of microorganisms.

This complete procedure, illustrated in Figure 10.1, is to be followed each time a culture transfer is performed.

1. Inoculating needles or loops must always be sterilized by holding them in the hottest portion of the Bunsen burner flame until the entire wires become red hot. Once flamed, the loop is never put down but is held in the hand and allowed to cool for 10 to 20 seconds. The stock culture tube and the tube to be inoculated are held in the palm of the other hand and secured with the thumb. The two tubes are then separated to form a V in the hand.

2. The tubes are uncapped by grasping the first cap with the little finger and the second cap with the next finger and lifting the closures upward. Once removed, these caps must be kept in the hand that holds the sterile inoculating loop or needle. They must never be placed on the laboratory bench as this would compromise the sterile procedure. Following removal of the closures, the necks of the tubes are briefly passed through the flame and the sterile transfer instrument is further cooled by touching it to the sterile inside wall of the culture tube before removing a small sample of inoculum.

3. Depending on the culture medium, a loop or needle is used for removal of the inoculum. Loops are commonly used to obtain a sample of culture from a broth culture. Either instrument can be used to obtain the inoculum from an agar slant culture by carefully touching the surface of the solid medium in an area exhibiting growth so as not to gouge into the agar. A straight needle is always used when transferring microorganisms to an agar deep tube from both solid and liquid cultures.

4. The cell-laden loop or needle is inserted into the subculture tube. In the case of a broth medium, the loop or needle is shaken slightly to dislodge the organisms; with an agar slant medium it is drawn lightly over the hardened surface in a straight or zigzag line. For inoculation of an agar deep tube, a straight needle is inserted to the bottom of the tube in a straight line and rapidly withdrawn along the line of insertion.

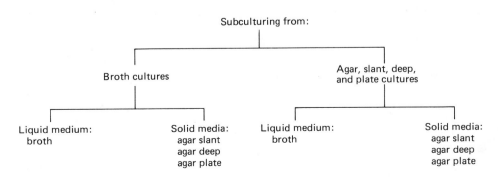

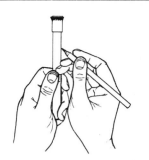

(a) Label the tube to be inoculated with the name of the organism and your initials.

(b) Place the tubes in the palm of your hand, secure with your thumb, and separate to form a V.

(c) Flame the needle or loop until the entire wire is red.

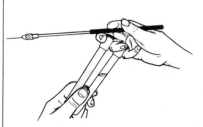

(d) With the sterile loop or needle in hand, uncap the tubes.

(e) Flame the necks of the tubes by rapidly passing them through the flame once.

(f) **Slant-to-broth transfer:** dislodge inoculum by slight agitation. **Broth-to-slant transfer:** following insertion to base of slant, withdraw the loop in a zigzag motion. **Slant-to-agar deep transfer:** insert the needle to the bottom of the tube and withdraw along the line of insertion.

(g) Flame the necks of the tubes by rapidly passing them through the flame.

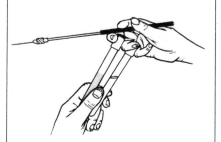

(h) Recap the tubes.

(i) Reflame the loop or needle.

FIGURE 10.1. Subculturing procedure

5. Following inoculation, the instrument is removed, the necks of the tubes are reflamed, and the caps are replaced in the order in which they were removed.

6. The needle or loop is again flamed to destroy existing organisms.

In this experiment students will master the manipulations required for aseptic transfer of microorganisms in broth-to-slant, slant-to-broth, and slant-to-agar deep transfers. The technique for transfer to and from agar plates is discussed in Experiment 11.

MATERIALS

Cultures

24-hour nutrient broth and nutrient agar slant cultures of *Serratia marcescens*.

Media

Per designated student group: One nutrient broth, one nutrient agar slant, and one nutrient agar deep tube.

Equipment

Bunsen burner, inoculating loop and needle, and wax pencil.

PROCEDURE

Following the procedure previously outlined and illustrated, perform the following transfers:

1. *S. marcescens* broth culture to a nutrient agar slant.

2. *S. marcescens* agar slant culture to a nutrient broth and a nutrient agar deep tube.

Techniques for Isolation of Pure Cultures

In nature, microbial populations do not segregate themselves by species but exist with a mixture of many other cell types. In the laboratory, these populations can be separated into **pure cultures.** These cultures contain only one type of organism and are suitable for the study of their cultural, morphological, and biochemical properties.

In this experiment, students will first use one of the techniques designed to produce discrete **colonies.** Colonies are individual, macroscopically visible masses of microbial growth on a solid medium surface, each representing the multiplication of a single organism. Once these discrete colonies are obtained, aseptic transfer will be made onto nutrient agar slants for the isolation of pure cultures.

PART A: Isolation of Discrete Colonies from a Mixed Culture

PURPOSE

To perform the spread-plate and/or the streak-plate inoculation procedure for the separation of a mixed culture so that discrete colonies can be isolated.

PRINCIPLE

The techniques commonly used for isolation of discrete colonies initially require that the number of organisms in the inoculum be reduced. The resulting diminution of the population size ensures that, following inoculation, individual cells will be sufficiently far apart on the surface of the agar medium to effect a separation of the different species present. The following are available techniques that accomplish this necessary dilution:

1. The **streak-plate** method is a rapid qualitative isolation method. It is essentially a dilution technique that involves spreading a loopful of culture over the surface of an agar plate. Although many types of procedures are performed, the four-way or quadrant streak is described. Refer to Figure 11.1, which schematically illustrates this procedure.

a. Place a loopful of culture on the agar surface in area 1. Flame and cool the loop, and drag it rapidly several times across the surface of area 1.

b. Reflame and cool the loop, and turn the Petri dish 90 degrees. Then touch the loop to a corner of the culture in area 1, and drag it several times across the agar in area 2. The loop should never enter area 1 again.

c. Reflame and cool the loop, and again turn the dish 90 degrees. Streak area 3 in the same manner as area 2.

d. Without reflaming the loop, again turn the dish 90 degrees, and now drag the culture from a corner of area 3 across area 4, using a wider streak. The loop must not touch any of the previously streaked areas. The flaming of the loop at the points indicated is to effect the dilution of the culture so that fewer organisms are streaked in each area, resulting in the final desired separation.

2. The **spread-plate** technique requires that a previously diluted mixture of microorganisms be used. During inoculation, the cells are spread over the surface of a solid agar medium with a sterile, L-shaped bent rod while the Petri dish is spun on

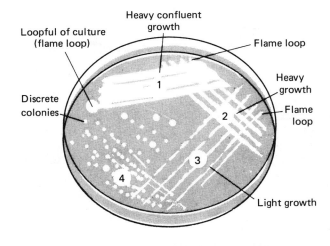

FIGURE 11.1. Four-way streak plate inoculation

a ''lazy-Susan'' turntable (Figure 11.2). The step-by-step procedure for this technique is as follows:

a. Place the bent glass rod into the beaker and add a sufficient amount of 95 percent ethyl alcohol to cover the lower, bent portion.

b. With a sterile loop, place a loopful of *M. luteus* culture in the center of the appropriately labeled nutrient agar plate that has been placed on the turntable. Replace the cover.

c. Remove the glass rod from the beaker and pass it through the Bunsen burner flame. With the bent portion of the rod pointing downward to prevent the burning alcohol from running down your arm, allow the alcohol to burn off the rod completely. Cool the rod for 10 to 15 seconds.

d. Remove the Petri dish cover and spin the turntable.

e. While the turntable is spinning, lightly touch the sterile bent rod to the surface of the agar and move it back and forth. This will spread the culture over the agar surface.

f. When the turntable comes to a stop, replace the cover. Immerse the rod in alcohol and re-flame.

3. The **pour-plate** technique requires a serial dilution of the mixed culture by means of a loop or pipette. The diluted inoculum is then added to a molten agar medium in a Petri dish, mixed, and allowed to solidify. The serial dilution and pour plate procedures are outlined in Experiment 13.

MATERIALS

Cultures

24-hour to 48-hour nutrient broth cultures of a mixture of one part *Serratia marcescens* and three parts *Escherichia coli* and a mixture of one part *Escherichia coli* and ten parts *Micrococcus luteus*. For the spread-plate procedure, adjust the cultures to an O.D. of 0.1 at 600 mμ.

Media

Two nutrient agar plates per designated student group for each inoculation technique to be performed.

Equipment

Bunsen burner, inoculating loop, turntable, 95 percent ethyl alcohol, 500-ml beaker, L-shaped bent glass rod, and wax pencil.

PROCEDURE

1. With a wax pencil, label all plates with the name of the organisms and the inoculation technique (spread plate or streak plate). Identify each with your initials.

2. Following the procedures previously described, prepare a spread-plate and/or streak-plate inoculation of each test culture.

3. Incubate all plates in an inverted position for 48 to 72 hours at 25 degrees C.

PART B: Isolation of Pure Cultures from a Spread-Plate or Streak-Plate Preparation

PURPOSE

To prepare a stock culture of an organism using isolates from the mixed agar streak-plate and/or spread-plate cultures prepared in Part A of this experiment.

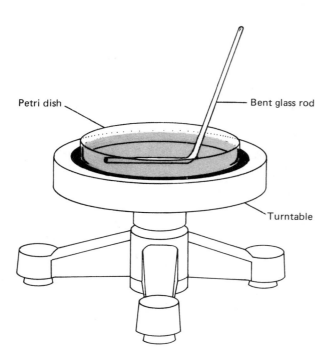

Petri dish

Bent glass rod

Turntable

FIGURE 11.2. Petri dish turntable

PRINCIPLE

Once discrete, well-separated colonies develop on the surface of a nutrient agar plate culture, each may be picked up with a sterile needle and transferred to separate nutrient agar slants. Each of these new slant cultures represents the growth of a single bacterial species and is designated as a **pure or stock culture.**

MATERIALS

Cultures

Mixed nutrient agar streak-plate and/or spread-plate preparations of *Serratia marcescens* and *Escherichia coli* and of *Micrococcus luteus* and *Escherichia coli* from Part A.

Media

Three nutrient agar slants per designated student group.

Reagents

Crystal violet, Gram's iodine, 95 percent ethyl alcohol, and safranin.

Equipment

Bunsen burner, inoculating needle, glass slides, lens paper, bibulous paper, staining tray, wax pencil, and microscope.

PROCEDURE

1. With a wax pencil, label the nutrient agar slants as *M. luteus, S. marcescens,* and *E. coli.* Identify each with your initials.

2. Aseptically transfer the yellow *M. luteus,* the white *E. coli,* and the red *S. marcescens* to the appropriately labeled agar slants as shown in Figure 11.3 on page 66.

3. Incubate the cultures for 48 to 72 hours at 25 degrees C.

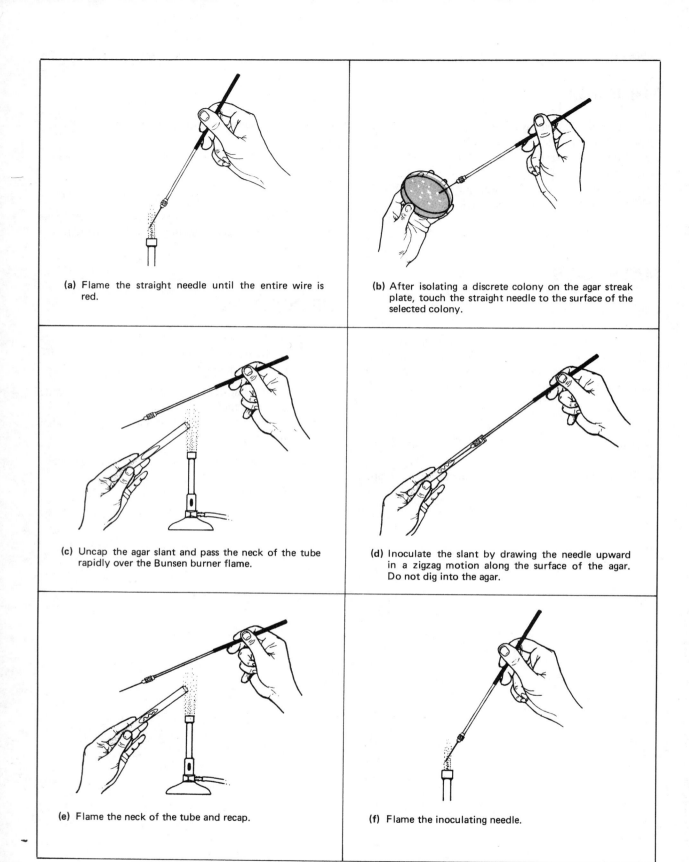

(a) Flame the straight needle until the entire wire is red.

(b) After isolating a discrete colony on the agar streak plate, touch the straight needle to the surface of the selected colony.

(c) Uncap the agar slant and pass the neck of the tube rapidly over the Bunsen burner flame.

(d) Inoculate the slant by drawing the needle upward in a zigzag motion along the surface of the agar. Do not dig into the agar.

(e) Flame the neck of the tube and recap.

(f) Flame the inoculating needle.

FIGURE 11.3. Procedure for the preparation of a pure culture

Cultural Characteristics of Microorganisms

PURPOSE

To determine the cultural characteristics of microorganisms as an aid in identifying and classifying organisms into taxonomic groups.

PRINCIPLE

When grown on a variety of media, microorganisms will exhibit differences in the macroscopic appearance of their growth. These differences, called **cultural characteristics,** are used as bases for separating microorganisms into taxonomic groups. The cultural characteristics for all known microorganisms are contained in *Bergey's Manual of Determinative Bacteriology,* 8th ed. They are determined by culturing the organisms on nutrient agar slants and plates, nutrient broth, and nutrient gelatin. The patterns of growth to be considered in each of these media are described below and some are illustrated in Figure 12.1 on page 70.

NUTRIENT AGAR SLANTS These have a single straight line of inoculation on the surface, and are evaluated in the following manner:

1. **Abundance of growth:** The amount of growth is designated as none, slight, moderate, or large.

2. **Pigmentation:** Chromogenic microorganisms may produce intracellular pigments that are responsible for the coloration of the organism as seen in surface colonies. Other organisms produce extracellular soluble pigments that are excreted into the medium and also produce a color. Most organisms, however, are nonchromogenic and will appear white to gray.

3. **Optical characteristics:** Optical characteristics may be evaluated on the basis of the amount of light transmitted through the growth. These characteristics are described as **opaque** (no light transmission), **translucent** (partial transmission), or **transparent** (full transmission).

4. **Form:** The appearance of the single line streak of growth on the agar surface is designated as:

    a. **Filiform:** Continuous, threadlike growth with smooth edges.

    b. **Echinulate:** Continuous, threadlike growth with irregular edges.

    c. **Beaded:** Nonconfluent to semiconfluent colonies.

    d. **Effuse:** Thin, spreading growth.

    e. **Arborescent:** Treelike growth.

    f. **Rhizoid:** Rootlike growth.

NUTRIENT AGAR PLATES These demonstrate well-isolated colonies and are evaluated in the following manner:

1. **Size:** Pinpoint, small, moderate, or large.

2. **Pigmentation:** Color of colony.

3. **Form:** The shape of the colony is described as follows:

    a. **Circular:** Unbroken peripheral edge.

    b. **Irregular:** Indented peripheral edge.

    c. **Rhizoid:** Rootlike spreading growth.

4. **Margin:** The appearance of the outer edge of the colony is described as follows:

    a. **Entire:** Sharply defined, even.

    b. **Lobate:** Marked indentations.

    c. **Undulate:** Wavy indentations.

    d. **Serrate:** Toothlike appearance.

    e. **Filamentous:** Threadlike, spreading edge.

5. **Elevation:** The degree to which colony growth is raised on the agar surface is described as follows:

    a. **Flat:** Elevation not discernible.

    b. **Raised:** Slightly elevated.

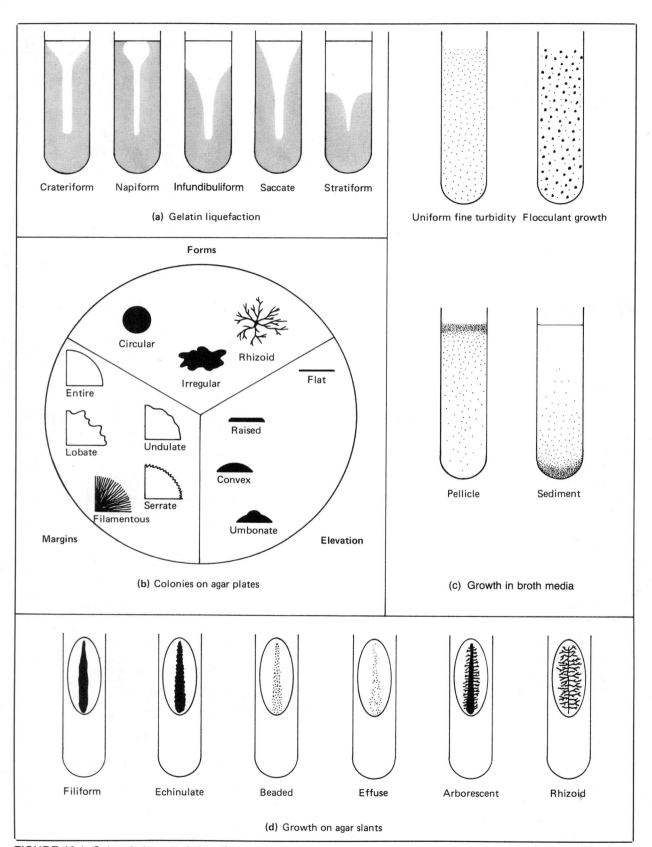

Crateriform Napiform Infundibuliform Saccate Stratiform

(a) Gelatin liquefaction

Uniform fine turbidity Flocculant growth

Forms

Circular

Rhizoid

Entire

Irregular

Flat

Lobate Undulate

Raised

Filamentous Serrate

Convex

Margins

Umbonate

Elevation

(b) Colonies on agar plates

Pellicle Sediment

(c) Growth in broth media

Filiform Echinulate Beaded Effuse Arborescent Rhizoid

(d) Growth on agar slants

FIGURE 12.1 Cultural characteristics of bacteria

70

    c. **Convex:** Dome-shaped elevation.

    d. **Umbonate:** Raised, with elevated convex central region.

NUTRIENT BROTH CULTURES These are evaluated as to the distribution and appearance of the growth as follows:

1. **Uniform fine turbidity:** Finely dispersed growth throughout.

2. **Flocculant:** Flaky aggregates dispersed throughout.

3. **Pellicle:** Thick, padlike growth on surface.

4. **Sediment:** Concentration of growth at the bottom of broth culture may be granular, flaky, or flocculant.

NUTRIENT GELATIN This solid medium may be liquefied through the enzymatic action of gelatinase. Liquefaction occurs in a variety of patterns:

1. **Crateriform:** Liquefied surface area is saucer-shaped.

2. **Napiform:** Bulbous-shaped liquefaction at surface.

3. **Infundibuliform:** Funnel-shaped.

4. **Saccate:** Elongated, tubular.

5. **Stratiform:** Complete liquefaction of the upper half of the medium.

MATERIALS

Cultures

24-hour nutrient broth cultures of *Pseudomonas aeruginosa, Bacillus cereus, Micrococcus luteus, Escherichia coli,* and *Mycobacterium smegmatis.*

Media

Per designated student group: five each of nutrient agar plates, nutrient agar slants, nutrient broth tubes, and nutrient gelatin tubes.

Equipment

Bunsen burner, inoculating loop, and needle.

PROCEDURE

1. With a wax pencil, label each plate and tube with the name of the organism to be inoculated and identify each with your initials.

2. Using sterile technique, inoculate each of the media below in the following manner:

    a. Nutrient agar plates: With a sterile loop, prepare a streak-plate inoculation of each of the cultures for the isolation of discrete colonies.

    b. Nutrient agar slants: With a sterile needle, make a single line streak of each of the cultures provided, starting at the butt and drawing the needle up the center of the slanted agar surface.

    c. Nutrient broth: Using a sterile loop, inoculate each organism into a tube of nutrient broth. Shake the loop a few times to dislodge the inoculum.

    d. Nutrient gelatin: Using a sterile needle, prepare a stab inoculation of each of the cultures provided.

3. Incubate all cultures at 37 degrees C for 24 to 48 hours.

Observations and Results   EXPERIMENT **12**

Refer to Figure 12.1 and the descriptions presented in the introductory section of this exercise while making the following observations:

1. Place all gelatin cultures in refrigerator for 30 minutes to determine whether liquefaction of the medium has developed. Record your observations in the chart below according to the presence or absence of liquefaction and its type, if it has occurred.

Nutrient Gelatin Cultures

	M. luteus	*P. aeruginosa*	*M. smegmatis*	*E. coli*	*B. cereus*
Draw pattern of liquefaction					
Liquefaction (+) or (−): Type of liquefaction:	_____ _____	_____ _____	_____ _____	_____ _____	_____ _____

2. Observe a single, well-isolated colony on each of the nutrient agar plate cultures and identify its size, elevation, margin, consistency, and pigmentation. Record your observations in the chart below.

Nutrient Agar Plate Cultures

	M. luteus	*P. aeruginosa*	*M. smegmatis*	*E. coli*	*B. cereus*
Draw distribution of colonies					
Size: Elevation: Margin: Consistency: Pigmentation:	_____ _____ _____ _____ _____	_____ _____ _____ _____ _____	_____ _____ _____ _____ _____	_____ _____ _____ _____ _____	_____ _____ _____ _____ _____

3. Observe each of the nutrient agar slant cultures as to the amount, pigmentation, consistency and form of the growth. Record your observations in the chart below.

Nutrient Agar Slant Cultures

	M. luteus	*P. aeruginosa*	*M. smegmatis*	*E. coli*	*B. cereus*
Draw distribution of growth on slant surface	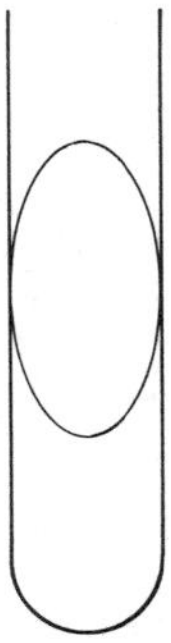		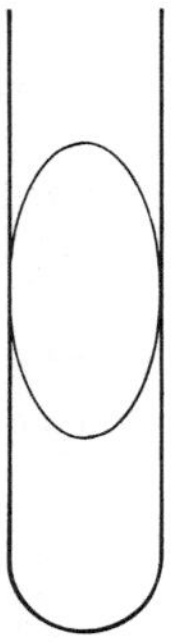	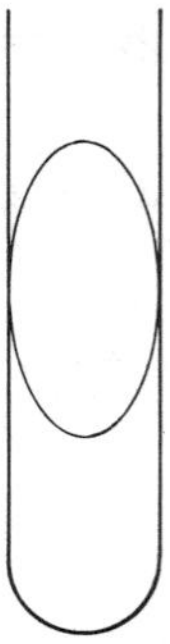	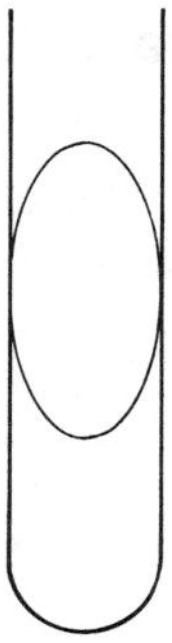
Amount of growth:	_____	_____	_____	_____	_____
Pigmentation:	_____	_____	_____	_____	_____
Consistency:	_____	_____	_____	_____	_____
Form:	_____	_____	_____	_____	_____

4. Observe each of the nutrient broth cultures as to the appearance of the growth (flocculation, turbidity, sediment, or pellicle). Record your results in the chart below.

Nutrient Broth Cultures

	M. luteus	*P. aeruginosa*	*M. smegmatis*	*E. coli*	*B. cereus*
Draw distribution of growth	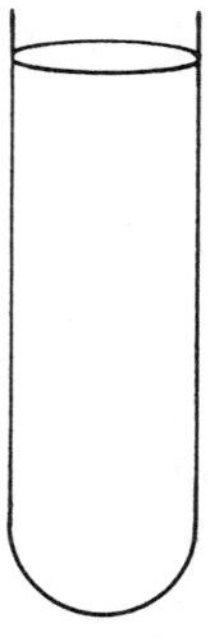	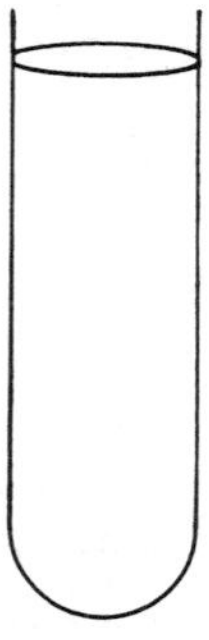	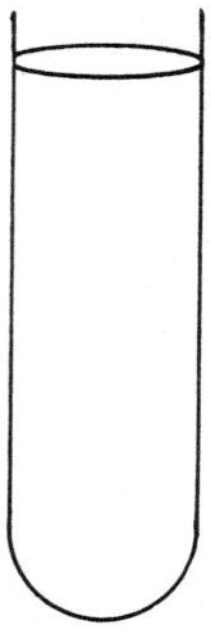	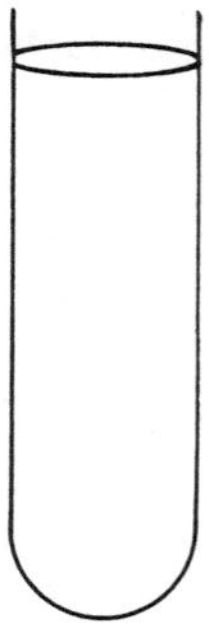	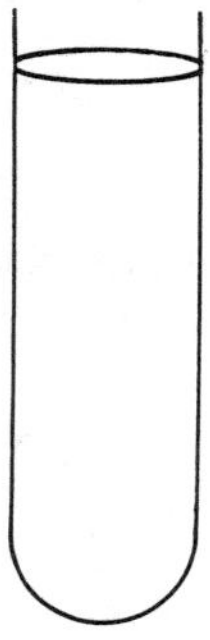
Appearance of growth:	_____	_____	_____	_____	_____

Serial Dilution-Agar Plating Procedure to Quantitate Viable Cells

PURPOSE

To determine quantitatively the number of viable cells in a bacterial culture.

PRINCIPLE

Studies involving the analysis of materials such as food, water, milk, and in some cases air require quantitative enumeration of microorganisms in the substances. Many methods have been devised to accomplish this, including direct microscopic counts, Breed smears, an electronic cell counter such as the Coulter counter, chemical methods for estimating cell mass or cellular constituents, turbidimetric measurements for increases in cell mass, and the serial dilution-agar plate method.

Direct Microscopic Counts

These require the use of a specialized counting chamber called the **Petroff-Hauser chamber,** in which an aliquot of a cell suspension is counted and the total cell number is determined mathematically. Although rapid, it has the disadvantages that both living and dead cells are counted and the method is not sensitive to populations of fewer than one million cells.

Breed smears are used mainly to quantitate bacterial cells in milk. Using stained smears confined to a 1-mm^2 ruled area of the slide, the total population is determined mathematically. This method also fails to discriminate between viable and dead cells.

Electronic Cell Counters

The **Coulter counter** is an example of an instrument capable of rapidly counting the number of cells suspended in a conducting fluid that passes through a minute orifice through which an electric current is flowing. Cells, which are nonconductors, increase the electrical resistance of the conducting fluid and the resistance is electronically recorded, enumerating the number of organisms flowing through the orifice. In addition to inability to distinguish between living and dead cells, the apparatus is also unable to differentiate inert particulate matter from cellular material.

Chemical Methods

While not considered as means of direct quantitative analysis, they may be used to measure indirectly increases both in protein concentration and in DNA production. In addition, cell mass can be estimated by dry weight determination of a specific aliquot of the culture. Measurement of certain metabolic parameters may also be used to quantitate bacterial populations. The amount of oxygen consumed (oxygen uptake) is directly proportional to the increasing number of vigorously growing aerobic cells, while the rate of carbon dioxide production is related to increased growth of anaerobic organisms.

Spectrophotometric Analysis

Increased turbidity in a culture is another index of growth. With these instruments, the amount of transmitted light decreases as the cell population increases, and the decrease in radiant energy is converted to electrical energy and indicated on a galvanometer. This method is rapid but limited because sensitivity is restricted to microbial suspensions of 10 million cells or greater.

While all of these methods may be used to enumerate the number of cells in a bacterial culture, the major disadvantage common to all is that the total count includes dead as well as living cells. Sanitary and medical microbiology at times require determination of viable cells. To accomplish this, the serial dilution-agar plating technique is used. Briefly, this method involves serial dilution of a bacterial suspension in sterile water blanks, which serve as a diluent of known volume. Once diluted, the suspensions are plated on suitable nutrient media. The **pour-plate technique,** illustrated

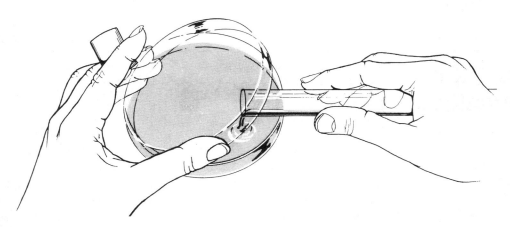

FIGURE 13.1. Pour-plate technique

in Figure 13.1, is the procedure usually employed. Molten agar, cooled to 45 degrees C, is poured into a Petri dish containing a specified amount of the diluted sample. Following addition of the molten and cooled agar, the cover is replaced, and the plate is gently rotated in a circular motion to achieve uniform distribution of microorganisms. This procedure is repeated for all dilutions to be plated. Dilutions should be plated in duplicate for greater accuracy, incubated overnight, and counted on a **Quebec colony counter** either by hand or by an electronically modified version of this instrument.

Plates suitable for counting must contain not fewer than 30 or more than 300 colonies. The total count of the suspension is obtained by multiplying the number of cells per plate by the dilution factor, which is the reciprocal of the dilution.

Advantages of the serial dilution-agar plating technique are:

1. Only viable cells are counted.

2. It allows isolation of discrete colonies that can be subcultured into pure cultures, which may then be easily studied and identified.

Disadvantages of this method are:

1. Overnight incubation is necessary before colonies develop on the agar surface.

2. It is necessary to use more glassware in this procedure.

3. The need for greater manipulation may result in erroneous counts due to errors in dilution or plating.

The following experiment uses the pour-plate technique for plating serially diluted culture samples. The procedure to be followed is illustrated in Figure 13.2.

MATERIALS

Culture

24-hour to 48-hour nutrient broth culture of *Escherichia coli*.

Media

Per designated student group: six 20-ml nutrient agar deep tubes and seven sterile 9-ml water blanks.

Equipment

Hot plate, water bath, thermometer, test tube rack, Bunsen burner, sterile 1-ml serological pipettes, sterile Petri dishes, Quebec colony counter, manual hand counter, disinfectant solution in a 500-ml beaker, and wax pencil.

PROCEDURE

1. Place the six agar deep tubes in a water bath and place on a hot plate. Bring to 100 degrees C and continue heating until all agar is liquefied. Remove the water bath from the hot plate and cool to 45 degrees C by adding cold water. Check the temperature with the thermometer. Place the water bath on the hot plate and maintain at 45 degrees C with low heat.

2. With a wax pencil, label the *E. coli* culture tube as number 1 and the seven 9-ml water blanks as numbers 2 through 8. Place the labeled tubes in a

test-tube rack. Label the Petri dishes as follows: 1A, 1B, 2A, 2B, 3A, and 3B.

3. Mix the *E. coli* culture (tube number 1) by rolling the tube between the palms of your hands to ensure even dispersal of cells in the culture.

4. With a sterile pipette, aseptically transfer 1 ml from the bacterial suspension tube number 1 to water blank tube number 2. Discard the pipette in the

beaker of disinfectant. The culture has been diluted 10 times to 10^{-1}.

5. Mix tube number 2 and with a fresh pipette transfer 1 ml to tube number 3. Discard the pipette. The culture has been diluted 100 times to 10^{-2}.

6. Mix tube number 3 and with a fresh pipette transfer 1 ml to tube number 4. Discard the pipette. The culture has been diluted 1000 times to 10^{-3}.

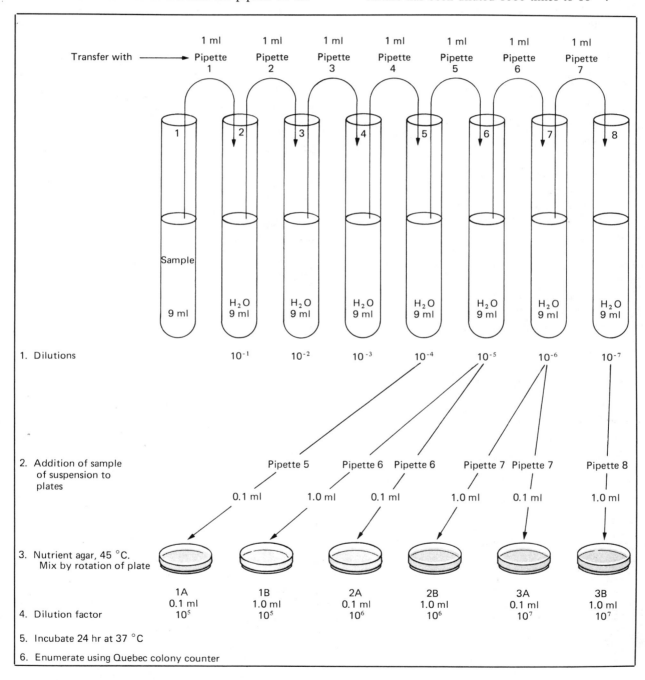

FIGURE 13.2. Serial dilution-agar plate procedure

7. Mix tube number 4 and with a fresh pipette transfer 1 ml to tube number 5. Discard the pipette. The culture has been diluted 10,000 times to 10^{-4}.

8. Mix tube number 5 and with a fresh pipette transfer 0.1 ml of this suspension to plate 1A. Return the pipette to tube number 5 and transfer 1 ml to tube number 6. Discard the pipette. The culture has been diluted 100,000 times to 10^{-5}.

9. Mix tube number 6 and with a fresh pipette transfer 1 ml of this suspension to plate 1B. Return the pipette to tube number 6 and transfer 0.1 ml to plate 2A. Return the pipette to tube number 6 and transfer 1 ml to tube number 7. Discard the pipette. The culture has been diluted 1,000,000 times to 10^{-6}.

10. Mix tube number 7 and with a fresh pipette transfer 1 ml of this suspension to plate 2B. Return the pipette to tube number 7 and transfer 0.1 ml to plate 3A. Return the pipette to tube number 7 and transfer 1 ml to tube number 8. Discard the pipette. The culture has been diluted 10,000,000 to 10^{-7}.

11. Mix tube number 8 and with a fresh pipette transfer 1 ml of this suspension to plate 3B. Discard the pipette. The dilution procedure is now complete.

12. Check the temperature of the molten agar medium to be sure the temperature is 45 degrees C. Remove a tube from the water bath and wipe the outside surface dry with a paper towel. Using sterile technique, pour the agar into plate 1A as shown in Figure 13.1 and rotate the plate gently to ensure uniform distribution of the cells in the medium.

13. Repeat step 12 for the addition of molten nutrient agar to plates 1B, 2A, 2B, 3A, and 3B.

14. Once the agar has solidified, incubate the plates in an inverted position for 24 hours at 37 degrees C.

Observations and Results   EXPERIMENT **13**

1. Using a Quebec colony counter and a mechanical hand counter, observe all colonies on plates. Plates with more than 300 colonies cannot be counted and are designated as **too numerous to count—TNTC;** plates with fewer than 30 colonies are designated as **too few to count—TFTC.** Count only plates containing between 30 and 300 colonies. Remember to count all subsurface as well as surface colonies.

2. The number of organisms per ml of original culture is calculated by multiplying the number of colonies counted, by the dilution factor:

Number of cells per ml = Number of colonies × Dilution factor

Examples:
1. Colonies per plate = 50
   Dilution factor = 1:1 × 10^6 (1:1,000,000)
   Volume of dilution added to plate = 1 ml
   50 × 1,000,000 = 50,000,000 cells/ml
   $\qquad\qquad\qquad$ (5 × 10^7)

2. Colonies per plate = 50
   Dilution factor = 1:1 × 10^5 (1:100,000)
   Volume of dilution added to plate = 0.1 ml
   50 × 100,000 = 5,000,000 cells/0.1 ml
   $\qquad\qquad\qquad$ (5 × 10^6)
   5,000,000 × 10 = 50,000,000 cells/ml
   (5 × 10^6) $\qquad\qquad$ (5 × 10^7)

3. Record your observations and calculated bacterial counts per ml of sample in the chart.

4. Since the dilutions plated are replicates of each other, determine the average of the duplicate bacterial counts per ml of sample and record in the chart.

Plate	Dilution	Ml of dilution plated	Number of colonies	Bacterial count per ml of sample	Average count per ml of sample
1A					
1B					
2A					
2B					
3A					
3B					

REVIEW QUESTIONS

1. What is the major disadvantage of microbial counts performed by methods other than the serial dilution-agar plating procedure?

2. Distinguish between dilution and dilution factor.

3. What are the advantages and disadvantages of the serial dilution-agar plating procedure?

4. If 0.1 ml of a 1×10^{-6} dilution plate contains 56 colonies, calculate the number of cells per ml of the original culture.

5. How would you record your observation of a plate containing 305 colonies? _____ .
   A plate with 15 colonies? _____ .

6. Explain the chemical methods for measuring cell growth.

1. **CARBON** is the most essential and central atom c
   function. Among microbial cells, two carbon-depende

   a. **Autotrophs:** These organisms can be cultivated i
      ganic compounds and they specifically use inorgar
      ide.

   b. **Heterotrophs:** These are organisms that cannot
      solely of inorganic compounds, but must be supp
      glucose.

2. **NITROGEN** is an essential atom in many cellular mac
   nucleic acids. Proteins serve as the structural molecul
   cell and as functional molecules, enzymes, that are re
   of the cell. Nucleic acids include DNA, the genetic ba
   an active role in protein synthesis within the cell. Nu
   microbial types use atmospheric nitrogen, others rely
   monium or nitrate salts, and still others require nitroge
   as amino acids.

3. **NONMETALLIC ELEMENTS:** The major nonmetal

   a. **Sulfur:** This is integral to some amino acids and i
      and functional proteins. Its sources include organic
      amino acids, inorganic compounds such as sulfate:

   b. **Phosphorus:** This is necessary for the formation
      the nucleic acids DNA and RNA, and also for
      compound adenosine triphosphate, ATP. Phosphor
      salts for use by all microbial cells.

4. **METALLIC ELEMENTS:** Ca^{++}, Zn^{++}, Na^+, K^+,
   are some of the metallic ions necessary for continued e
   activities. Some of these activities are osmoregulation
   electron transport during biooxidation. It must be rem
   trients and are required in trace concentrations only. Ir

5. **VITAMINS** are organic substances that contribute to
   minute concentrations for cell activities. They are al
   required for the formation of active enzyme systems.
   be supplied in a preformed state for normal metabolic ac
   extensive vitamin synthesizing pathways, whereas other
   from other compounds present in the medium.

6. **WATER:** All cells require water in the medium so th
   can cross the cell membrane.

7. **ENERGY:** Active transport, biosynthesis, and biodeg
   metabolic activities of cellular life. These activities can
   ability of energy within the cell. Two bioenergetic type

   a. **Phototrophs:** These use radiant energy as their sol

   b. **Chemotrophs:** They depend on oxidation of orga
      inorganic compounds such as H_2S or $NaNO_2$ as en

   Three of the most important physical factors that influe
   are temperature, pH, and gaseous requirements. An unders
   metabolism is essential.

mmon to all cellular structures and
t types are noted:

a medium consisting solely of inor-
ic carbon in the form of carbon diox-

e cultivated in a medium consisting
ied with organic nutrients, especially

omolecules, particularly proteins and
s forming the so-called fabric of the
sponsible for the metabolic activities
is of cell life, and RNA, which plays
ritional diversity exists in that some
n inorganic compounds such as am-
-containing organic compounds such

ic ions used for cellular nutrition are:

therefore a component of structural
compounds such as sulfur-containing
, and elementary sulfur.

of essential macromolecules such as
ynthesis of the high-energy organic
is is supplied in the form of phosphate

Cu^{++}, Mn^{++}, Mg^{++}, and $Fe^{+2,+3}$
ficient performance of varied cellular
, regulation of enzyme activity, and
mbered that these ions are micronu-
organic salts supply these materials.

cellular growth and are essential in
o sources of coenzymes, which are
Cells vary as to which vitamins must
ivities. Some microbial types possess
can only synthesize a limited number

t the low-molecular-weight nutrients

radation of macromolecules are the
only be sustained by constant avail-
of microorganisms exist.

energy source.

ic compounds such as glucose, or
rgy sources.

ce the growth and survival of cells
anding of the role they play in cell

Temperature is important because of its influence on the rate of chemical reactions through its action on cellular enzymes. In general, the optimum temperature for enzymatic activities in all cell types is 20 degrees to 40 degrees C. Low temperatures slow down or inhibit enzyme activity, thereby slowing down or inhibiting cell metabolism and consequently, cell growth. High temperatures coagulate and thus irreversibly denature thermolabile enzymes. Although enzymes differ in their degree of heat sensitivity, generally temperatures in the range of 70 degrees C will destroy most essential enzymes and cause cell death.

The pH of the extracellular environment greatly affects cells' enzymatic activities. The optimum pH for cell metabolism is in the neutral range of 7. An increase in the hydrogen ion concentration resulting in an acidic pH of below 7, or a decrease in the hydrogen ion concentration resulting in an alkaline pH of above 7, is often detrimental. Either increase or decrease will slow down the rate of chemical reactions due to the destruction of cellular enzymes, thereby affecting the rate of growth and ultimately, survival.

The gaseous requirement in most cells is atmospheric oxygen, which is necessary for the biooxidative process of respiration. Atmospheric oxygen plays a vital role in adenosine triphosphate (ATP) formation and the availability of energy in a utilizable form for cell activities. Other cell types, however, lack the enzyme systems for respiration in the presence of oxygen and therefore must use an anaerobic form of respiration, namely fermentation.

The following exercises will demonstrate the diversity of microorganisms as to nutritional and environmental requirements.

Nutritional Requirements: Media for the Routine Cultivation of Bacteria

PURPOSES

1. To evaluate several types of media as to their ability to support the growth of different bacterial species.

2. To evaluate the nutritional needs of the bacteria under study.

PRINCIPLE

To satisfy the diverse nutritional needs of bacteria, bacteriologists employ two major categories of media for routine cultivation.

Chemically Defined Media

These are composed of known quantities of chemically pure, specific organic, and/or inorganic compounds. Their use requires knowledge of the organism's specific nutritional needs. The following two chemically defined media are used in this exercise:

1. **Inorganic synthetic broth:** This completely inorganic medium is prepared by incorporating the following salts per 1000 ml of water:

Sodium chloride (NaCl)	5 gm
Magnesium sulfate ($MgSO_4$)	0.2 gm
Ammonium dihydrogen phosphate ($NH_4H_2PO_4$)	1 gm
Dipotassium hydrogen phosphate (K_2HPO_4)	1 gm
Atmospheric CO_2	

2. **Glucose salts broth:** This medium is composed of salts incorporated into the inorganic synthetic broth medium plus **glucose,** 5 gm per liter, which serves as the sole organic carbon source.

Artificial Media

They are composed of a limited number of complex substances, plant and animal extracts, whose exact chemical composition is not known. They are capable of supporting the growth of most heterotrophs. The following two artificial media are used in this exercise:

1. **Nutrient broth:** This basic artificial medium is prepared by incorporating the following ingredients per 1000 ml of water:

Peptone:	5 gm
Beef extract:	3 gm

   Peptone, a semidigested protein, is primarily a nitrogen source. The **beef extract,** a beef derivative, is a source of organic carbon, nitrogen, vitamins, and inorganic salts.

2. **Yeast extract broth:** It is composed of the basic artificial medium ingredients used in the nutrient broth plus yeast extract, 5 gm per liter, which is a rich source of vitamin B and provides additional organic nitrogen and carbon compounds.

The yeast extract broth is an example of an **enriched medium** and is used for the cultivation of **fastidious** microorganisms, organisms that have highly elaborate and specific nutritional needs. These bacteria do not grow or grow poorly on a basic artificial medium and require the addition of one or more growth-supporting substances, enrichments such as additional plant or animal extracts, vitamins, or blood.

MATERIALS

Cultures

Saline suspension of 24-hour trypticase soy broth cultures adjusted to 0.05 optical density at a wavelength of 600 mμ of *Escherichia coli, Alcaligenes faecalis,* and *Streptococcus mitis*.

Media

Per designated student group three of all of the following contained in 13 × 100 mm test tubes: inorganic synthetic broth, glucose salt broth, nutrient broth, and yeast extract broth.

Equipment

Bunsen burner, sterile 1-ml serological pipettes, Bausch & Lomb Spectronic 20 spectrophotometer.

PROCEDURE

1. Placing the label toward the top of the test tube, label each of the tubes to be inoculated with the name of the medium and the name of the organism to be inoculated.

2. Using a sterile 1-ml pipette, add 0.1 ml of the *E. coli* culture to one test tube of each of the media.

3. Repeat step 2 for inoculation with *A. faecalis* and *S. mitis*.

4. Incubate the test cultures for 24 to 48 hours at 37 degrees C.

Nutritional Requirements:
Use of Differential and Selective Media

PURPOSE

To acquaint students with the use and function of specialized. media for selection and differentiation of microorganisms.

PRINCIPLE

Numerous special-purpose media are available for functions such as:

1. Isolation of bacterial types from a mixed population of organisms.

2. Differentiation among closely related groups of bacteria on the basis of macroscopic appearance of the colonies and biochemical reactions within the medium.

3. Enumeration of bacteria in sanitary microbiology such as water and sewage, and also in food and dairy products.

4. Assay of naturally occurring substances such as antibiotics, vitamins, and products of industrial fermentation.

5. Characterization and identification of bacteria by their ability to produce chemical changes in different media.

In addition to nutrients necessary for the growth of all bacteria, special-purpose media contain one or more chemical compounds that are essential for their functional specificity. In this exercise, the following two types of media will be studied and evaluated:

Selective Media

They are used to isolate specific groups of bacteria. These media incorporate chemical substances that inhibit the growth of one type of bacteria while permitting growth of another, thus facilitating bacterial isolation.

Differential Media

These can distinguish among morphologically and biochemically related groups of organisms. They incorporate chemical compounds that, following inoculation and incubation, produce a characteristic change in the appearance of bacterial growth and/or the medium surrounding the colonies, which permits differentiation.

The following media, which are representative of these two types, will be investigated in this exercise:

1. **Mannitol salt agar:** This medium contains a high salt concentration, 7.5 percent NaCl, which is inhibitory to the growth of most bacteria other than the staphylococci. The medium also performs a differential function as it contains the carbohydrate mannitol, which some staphylococci are capable of fermenting, and phenol red, a pH indicator for detecting acid produced by mannitol-fermenting staphylococci. These staphylococci exhibit a yellow zone surrounding their growth; staphylococci that do not ferment mannitol will not produce a change in coloration.

2. **Blood agar:** Blood that is incorporated into this medium is an enrichment ingredient for the cultivation of fastidious organisms such as the *Streptococcus* sp. The blood also permits demonstration of the following hemolytic properties of streptococci:

   a. **Gamma hemolysis:** No lysis of red blood cells results in no significant change in the appearance of the medium surrounding the colonies.

   b. **Alpha hemolysis:** Incomplete lysis of red blood cells with reduction of hemoglobin to methemoglobin results in a greenish halo around the bacterial growth.

91

c. **Beta hemolysis:** Lysis of red blood cells with complete destruction and use of hemoglobin by the organism results in a clear zone surrounding the colonies. This hemolysis is produced by two types of beta hemolysins, namely **streptolysin O,** an antigenic, oxygen-labile enzyme, and **streptolysin S,** a nonantigenic, oxygen-stable lysin. The hemolytic reaction is enhanced when blood agar plates are streaked and simultaneously stabbed to show subsurface hemolysis by streptolysin O in an environment with reduced oxygen tension.

Complete lysing

Test TA

3. **MacConkey agar:** The inhibitory action of crystal violet on the growth of gram-positive organisms allows for the isolation of gram-negative bacteria. Incorporation of the carbohydrate lactose, bile salts, and the pH indicator neutral red permits differentiation of enteric bacteria on the basis of their ability to ferment lactose. On this basis enteric bacteria are separated into two groups:

a. **Coliform bacilli** produce acid as a result of lactose fermentation. The bacteria exhibit a red coloration on their surface. *E. coli* produce greater quantities of acid from lactose than other coliform species. When this occurs the medium surrounding the growth also becomes red due to the action of the acid that precipitates the bile salts and is followed by absorption of the neutral red.

b. **Dysentery, typhoid, and paratyphoid bacilli** are not lactose fermenters and therefore do not produce acid. The colonies appear uncolored and frequently transparent.

4. **Eosin-methylene blue agar (Levine):** Lactose and the dyes eosin and methylene blue permit differentiation between enteric lactose fermenters and nonfermenters as well as identification of the colon bacillus *Escherichia coli*. The *E. coli* colonies are blue-black with a metallic green sheen caused by the large quantity of acid that is produced and that precipitates out the dyes onto the growth's surface. Other coliform bacteria such as *Enterobacter aerogenes* produce thick, mucoid, pink colonies on this medium. Enteric bacteria that do not ferment lactose produce colorless colonies, which because of their transparency, appear to take on the purple color of the medium. This medium is also partially inhibitory to the growth of gram-positive organisms, and thus gram-negative growth is more abundant.

MATERIALS

Cultures

24-hour to 48-hour trypticase soy broth cultures of *Enterobacter aerogenes, Escherichia coli, Streptococcus* var. Lancefield Group E, *Streptococcus mitis, Streptococcus faecalis, Staphylococcus aureus, Staphylococcus epidermidis,* and *Salmonella typhimurium.*

Media

Per designated student group: one each of mannitol salts agar plate, blood agar plate, MacConkey agar plate, and eosin-methylene blue agar plate.

Equipment

Bunsen burner, inoculating loop, and wax pencil.

PROCEDURE

1. Using the bacterial organisms listed in step 2, prepare and inoculate each of the plates in the following manner:

   a. Write the name of the medium and your initials on the cover of each plate.

   b. Divide each of the Petri dishes into the required number of sections (one section for each different organism) by marking the **bottom of the dish** with a wax pencil. Label each section with the name of the organism to be inoculated as illustrated in Figure 15.1.

   c. Using sterile technique, inoculate all plates except the blood agar plate with the designated organisms by making a single line of inoculation of each organism in its appropriate section. Be sure to close the Petri dish and flame the inoculating needle between inoculation of the different organisms. Refer to Figure 15.1 for an illustration of the procedure.

   d. Using sterile technique, inoculate the blood agar plate as described in step 1c. Upon completion of each single line of inoculation, use the inoculating loop and proceed to make three to four stabs at a 45-degree angle across the streak.

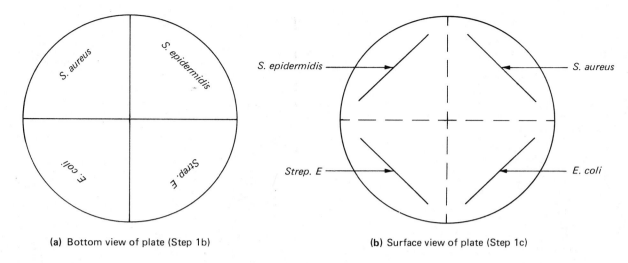

(a) Bottom view of plate (Step 1b)   (b) Surface view of plate (Step 1c)

FIGURE 15.1. Mannitol salts agar plate preparation and inoculation procedure

2. Inoculate each of the different media with the following:

   a. Mannitol salts agar: *S. aureus, S. epidermidis, Streptococcus* var. Lancefield Group E, and *E. coli.*

   b. Blood agar: *S. faecalis, S. mitis,* and *Strep-*

   tococcus var. Lancefield Group E.

   c. MacConkey agar and eosin-methylene blue agar: *E. coli, E. aerogenes, S. typhimurium,* and *S. aureus.*

3. Incubate all plates in an inverted position for 24 to 48 hours at 37 degrees C.

1. Support the growth of all of organism

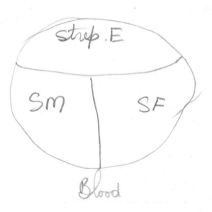

Physical Factors: Temperature

PURPOSE

To acquaint students with the diverse growth temperature requirements of bacteria.

PRINCIPLE

Bacteria, as a group of living organisms, are capable of growth within an overall temperature range of minus 5 degrees C to 80 degrees C. Each species, however, requires a narrower range that is determined by the heat sensitivity of its enzyme systems. Specific temperature ranges consist of the following **cardinal temperature points:**

1. **Minimum growth temperature:** the lowest temperature at which growth will occur. Below this temperature enzyme activity is inhibited and the cells are metabolically inactive so that growth is negligible or absent.

2. **Maximum growth temperature:** the highest temperature at which growth will occur. Above this temperature most cell enzymes are destroyed and the organism dies.

3. **Optimum growth temperature:** the temperature at which the rate of reproduction is most rapid; however, it is not necessarily optimum or ideal for other cell activities.

All bacteria can be classified into one of three major groups depending on their temperature requirements.

1. **Psychrophiles:** Bacterial species that will grow within a temperature range of minus 5 degrees C to 20 degrees C. The distinguishing characteristic of all psychrophiles is that they will grow between 0 and 5 degrees C.

2. **Mesophiles:** Bacterial species that will grow within a temperature range of 20 degrees C to 45 degrees C. The distinguishing characteristics of all mesophiles are their ability to grow at human body temperature (37 degrees C) and their inability to grow at temperatures above 45 degrees C. Included among the mesophiles are two distinct groups:

   a. Those whose optimum growth temperature is in the range of 20 degrees C to 30 degrees C are plant saprophytes.

   b. Those whose optimum growth temperature is in the range of 35 degrees C to 40 degrees C are organisms that prefer to grow in the bodies of warm-blooded hosts.

3. **Thermophiles:** Bacterial species that will grow at 35 degrees C and above. Two groups of thermophiles exist:

   a. **Facultative thermophiles:** Organisms that will grow at 37 degrees C, with an optimum growth temperature of 45 degrees C to 60 degrees C.

   b. **Obligate thermophiles:** Organisms that will only grow at temperatures above 50 degrees C, with optimum growth temperatures above 60 degrees C.

MATERIALS

Cultures

24-hour to 48-hour nutrient broth cultures of *Escherichia coli, Bacillus stearothermophilus,* and *Pseudomonas savastanoi.*

Media

Twelve nutrient agar slants per designated student group.

Equipment

Bunsen burner, inoculating loop, refrigerator set at 4 degrees C, and two incubators set at 40 degrees C and 60 degrees C.

PROCEDURE

1. Label each of the nutrient agar slants with the name of the test organism to be inoculated, the temperature of incubation (4, 20, 40, or 60 degrees C), and your initials.

2. Aseptically inoculate four nutrient agar slants labeled 4, 20, 40, and 60 degrees C by streaking the surface with *E. coli*.

3. Repeat step 2 for inoculation of *B. stearothermophilus* and *P. savastanoi*.

4. Incubate the cultures at each of the four experimental temperatures for 24 to 48 hours.

Physical Factors:
pH of the Extracellular Environment

PURPOSE

To acquaint students with the pH requirements of microorganisms.

PRINCIPLE

Growth and survival of microorganisms is greatly influenced by the pH of the environment, and all bacteria and other microorganisms differ as to their requirements. Each species has the ability to grow within a specific pH range, which may be broad or limited, with the most rapid growth occurring within a narrow optimum range. These specific pH needs reflect the organisms' adaptation to their natural environment. For example, enteric bacteria are capable of survival within a broad pH range, which is characteristic of their natural habitat, the digestive system. Bacterial blood parasites, on the other hand, can only tolerate a narrow range, as the pH of the circulatory system remains fairly constant at approximately 7.3.

Despite this diversity and the fact that certain organisms can grow at extremes of the pH scale, generalities can be made. The specific range for bacteria is between 4 and 9, with the optimum being 6.5 to 7.5. Fungi, molds, and yeasts prefer an acidic environment, with optimum activities at a pH of 4 to 6.

As a neutral or nearly neutral environment is generally advantageous to the growth of microorganisms, the pH of the laboratory medium is frequently adjusted to approximately 7. Metabolic activities of the microorganisms will result in the production of wastes, such as acids from carbohydrate degradation and alkali from protein breakdown, and these will cause shifts in pH that can be detrimental to growth.

To retard this shift, chemical substances that act as **buffers** are frequently incorporated when the medium is prepared. A commonly used **buffering** system involves the addition of equimolar concentrations of K_2HPO_4, a salt of a weak base, and KH_2PO_4, a salt of a weak acid. In a medium that has become acidic, the

K_2HPO_4 absorbs excess H^+ to form a weakly acidic salt and a potassium salt with the anion of the strong acid.

$$K_2HPO_4 \; + \; HCl \; \longrightarrow \; KH_2PO_4 \; + \; KCl$$

| **Salt of a weak base** | **Strong acid** | **Salt of a weak acid** | **Potassium chloride salt** |

In a medium that has become alkaline, KH_2PO_4 releases H^+ to form water by combining with the excess OH^-, and the remaining anionic portion of the weakly acidic salt combines with the cation of the alkali.

$$KH_2PO_4 \; + \; KOH \; \rightarrow \; K_2HPO_4 \; + \; H_2O$$

| **Salt of a weak acid** | **Strong base** | **Salt of a weak base** | **Water** |

Most media contain amino acids, peptones, and proteins, which because of their amphoteric nature, can act as natural buffers. For example, amino acids are zwitterions, molecules in which the amino group and the carboxyl group ionize to form dipolar ions. These behave in the following manner:

MATERIALS

Cultures

Saline suspensions of 24-hour nutrient broth cultures, adjusted to an O.D. of 0.05 at a wavelength of 600 mμ, of *Alcaligenes faecalis*, *Escherichia coli*, and *Saccharomyces cerevisiae*.

Media

Per designated student group: twelve trypticase soy broth, three at each of the following pH designations: 3, 5, 7, and 9. The pH is adjusted with 1N sodium hydroxide or 1N hydrochloric acid.

Equipment

Bunsen burner, sterile 1-ml pipettes, Bausch and Lomb Spectronic 20 spectrophotometer.

PROCEDURE

1. Label each of the trypticase soy broth tubes with pH of the medium, name of the organism to be inoculated, and your initials.

2. Using a sterile pipette, inoculate a series of the tubes, pH values of 3, 5, 7, and 9, with *E. coli* by adding 0.1 ml of the saline culture to each.

3. Repeat step 2 for the inoculation of *A. faecalis* and *S. cerevisiae*.

4. Incubate the *A. faecalis* and *E. coli* cultures for 24 to 48 hours at 37 degrees C, and the *S. cerevisiae* cultures for 48 to 72 hours at 25 degrees C.

Observations and Results

1. Using the spectrophotometer as described in Experiment 14, determine the optical density of all cultures and record the readings in the chart.

Microbial species	Optical density readings			
	pH 3	pH 5	pH 7	pH 9
E. coli				
A. faecalis				
S. cerevisiae				

2. Summarize your results as to the overall range and optimum pH of each organism studied in the chart.

Microbial species	pH range	Optimum pH
E. coli		
A. faecalis		
S. cerevisiae		

REVIEW QUESTIONS

1. Explain the mechanism by which buffers prevent radical shifts in pH.

2. Explain why it is necessary to incorporate buffers into media in which microorganisms are grown.

3. Why are proteins and amino acids considered to be natural buffers?

4. Explain why microorganisms differ in their pH requirements.

5. Will all microorganisms grow optimally at a neutral pH? _____ Explain.

Physical Factors:
Atmospheric Oxygen Requirements

PURPOSE

To acquaint students with the diverse atmospheric oxygen requirements of microorganisms.

PRINCIPLE

Microorganisms exhibit great diversity as to their ability to use free oxygen (O_2) for cellular respiration. These variations in O_2 requirements reflect the differences in biooxidative enzyme systems present in the various species. As such, they can be classified into one of four major groups according to their O_2 needs.

1. **Aerobes:** These require the presence of atmospheric oxygen for growth. Their enzyme system necessitates use of O_2 as the final hydrogen (electron) acceptor in the complete oxidative degradation of high-energy molecules such as glucose.

2. **Anaerobes:** They require the absence of free oxygen for growth, as their oxidative enzyme system requires the presence of molecules other than O_2 to act as the final hydrogen (electron) acceptor. In these organisms, the presence of atmospheric oxygen results in formation of hydrogen peroxide, H_2O_2, which is highly toxic and results in their death. This is a result of the sensitivity of some of their essential enzymes to H_2O_2 and absence of the enzyme catalase that decomposes H_2O_2 to water and oxygen.

3. **Facultative anaerobes:** These organisms can grow in the presence or absence of free oxygen. Their enzyme systems completely oxidize high-energy substrates in the presence of oxygen, and incompletely oxidize substrates in the absence of O_2.

4. **Microaerophiles:** They require limited amounts of atmospheric oxygen for growth. Oxygen in excess of the required amount appears to block the activities of their oxidative enzymes and results in death.

Some O_2 will grow below the surface

Oxygen needs of microorganisms can be determined by noting their growth distribution following a **shake-tube inoculation.** This procedure requires introduction of the inoculum into a melted agar medium, shaking of the test tube to disperse the microorganisms throughout the agar, and rapid solidification of the medium to ensure that the cells remain dispersed. Following incubation, the growth distribution indicates the organisms' oxygen requirement. Aerobes exhibit surface growth, whereas anaerobic growth is limited to the bottom of the deep tube. Facultative anaerobes, because of their indifference to the presence or absence of oxygen, exhibit growth throughout the medium. Microaerophiles grow in a zone slightly below the surface. Figure 18.1 on page 106 illustrates the shake-tube inoculation procedure and the distribution of growth following an appropriate incubation period.

MATERIALS

Cultures

24-hour to 48-hour nutrient broth cultures of *Staphylococcus aureus* and *Corynebacterium xerosis*; 48-hour to 72-hour Sabouraud broth cultures of *Saccharomyces cerevisiae,* and *Aspergillus niger;* and 48-hour thioglycollate broth cultures of *Peptostreptococcus anaerobius*.

Media

Five brain-heart infusion agar deep tubes per designated student group.

Equipment

Bunsen burner, water bath, ice-water bath, thermometer, and sterile Pasteur pipettes.

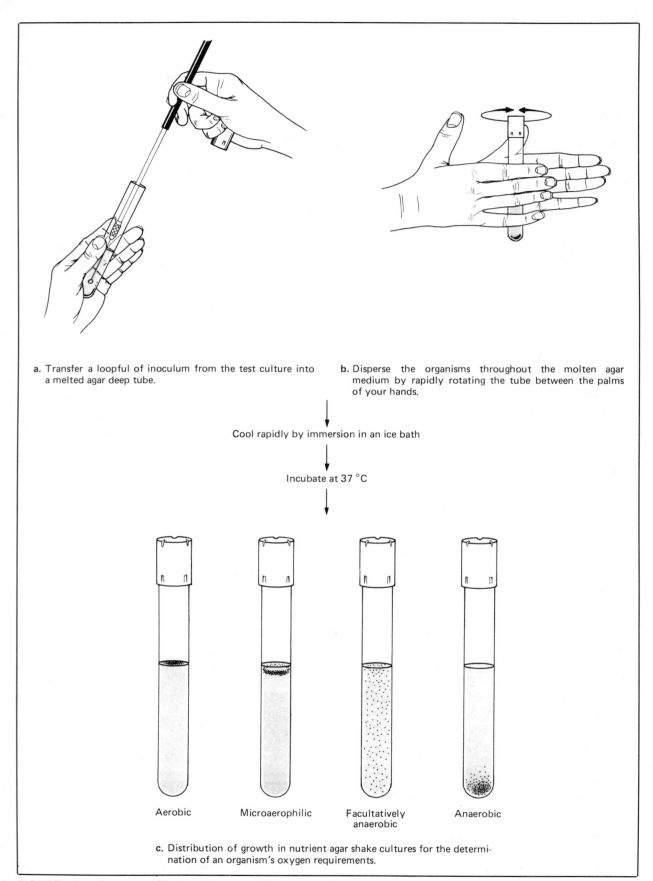

a. Transfer a loopful of inoculum from the test culture into a melted agar deep tube.

b. Disperse the organisms throughout the molten agar medium by rapidly rotating the tube between the palms of your hands.

Cool rapidly by immersion in an ice bath

Incubate at 37 °C

Aerobic   Microaerophilic   Facultatively anaerobic   Anaerobic

c. Distribution of growth in nutrient agar shake cultures for the determination of an organism's oxygen requirements.

FIGURE 18.1. Procedure for determination of oxygen requirements

PROCEDURE

1. Label the infusion agar tubes of media with the name of the organism to be inoculated and identify each with your initials.

2. Liquefy the sterile infusion agar by boiling in a water bath at 100 degrees C.

3. Cool molten agar to 45 degrees C by checking with a thermometer inserted into the water bath.

4. Using sterile technique, inoculate each experimental organism by introducing one drop of the culture from a sterile Pasteur pipette into the appropriately labeled tubes of molten agar.

5. Vigorously rotate the freshly inoculated molten infusion agar between the palms of the hands to distribute the organisms.

6. Place inoculated test tubes in an upright position into the ice-water bath to solidify the medium rapidly.

7. Incubate the *S. aureus, C. xerosis* and *P. anaerobius* cultures for 24 to 48 hours at 37 degrees C. and the *A. niger* and *S. cerevisiae* cultures for 48 to 72 hours at 25 degrees C.

Techniques for the Cultivation of Anaerobic Microorganisms

PURPOSE

To acquaint students with the methods for cultivation of anaerobic organisms.

PRINCIPLE

Microorganisms differ in their ability to use oxygen for cellular respiration. **Respiration** involves the oxidation of substrates for energy necessary to life. A substrate is **oxidized** when it loses a hydrogen ion and its electron H_e^+). Since the H_e^+ cannot remain free in the cell, it must immediately be picked up by an electron acceptor, which becomes **reduced**. Therefore reduction is the gain of the H_e^+. These are termed **oxidation-reduction (redox)** reactions. Some microorganisms have enzyme systems in which oxygen can serve

as an electron acceptor, thereby being reduced to water. These cells have high oxidation-reduction potentials; others have low potentials and must use other substances as electron acceptors.

The enzymatic differences in microorganisms are explained more fully in the section dealing with metabolism (see Part V). This discussion is limited to cultivation of the strict anaerobes, which cannot be cultivated in the presence of atmospheric oxygen (Fig. 19.1). The procedure is somewhat more difficult because it involves sophisticated equipment and media enriched with substances that lower the redox potential. Table 19.1 on page 112 shows some of the methods available for anaerobic cultivation.

The following experiment uses the pyrogallic acid-sodium hydroxide technique and the GasPak anaerobic system.

MATERIALS

Cultures

24-hour to 48-hour nutrient broth cultures of *Bacillus cereus, Escherichia coli,* and *Micrococcus luteus;* and 48-hour thioglycollate broth culture of *Clostridium butyricum.*

Media

Four cotton-plugged nutrient agar slants and four nutrient agar plates per designated student group.

Equipment

Bunsen burner, inoculating loop, rubber stoppers, glass rod, sterile Pasteur pipettes, pyrogallic acid, 4 percent sodium hydroxide, GasPak anaerobic system, wax pencil, and 10-ml pipette.

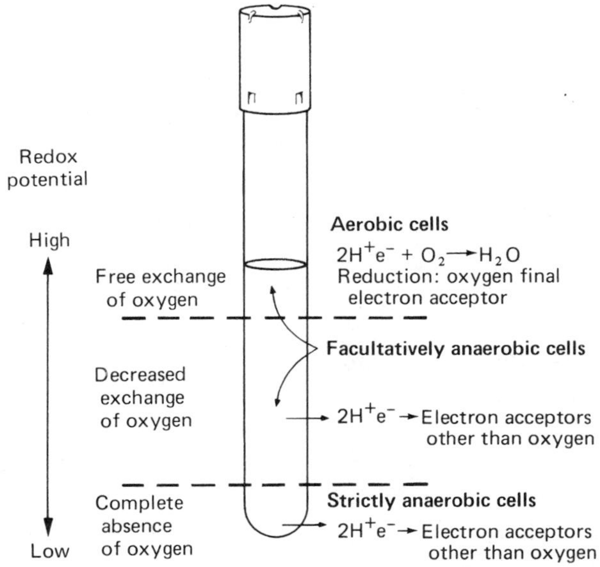

FIGURE 19.1. Illustration of redox potentials in a deep agar tube

Redox potential

High

Free exchange of oxygen

Decreased exchange of oxygen

Complete absence of oxygen

Low

Aerobic cells
$2H^+e^- + O_2 \rightarrow H_2O$
Reduction: oxygen final electron acceptor

Facultatively anaerobic cells

$2H^+e^- \rightarrow$ Electron acceptors other than oxygen

Strictly anaerobic cells
$2H^+e^- \rightarrow$ Electron acceptors other than oxygen

TABLE 19.1.   Methods for the cultivation of anaerobic microorganisms

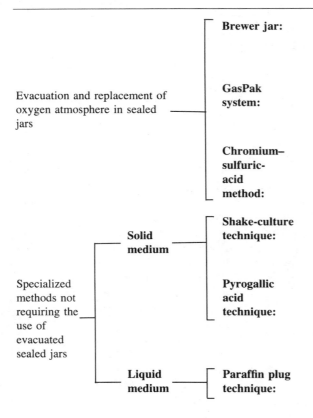

	Brewer jar:	High-vacuum pump evacuates O_2, which is replaced with a mixture of 95 percent $N_2 \uparrow$ + 5 percent $CO_2 \uparrow$. Platinum catalyst in jar lid results in binding of residual $O_2 \uparrow$ with $H_2 \uparrow$, causing formation of H_2O.
Evacuation and replacement of oxygen atmosphere in sealed jars	**GasPak system:**	Disposable $H_2 \uparrow$ + $CO_2 \uparrow$ envelope generator. Requires no evacuation of jar, no high-vacuum pumping equipment. Room-temperature catalyst that requires no electrical activation is used. Evolved $H_2 \uparrow$ reacts with O_2 to yield H_2O. (See Figure 19.2.)
	Chromium–sulfuric-acid method:	$H_2 \uparrow$ is generated in a desiccator jar following the reaction of 15 percent H_2SO_4 with chromium powder. $H_2SO_4 + Cr^{++} \longrightarrow CrSO_4 + H_2 \uparrow$. As $H_2 \uparrow$ is evolved, O_2 is forced out of desiccator jar and replaced with $H_2 \uparrow$.
Specialized methods not requiring the use of evacuated sealed jars — Solid medium	**Shake-culture technique:**	Molten and cooled nutrient agar is inoculated with loopful of organism. The tube is shaken, cooled rapidly, and incubated. Position of growth in tube is an index as to gaseous requirement of organism. (See Figure 18.1.)
	Pyrogallic acid technique:	Streak cultures on nutrient agar slants. Push a cotton plug into tube until it nearly touches slant. Fill space above cotton with pyrogallic acid crystals and add sodium hydroxide. Stopper tightly. Invert and incubate. Chemicals absorb O_2, producing anaerobic environment. (See Figure 19.3.)
Liquid medium	**Paraffin plug technique:**	Any medium containing reducing substances such as thioglycollate, brain-heart infusion, liver-veal, cystine, or ascorbic acid may be used. The medium is heated to drive off O_2, rapidly cooled, and inoculated with a loopful of culture. This is immediately sealed with a half-inch of melted paraffin and incubated.

PROCEDURE

Pyrogallic Acid-Sodium Hydroxide Technique

1. Label four cotton-plugged nutrient agar slants with the name of the organism to be inoculated and identify each with your initials.

2. Using sterile technique, inoculate each experimental organism by streaking the surface of each of the appropriately labeled agar slants.

3. Replace cotton plug and ignite it by passing it through the Bunsen burner. While cotton is burning, push it into tube with a glass rod until it nearly touches the slant. This will extinguish the flame.

4. Add sufficient pyrogallic acid crystals to fill the space between the cotton and the top of tube.

5. With a Pasteur pipette, add approximately 2 ml of 4 percent sodium hydroxide to the pyrogallic acid crystals and immediately stopper tightly with a rubber stopper.

6. Invert tubes and incubate for 24 to 48 hours at 37 degrees C.

GasPak Anaerobic Technique

1. With a wax pencil, divide the bottom of each nutrient agar plate into two sections.

2. Label each section on two plates with the name of the organism to be inoculated and identify plates with your initials.

3. Repeat step 2 to prepare a duplicate set of cultures.

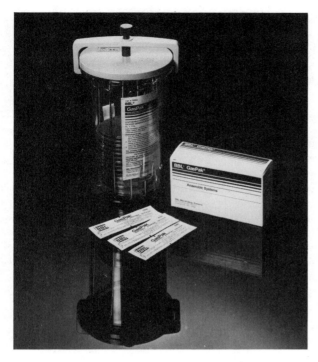

FIGURE 19.2. The GasPak system (Courtesy of BBL Microbiology Systems, Division of Becton Dickinson and Co.)

4. Using sterile technique, make a single line streak inoculation of each test organism in its respectively labeled section on both sets of plates.

5. Tear off the corner of the hydrogen and carbon dioxide gas generator and insert this inside the GasPak jar.

6. Place one set of plate cultures in an inverted position inside the GasPak chamber.

7. Expose the anaerobic indicator strip and place it inside the anaerobic jar so that the wick is visible from the outside.

8. With a pipette, add the required 10 ml of water to the gas generator and quickly seal the chamber with its lid.

9. Place the sealed jar in an incubator at 37 degrees C for 24 to 48 hours. After several hours of incubation observe the indicator strip for a color change to colorless, which is indicative of anaerobic conditions.

10. Incubate the duplicate set of plates in an inverted position for 24 to 48 hours at 37 degrees C under aerobic conditions.

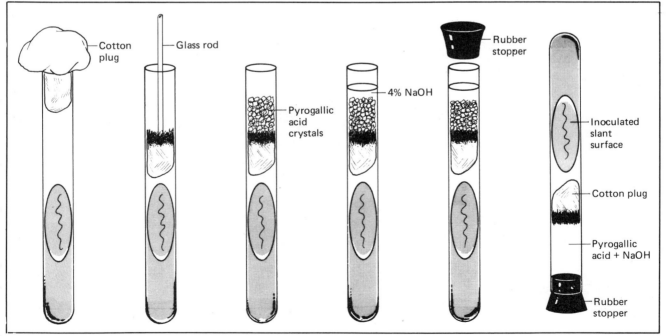

FIGURE 19.3. Pyrogallic acid-sodium hydroxide method

Observations and Results   EXPERIMENT **19**

1. Observe the pyrogallic acid-sodium hydroxide, GasPak system, and aerobically incubated plate cultures for the presence or absence of growth. Record your results in the chart.

2. Based on your observations, indicate the oxygen requirement classification of each test organism as strict anaerobe, facultative anaerobe, or aerobe.

Bacterial species	Pyrogallic acid-sodium hydroxide	GasPak anaerobic incubation	Incubator aerobic incubation	Oxygen requirement classification
M. luteus				
B. cereus				
E. coli				
C. butyricum				

REVIEW QUESTIONS

1. Why do anaerobic spores germinate quickly in the tissues of deep puncture wounds?

2. What is the function of the pyrogallic acid and sodium hydroxide in establishing anaerobic conditions?

3. Why can media such as brain-heart infusion and thioglycollate be used for the cultivation of anaerobes?

4. What is the purpose of the indicator strip and the gas generator in the GasPak system?

The Bacterial Growth Curve

PURPOSES

1. To acquaint students with the population growth dynamics of bacterial cultures.

2. To plot a growth curve and determine the generation time of bacterial cultures.

PRINCIPLE

Bacterial population growth studies require inoculation of viable cells into a sterile broth medium and incubation of the culture under optimum temperature, pH, and gaseous conditions. Under these conditions, the cells will reproduce rapidly and the dynamics of the microbial growth can be charted by means of a population growth curve, which is constructed by plotting the increase in cell numbers versus time of incubation. The curve can be used to delineate stages of the growth cycle. It also facilitates measurement of cell numbers and the rate of growth of a particular organism under standardized conditions as expressed by its **generation time,** the time required for a microbial population to double.

The stages of a typical growth curve (Figure 20.1) are:

1. **Lag phase:** During this stage the cells are adjusting to their new environment. Cellular metabolism is accelerated, resulting in rapid biosynthesis of cellular macromolecules, primarily enzymes, in preparation for the next phase of the cycle. Although the cells are increasing in size, there is no cell division and therefore no increase in numbers.

2. **Logarithmic (log) phase:** Under optimum nutritional and physical conditions the physiologically robust cells reproduce at a uniform and rapid rate by binary fission. Thus there is a rapid exponential increase in population, which doubles regularly until a maximum number of cells is reached. The time required for the population to double is the generation time. The length of the log phase varies, depending on the organisms and the composition of the medium. The average may be estimated to last 6 to 12 hours.

3. **Stationary phase:** During this stage the number of cells undergoing division is equal to the number of cells that are dying. Therefore there is no further increase in cell number and the population is maintained at its maximum level for a period of time. The primary factors responsible for this phase are the exhaustion of some essential metabolites and the accumulation of toxic acidic or alkaline end products in the medium.

4. **Decline or death phase:** Because of the continuing depletion of nutrients and buildup of metabolic wastes, the microorganisms die at a rapid and uniform rate. The decrease in population closely parallels its increase during the log phase. Theoretically, the entire population should die during a time interval equal to that of the log phase. This does not occur, however, since a small number of highly resistant organisms persists for an indeterminate length of time.

Construction of a complete bacterial growth curve requires that aliquots of a 24-hour shake-flask culture be measured for population size at intervals during the incubation period. Such a procedure does not lend itself to a regular laboratory session. Therefore this experiment follows a modified procedure designed to demonstrate only the lag and log phases. The curve will be plotted on semilog paper by using two values for the

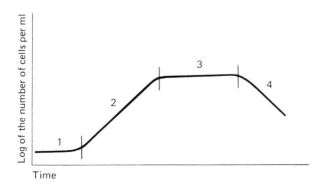

FIGURE 20.1. Population growth curve

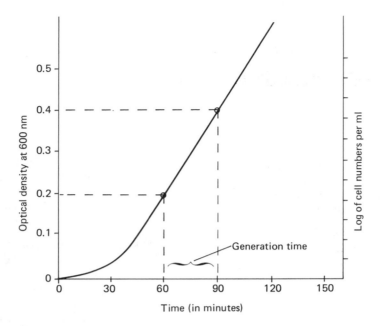

FIGURE 20.2. Indirect method of determining generation time

measurement of growth. The direct method requires enumeration of viable cells in serially diluted samples of the test culture taken at 30-minute intervals as described in Experiment 13. The indirect method uses spectrophotometric measurement of the developing turbidity at the same 30-minute intervals as an index of increasing cellular mass.

Students will determine generation time with indirect and direct methods by using data on the growth curve. Indirect determination is made by simple extrapolation from the log phase as illustrated in Figure 20.2. Select two points on the optical density scale, such as 0.2 and 0.4, that represent a doubling of turbidity. Using a ruler, extrapolate by drawing a line between each of the selected optical densities on the ordinate and the plotted line of the growth curve. Then draw perpendicular lines from these end points on the plotted line of the growth curve to their respective time intervals on the abscissa. With this information, determine the generation time as follows:

$$GT = t_{(O.D.\ 0.4)} - t_{(O.D.\ 0.2)}$$
$$GT = 90 \text{ minutes} - 60 \text{ minutes} = 30 \text{ minutes}$$

The direct method uses the log of cell number scale on the growth curve and the following formula:

$$g = \frac{t \log 2}{\log b - \log B}$$

g = generation time;
B = number of bacterial cells at the beginning of the log phase, or at some point during the log phase;
b = number of bacterial cells at the end of the log phase;
t = time in hours or minutes between B and b.

MATERIALS

Cultures

10-hour to 12-hour (log phase) brain-heart infusion broth cultures of *Escherichia coli*. Cultures may be maintained in log phase by immersion in an ice-water bath.

Media

Per designated student group: 100 ml of brain-heart infusion in a 250-ml Erlenmeyer flask; twenty-one 99-ml sterile water blanks; and three 100-ml bottles of nutrient agar.

Equipment

37 degrees C water bath shaker incubator, Bausch and Lomb Spectronic 20 spectrophotometer, 13 × 100 mm

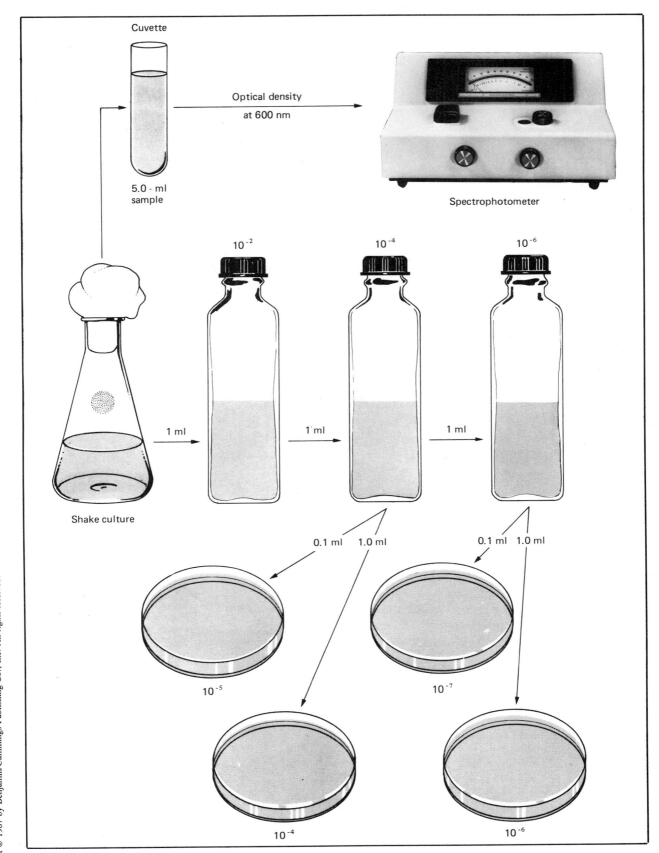

Cuvette

Optical density
at 600 nm

5.0 - ml
sample

Spectrophotometer

10^{-2} 10^{-4} 10^{-6}

1 ml 1 ml 1 ml

Shake culture

0.1 ml 1.0 ml   0.1 ml 1.0 ml

10^{-5} 10^{-7}

10^{-4} 10^{-6}

FIGURE 20.3. Spectrophotometric and dilution-plating procedure for use in bacterial growth curves

cuvettes, Quebec colony counter, twenty-eight Petri dishes, 1-ml and 10-ml sterile pipettes, 1000-ml beaker, and Bunsen burner.

PROCEDURE

1. Separate the 21 99-ml sterile water blanks into 7 sets of 3 water blanks each. Label each set as to time of inoculation (t_0, t_{30}, t_{60}, t_{90}, t_{120}, t_{150}, t_{180}) and the dilution to be effected in each water blank (10^{-2}, 10^{-4}, 10^{-6}).

2. Label seven sets of Petri dishes as to time of inoculation, dilution to be plated (10^{-4}, 10^{-5}, 10^{-6}, 10^{-7}) and identify each with your initials.

3. Liquefy the three bottles of nutrient agar in a water bath. Cool and maintain at 45 degrees C.

4. With a sterile pipette, add approximately 5 ml of the log phase $E.\ coli$ culture to the flask containing 100 ml of brain-heart infusion broth that has been labeled with your initials. The approximate initial O.D (t_0) should be 0.08 to 0.1 at 600 mμ. Refer to Experiment 14 for proper use of the spectro-photometer.

5. After the t_0 O.D has been determined, shake the culture flask and aseptically transfer 1 ml to the 99-ml water blank labeled t_0 10^{-2} and continue to dilute serially to 10^{-4} and 10^{-6}.

6. Place the culture flask in a water bath shaker set at 120 rpm at 37 degrees C and time for the required 30-minute intervals.

7. Plate the t_0 dilutions on the appropriately labeled t_0 plates as shown in Figure 20.3, on page 119. Aseptically pour 15 ml of the molten agar into each plate and mix by gentle rotation.

8. Thereafter, at each 30-minute interval, aseptically transfer a 5-ml aliquot of the culture to a cuvette and determine its optical density. Also, aseptically transfer a 1-ml aliquot of the culture into the 10^{-2} water blank of the set labeled with the appropriate time, complete the serial dilution, and plate in the respectively labeled Petri dishes.

9. When the pour-plate cultures harden, incubate them in an inverted position for 24 hours at 37 degrees C.

Observations and Results

1. Perform cell counts on all plates as described in Experiment 13.

2. Record the optical densities and corresponding cell counts in the chart.

Incubation time (minutes)	Optical density @ 600 mμ	Plate counts cells/ml	Log of cells/ml
0			
30			
60			
90			
120			
150			
180			

3. On the provided semilog paper, plot:

   a. Optical densities on the ordinate and incubation times on the abscissa.

   b. Log of the cell numbers on the ordinate and incubation times on the abscissa.
   On both graphs, with a ruler connect the points by drawing the best line between the plotted points. The log phase is represented by the straight line portion.

4. Calculate the generation time for this culture by the direct method using the mathematical formula, and by the indirect method extrapolating from the O.D. scale on the plotted curve. Show calculations and record the generation time.

   GENERATION TIME
      Direct method:

      Indirect method:

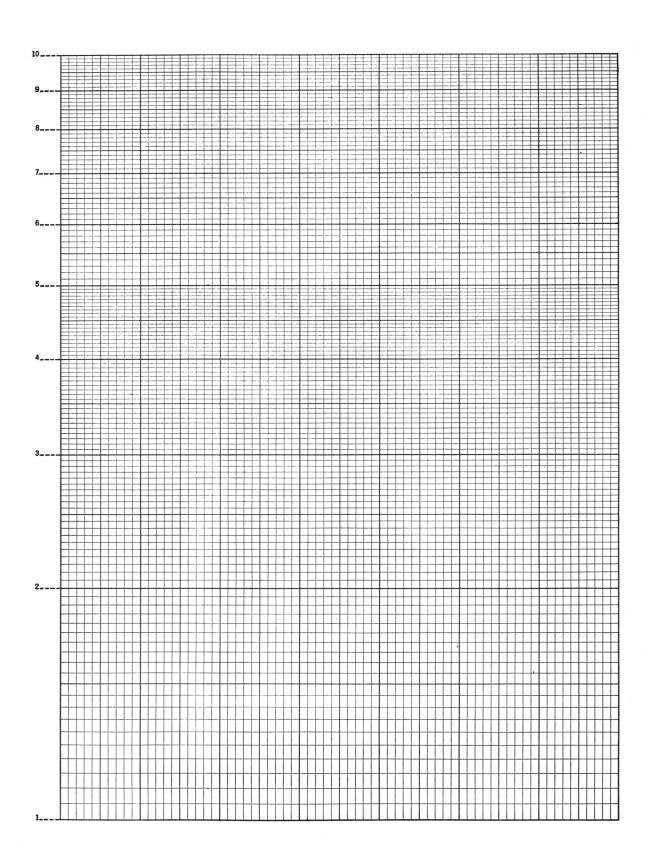

Optical Density versus Incubation Time

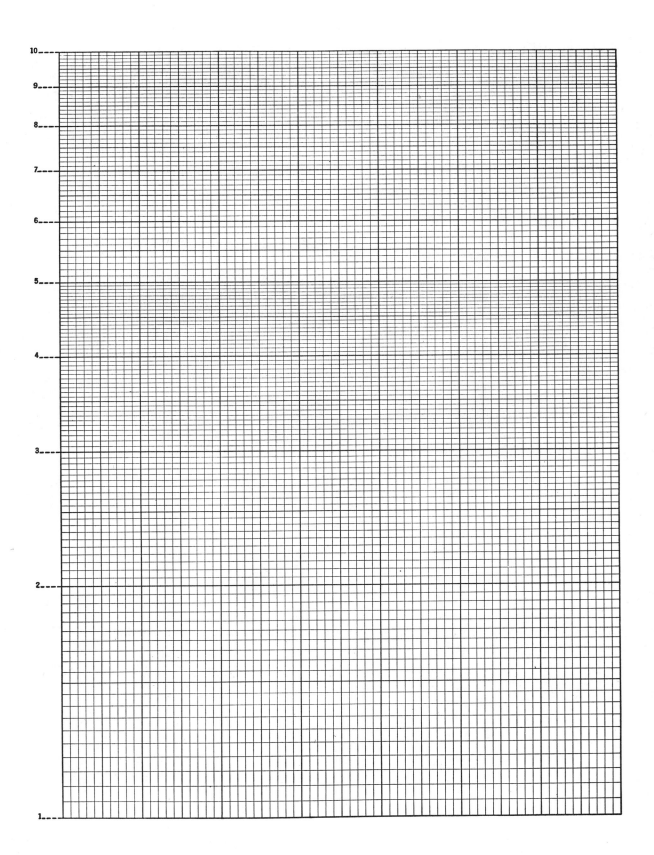

10 ——
9 ——
8 ——
7 ——
6 ——
5 ——
4 ——
3 ——
2 ——
1 ——

Cell Numbers (CFUs) versus Incubation Time

REVIEW QUESTIONS

1. Explain why no increase in population size can be observed during the lag phase.

2. Cite the factors responsible for the stationary phase.

3. Is generation time a useful parameter to indicate the types of media best suited to support the growth of a specific organism? _____ Explain.

4. Why do variations in generation time exist among different species of microorganisms?

5. Why do variations in generation time exist within a single species?

6. Does the term *growth* convey the same meaning when applied to bacteria and to multicellular organisms? _____. Explain.

V

Biochemical Activities of Microorganisms

PURPOSES

This section is designed to instruct students in:

1. The nature and activities of exoenzymes and endoenzymes.

2. Experimental procedures for differentiation of enteric microorganisms.

3. Biochemical test procedures for identification of microorganisms.

INTRODUCTION

Microorganisms must be separated and identified for a wide variety of reasons, such as:

1. Determination of pathogens responsible for infectious diseases.

2. Selection and isolation of strains of fermentative microorganisms necessary for the industrial production of alcohols, solvents, vitamins, organic acids, antibiotics, and industrial enzymes.

3. Isolation and development of suitable microbial strains necessary for the manufacture and the enhancement of quality and flavor in certain food materials such as yogurt, cheeses, and milk products.

4. Comparison of biochemical activities for taxonomic purposes.

To accomplish these tasks, the microbiologist is assisted by the fact that, just as human beings possess a characteristic and specific set of fingerprints, microorganisms all have their own identifying biochemical characteristics. These so-called biochemical fingerprints are the properties controlled by the cells' enzymatic activity, and are responsible for bioenergetics, biosynthesis, and biodegradation.

The sum total of all these chemical reactions is defined as **cellular metabolism,** and the biochemical transformations that occur both outside and inside the cell are governed by biological catalysts called **enzymes.**

EXTRACELLULAR ENZYMES (exoenzymes) These act on substances outside of the cell. Most high-molecular-weight substances are not able to pass through cell membranes and therefore

125

these raw materials—foodstuffs such as polysaccharides, lipids, and proteins—must be degraded to low-molecular-weight materials—nutrients—before they can be transported into the cell. Because of the reactions involved, exoenzymes are mainly **hydrolytic enzymes** that reduce high-molecular-weight materials into their building blocks by introducing water into the molecule. This liberates smaller molecules, which may then be transported into the cell and assimilated.

INTRACELLULAR ENZYMES (endoenzymes) These enzymes function inside the cell and are mainly responsible for synthesis of new protoplasmic requirements and production of cellular energy from assimilated materials. The ability of cells to act on nutritional substrates permeating cell membranes indicates the presence of many endoenzymes capable of transforming the chemically specific substrates into essential materials.

This transformation is necessary for cellular survival and function, and is the basis of cellular metabolism. As a result of these metabolic processes metabolic products are formed and excreted by the cell into the environment. Assay of these end products not only aids in identification of specific enzyme systems, but also serves to identify, separate, and classify microorganisms. The following represents a simplified schema of experimental procedures used to acquaint students with the intracellular and extracellular enzymatic activities of microorganisms.

Biochemical Activities of Microorganisms

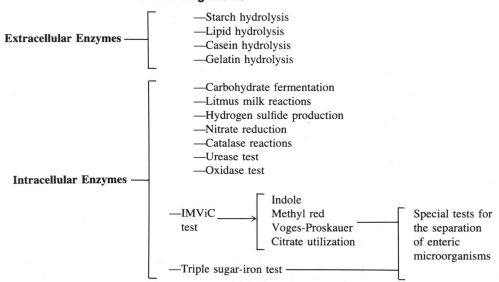

The experiments to be carried out in this section can be performed in either of two ways. A short version uses a limited number of organisms to illustrate the possible end product(s) that may result from enzyme action on a substrate. The organisms for this version are designated in the individual exercises.

The alternative, or long, version involves the use of thirteen microorganisms. This version provides a complete overview of the biochemical fingerprints of the organisms and supplies the format for their separation and identification. These organisms were chosen to serve as a basis for the identification of an unknown microorganism in Experiment 31. If this alternative version is selected, the following organisms are recommended for use:

Escherichia coli	*Proteus vulgaris*	*Staphylococcus aureus*
Enterobacter aerogenes	*Pseudomonas aeruginosa*	*Bacillus cereus*
Klebsiella pneumoniae	*Alcaligenes faecalis*	*Corynebacterium xerosis*
Shigella dysenteriae	*Micrococcus luteus*	
Salmonella typhimurium	*Streptococcus lactis*	

Extracellular Enzymatic Activities of Microorganisms

PURPOSE

To enable students to determine the ability of micro-organisms to excrete hydrolytic extracellular enzymes capable of degrading the polysaccharide starch, the lipid tributyrin, and the proteins casein and gelatin.

PRINCIPLE

Because of their large sizes, high-molecular-weight nutrients such as polysaccharides, lipids, and proteins are not capable of permeating the cell membrane. These macromolecules must first be hydrolyzed by specific extracellular enzymes into their respective basic building blocks. These low-molecular-weight substances can then be transported into the cells and used for the synthesis of protoplasmic requirements and energy production. The following procedures are designed to investigate the exoenzymatic activities of different microorganisms.

Starch Hydrolysis

Starch is a high-molecular-weight, branching polymer composed of **glucose** molecules linked together by **glycosidic bonds.** The degradation of this macromolecule first requires the presence of the extracellular enzyme **amylase** for its hydrolysis into shorter polysaccharides, namely **dextrins**, and ultimately into **maltose** molecules. The final hydrolysis of this disaccharide, which is catalyzed by **maltase**, yields low-molecular-weight, soluble **glucose** molecules that can be transported into the cell and used for energy production through the process of glycolysis.

In this experimental procedure, starch agar is used to demonstrate the hydrolytic activities of these exoenzymes. The medium is composed of nutrient agar supplemented with starch, which serves as the polysaccharide substrate. The detection of the hydrolytic activity following the growth period is made by performing the starch test to determine the presence or absence of starch

in the medium. Starch in the presence of iodine will impart a blue-black color to the medium, indicating the absence of starch-splitting enzymes and representing a negative result. If the starch has been hydrolyzed, a clear zone of hydrolysis will surround the growth of the organism. This is a positive result.

Lipid Hydrolysis

Lipids are high-molecular-weight compounds possessing large amounts of energy. The degradation of lipids such as **triglycerides** is accomplished by hydrolyzing extracellular enzymes, called **lipases**, (esterases) that cleave the **ester bonds** in this molecule by the addition of water to form the building blocks **glycerol** and **fatty acids**. The following illustration shows this reaction:

$$
\begin{array}{l}
\text{CH}_2\text{—O—}\overset{\displaystyle\overset{O}{\|}}{\text{C}}\text{—R} \\[6pt]
\text{CH—O—}\overset{\displaystyle\overset{O}{\|}}{\text{C}}\text{—R}' \quad + 3\text{H}_2\text{O} \\[6pt]
\text{CH}_2\text{—O—}\overset{\displaystyle\overset{O}{\|}}{\text{C}}\text{—R}''
\end{array}
\xrightarrow{\text{Lipase}}
\begin{array}{l}
\text{CH}_2\text{OH} + \text{RCOOH} \\[6pt]
\text{CHOH} + \text{R}^1\text{COOH} \\[6pt]
\text{CH}_2\text{OH} + \text{R}''\text{COOH}
\end{array}
$$

Triglyceride   **Glycerol**   **Fatty acids**

Once assimilated into the cell, these basic lipid components can be further metabolized through the Krebs cycle, a series of reactions in which aerobic respiration produces cellular energy that can be used in the form of adenosine triphosphate (ATP). The components may also enter other metabolic pathways for the synthesis of the cell's protoplasmic requirements.

In this experimental procedure, tributyrin agar is used to demonstrate the hydrolytic activities of these exoenzymes. The medium is composed of nutrient agar supplemented with the triglyceride tributyrin as the lipid substrate. Tributyrin forms an emulsion when dispensed in the agar, producing an opaque medium that is necessary for observing exoenzymatic activity.

Following inoculation and incubation of the agar plate cultures, organisms excreting lipase will show a zone of **lipolysis,** which is demonstrated by a clear area surrounding the bacterial growth. This loss of opacity is the result of the hydrolytic reaction yielding soluble glycerol and fatty acids, and represents a positive reaction for lipid hydrolysis. In the absence of lipolytic enzymes, the medium retains its opacity. This is a negative reaction.

Casein Hydrolysis

Casein, the major milk protein, is a macromolecule composed of **amino acid** subunits linked together by **peptide bonds** (CO-NH). Before their assimilation into the cell, proteins must undergo step-by-step degradation into **peptones, polypeptides, dipeptides,** and ultimately into their building blocks, **amino acids.** This process is called peptonization, or **proteolysis,** and is mediated by extracellular enzymes called **proteases.** The function of these proteases is to cleave the peptide bond CO-NH by introducing water into the molecule. The reaction then liberates the amino acids, as illustrated:

Polypeptide chain

$$NH_2-RCH-CO-NH-RCH-CO-NH-RCH-etc...$$

Peptide bond

Protease + HOH

HOH

OH⁻   H⁺

$$NH_2-RCH-COOH + NH_2-RCH-COOH$$

The liberated low-molecular-weight, soluble amino acids can now be transported through the cell membrane into the intracellular amino acid pool for use in the synthesis of structural and functional cellular proteins.

In this experimental procedure, milk agar is used to demonstrate the hydrolytic activity of these exoenzymes. The medium is composed of nutrient agar supplemented with milk that contains the protein substrate casein. Similar to other proteins, milk protein is a colloidal suspension that gives the medium its color and opacity, as it deflects light rays rather than transmitting them.

Following inoculation and incubation of the agar plate cultures, organisms secreting proteases will exhibit a zone of proteolysis, which is demonstrated by a clear area surrounding the bacterial growth. This loss of opacity is the result of a hydrolytic reaction yielding soluble, noncolloidal amino acids, and represents a positive reaction. In the absence of protease activity, the medium surrounding the growth of the organism remains opaque, which is a negative reaction.

Gelatin Hydrolysis

Although the value of gelatin as a nutritional source is questionable (it is an incomplete protein, lacking the essential amino acid tryptophan), its value in identifying bacterial species is well established. Gelatin is a protein produced by hydrolysis of collagen, a major component of connective tissue and tendons in humans and other animals. Below temperatures of 25 degrees C gelatin will maintain its gel properties and exist as a solid; at temperatures above 25 degrees C gelatin is liquid.

Liquefaction is accomplished by some microorganisms capable of producing a proteolytic extracellular enzyme called **gelatinase,** which acts to hydrolyze this protein to **amino acids.** Once this degradation occurs, even very low temperatures of 4 degrees C will not restore the gel characteristic.

In this experimental procedure, nutrient gelatin deep tubes are used to demonstrate the hydrolytic activity of this enzyme. The medium consists of nutrient broth supplemented with 12 percent gelatin. This high gelatin concentration results in a stiff medium and also serves as the substrate for the activity of gelatinase.

Following inoculation and incubation, the cultures are placed in a refrigerator at 4 degrees C for 30 minutes. Cultures that remain liquefied produce gelatinase and demonstrate gelatin hydrolysis. Cultures that solidify on refrigeration lack gelatinase and give negative reactions.

MATERIALS

Cultures

24-hour to 48-hour trypticase soy broth cultures of *Escherichia coli, Bacillus cereus,* and *Pseudomonas aeruginosa* for the short version.

24-hour to 48-hour trypticase soy broth cultures of the 13 previously listed organisms for the long version.

Media

Short version: One plate each of starch agar, tributyrin agar, and milk agar and three nutrient gelatin deep tubes per designated student group. Long version: Four plates each of starch agar, tributyrin agar, and milk agar and fourteen nutrient gelatin deep tubes per designated student group.

Reagent

Gram's iodine solution.

Equipment

Bunsen burner, inoculating loop and needle, wax pencil, and refrigerator.

PROCEDURE

1. Prepare the starch agar, tributyrin agar, and milk agar plates for inoculation as follows:

   a. Short procedure: With a wax pencil divide the bottom of each Petri dish into three sections. Label the sections as *E. coli*, *B. cereus*, and *P. aeruginosa*, respectively. Place your initials on the cover of each dish.

   b. Long procedure: Repeat step 1a, dividing three plate bottoms into three sections and one plate bottom into four sections of each of the required media to accommodate the 13 test organisms.

2. Using sterile technique, make a single line streak inoculation of each test organism on the agar surface of its appropriately labeled section on each of the agar plates.

3. Label each of the nutrient gelatin deep tubes with the name of the bacterial organism to be inoculated and identify each one with your initials.

4. Using sterile technique, inoculate each experimental organism into its appropriately labeled deep tube by means of a stab inoculation.

5. Incubate all the tubes and inverted plates for 24 to 48 hours at 37 degrees C.

Observations and Results

Starch Hydrolysis

1. Flood the starch agar plate cultures with Gram's iodine solution, allow the iodine to remain in contact with the medium for 30 seconds, and pour off the excess.

2. Examine the cultures for the presence or absence of a blue-black color surrounding the growth of each test organism. Record your results in the chart.

3. Based on your observations, determine and record which organisms were capable of hydrolyzing the starch.

4. Refer to the color plate for illustrations of these reactions.

Lipid Hydrolysis

1. Examine the tributyrin agar plate cultures for the presence or absence of a clear area, or zone of lipolysis, surrounding the growth of each of the organisms. Record your results in the chart.

2. Based on your observations, determine and record which organisms were capable of hydrolyzing the lipid.

3. Refer to the color plate for illustrations of these reactions.

Bacterial species	Starch hydrolysis		Tributyrin Hydrolysis	
	Appearance of medium	Result (+) or (−)	Appearance of Medium	Result (+) or (−)
E. coli				
E. aerogenes				
K. pneumoniae				
S. dysenteriae				
S. typhimurium				
P. vulgaris				
P. aeruginosa				
A. faecalis				
M. luteus				
S. lactis				
S. aureus				
B. cereus				
C. xerosis				
Control				

Casein Hydrolysis

1. Examine the milk agar plate cultures for the presence or absence of a clear area or zone of proteolysis, surrounding the growth of each of the bacterial test organisms. Record your results in the chart.

2. Based on your observations, determine and record which of the organisms were capable of hydrolyzing the milk protein casein.

Gelatin Hydrolysis

1. Place all gelatin deep tube cultures into a refrigerator at 4 degrees C for 30 minutes.

2. Examine all the cultures to determine whether the medium is solid or liquid. Record your results in the chart.

3. Based on your observations, determine and record which of the organisms were capable of hydrolyzing gelatin.

Bacterial species	Casein hydrolysis		Gelatin Hydrolysis	
	Appearance of medium	Result (+) or (−)	Appearance of Medium	Result (+) or (−)
E. coli				
E. aerogenes				
K. pneumoniae				
S. dysenteriae				
S. typhimurium				
P. vulgaris				
P. aeruginosa				
A. faecalis				
M. luteus				
S. lactis				
S. aureus				
B. cereus				
C. xerosis				
Control				

Carbohydrate Fermentation

PURPOSE

To enable students to determine the ability of microorganisms to degrade and ferment carbohydrates with the production of an acid or acid and gas.

PRINCIPLE

Most microorganisms obtain their energy through a series of orderly and integrated enzymatic reactions leading to the biooxidation of a substrate, frequently a carbohydrate. The major pathways by which this is accomplished are shown opposite.

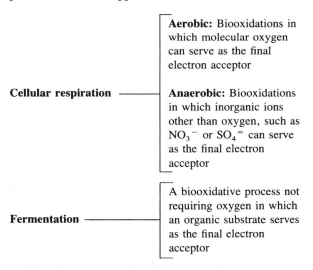

Cellular respiration —

Aerobic: Biooxidations in which molecular oxygen can serve as the final electron acceptor

Anaerobic: Biooxidations in which inorganic ions other than oxygen, such as NO_3^- or $SO_4^=$ can serve as the final electron acceptor

Fermentation —

A biooxidative process not requiring oxygen in which an organic substrate serves as the final electron acceptor

Organisms use carbohydrate differently depending on their enzyme complement. Some organisms are capable of fermenting sugars such as glucose anaerobically, while others use the aerobic pathway. Still others, facultative anaerobes, are enzymatically competent to use both aerobic and anaerobic pathways, and some organisms lack the ability to oxidize glucose by either. In this exercise the fermentative pathways are of prime concern.

In fermentation, substrates such as carbohydrates and alcohols undergo anaerobic dissimilation and produce an organic acid (for example, lactic, formic, or acetic acid) that may be accompanied by gases such as hydrogen or carbon dioxide. Facultative anaerobes are usually the so-called fermenters of carbohydrates. Fermentation is best described by considering the degradation of glucose by way of the **Embden-Meyerhof pathway,** also known as the **glycolytic pathway,** illustrated in Figure 22.1 on page 134.

As the diagram shows, one mole of glucose is converted to two moles of pyruvic acid, which is the major intermediate compound produced by glucose degradation. Subsequent metabolism of pyruvate may not be the same for all organisms, and a variety of end products may result that define their different fermentative capabilities. This can be seen in Figure 22.2 on page 134.

Fermentative degradation under anaerobic conditions is carried out in a fermentation broth containing a Durham tube, an inverted inner vial for the detection of gas production as illustrated. A typical carbohydrate fermentation medium contains:

1. Nutrient broth ingredients for the support of the growth of all organisms.

2. A specific carbohydrate that serves as the substrate for determining the organism's fermentative capabilities.

3. The pH indicator phenol red, which is red at a neutral pH (7) and changes to yellow at a slightly acidic pH of 6.8, suggesting that slight amounts of acid will cause a color change.

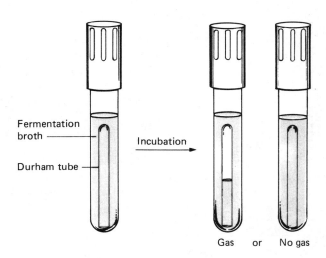

Fermentation broth

Durham tube

Incubation →

Gas   or   No gas

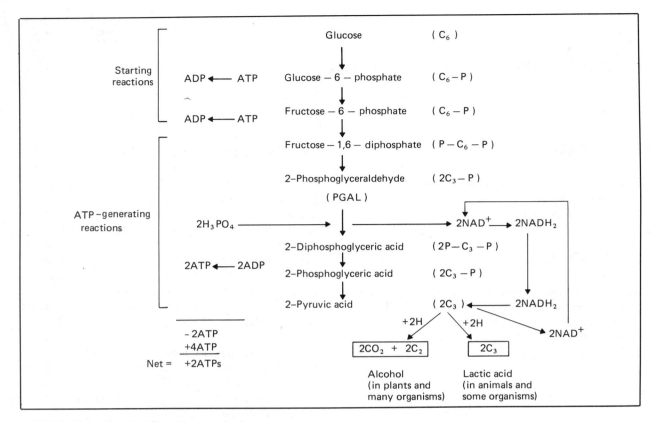

FIGURE 22.1. The Embden-Meyerhof scheme

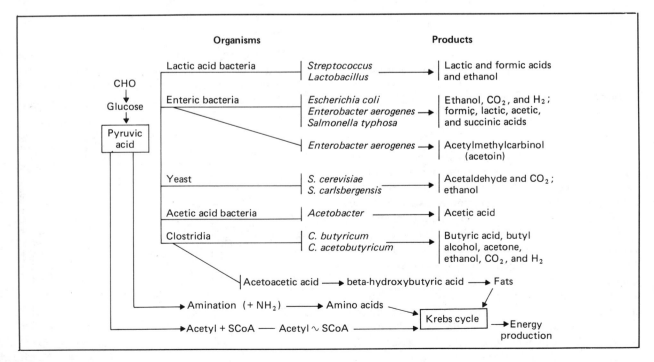

FIGURE 22.2. Variations in use of pyruvic acid

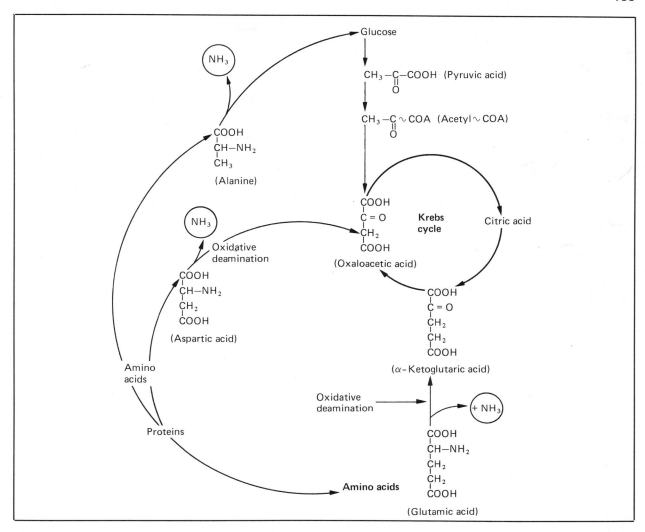

FIGURE 22.3 Proteins as energy sources for microbes

The critical nature of the fermentation reaction and the activity of the indicator make it imperative that all cultures should be observed within 48 hours. Extended incubation may mask acid-producing reactions by production of alkali because of enzymatic action on substrates other than the carbohydrate.

Following incubation, carbohydrates that have been fermented and produce acidic wastes will cause the phenol red to turn yellow, thereby indicating a positive reaction. In some cases, acid production is accompanied by the evolution of a gas (CO_2) that will be visible as a bubble in the inverted tube. Cultures that are not capable of fermenting a carbohydrate substrate will not change the indicator and the tubes will appear red; there will not be a concomitant evolution of gas. This is a negative reaction.

The lack of carbohydrate fermentation by some organisms should not be construed as absence of growth. The organisms use other nutrients in the medium as energy sources. Among these nutrients are peptones present in nutrient broth. Peptones can be degraded by microbial enzymes to amino acids that are in turn enzymatically converted by oxidative deamination to keto-amino acids. These are then metabolized through the Krebs cycle for energy production. These reactions liberate ammonia, which accumulates in the medium, forming ammonium hydroxide (NH_4OH) and producing an alkaline environment. When this occurs, the phenol red turns to a deep red in the now basic medium. This alternative pathway of aerobic respiration is illustrated in Figure 22.3.

MATERIALS

Cultures

24-hour to 48-hour trypticase soy broth cultures of *Escherichia coli, Alcaligenes faecalis, Salmonella ty-*

phimurium, and *Staphylococcus aureus* for the short version. 24-hour to 48-hour trypticase soy broth cultures of the 13 previously listed organisms for the long version.

Media

Phenol red lactose, dextrose, and sucrose broths per designated student group: five of each for short version, fourteen of each for long version.

Equipment

Bunsen burner and inoculating loop.

PROCEDURE

1. Label each of the specified tubes of media with the name of the bacterial organism to be inoculated and identify each with your initials.

2. Using sterile technique, inoculate each experimental organism into its appropriately labeled medium by means of loop inoculation. Care should be taken during this step not to shake the fermentation tube, as this may accidentally force a bubble of air into the inverted gas vial, displacing the medium and possibly rendering a false positive result.

3. Incubate all of the tubes for 24 to 48 hours at 37 degrees C.

Triple Sugar-Iron Agar Test

PURPOSE

To demonstrate a rapid screening procedure that will:

1. Differentiate among members of the Enterobacteriaceae.

2. Distinguish between the Enterobacteriaceae and other groups of intestinal bacilli.

PRINCIPLE

The **triple sugar-iron (TSI) agar test** is designed to differentiate among the different groups or genera of the Enterobacteriaceae, which are all gram-negative bacilli capable of fermenting glucose with the production of acid; and to distinguish the Enterobacteriaceae from other gram-negative intestinal bacilli. This differentiation is made on the basis of differences in carbohydrate fermentation patterns and hydrogen sulfide production by the various groups of intestinal organisms.

To observe carbohydrate utilization patterns, the TSI agar slants contain lactose and sucrose in 1 percent concentrations, and glucose in a concentration of 0.1 percent, which allows for detection of the utilization of this substrate only. The acid-base indicator phenol red is also incorporated to detect carbohydrate fermentation that is indicated by a change in color of the medium from orange-red to yellow in the presence of acids. Following a stab-and-streak inoculation and incubation of 18 to 24 hours, determination of carbohydrate fermentative activities is made as follows:

1. **Alkaline slant (red) and acid butt (yellow) with or without gas production (breaks in the agar butt):** Only glucose fermentation has occurred. The organisms preferentially degrade glucose first. Since this substrate is present in a minimal concentration, the small amount of acid on the slant surface

is oxidized rapidly, resulting in an alkaline reaction. Also, the peptones in the medium are used with the further production of alkali. In the butt the acid reaction is maintained due to reduced oxygen tension and slower growth of the organisms.

2. **Acid slant (yellow) and acid butt (yellow) with or without gas production.** Lactose and/or sucrose fermentation has occurred. Since these substances are present in higher concentrations they serve as substrates for continued fermentative activities with maintenance of an acid reaction in both slant and butt.

3. **Alkaline slant (red) and alkaline butt (red) or no change (orange-red) butt.** No carbohydrate fermentation has occurred. Instead, peptones are catabolized under anaerobic and/or aerobic conditions, resulting in an alkaline pH due to production of ammonia. If only aerobic degradation of peptones occurs, the alkaline reaction is evidenced only on the slant surface. If there is aerobic and anaerobic utilization of peptone, the alkaline reaction is present on the slant and the butt.

To obtain accurate results it is absolutely essential to observe the cultures within 18 to 24 hours following incubation to ensure that the carbohydrate substrates have not been depleted and that degradation of peptones yielding alkaline end products has not taken place.

The TSI agar medium also contains sodium thiosulfate, a substrate for hydrogen sulfide (H_2S) production, and ferrous sulfate for detection of this colorless end product. Following incubation, only cultures of organisms capable of producing H_2S will show an extensive blackening in the butt due to the precipitation of the insoluble ferrous sulfide. (Refer to Experiment 25 for a more detailed biochemical explanation of H_2S production.)

At the top of page 140 is a schema for the differentiation of intestinal bacilli on the basis of the TSI agar reactions.

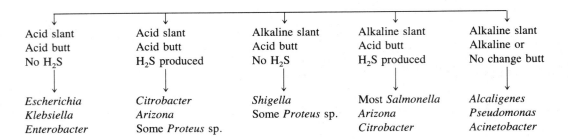

Acid slant	Acid slant	Alkaline slant	Alkaline slant	Alkaline slant
Acid butt	Acid butt	Acid butt	Acid butt	Alkaline or
No H$_2$S	H$_2$S produced	No H$_2$S	H$_2$S produced	No change butt
Escherichia	*Citrobacter*	*Shigella*	Most *Salmonella*	*Alcaligenes*
Klebsiella	*Arizona*	Some *Proteus* sp.	*Arizona*	*Pseudomonas*
Enterobacter	Some *Proteus* sp.		*Citrobacter*	*Acinetobacter*

MATERIALS

Cultures

24-hour trypticase soy broth cultures of *Pseudomonas aeruginosa, Escherichia coli, Salmonella typhimurium, Shigella dysenteriae, Proteus vulgaris,* and *Alcaligenes faecalis* for the short version.

24-hour trypticase soy broth cultures of the 13 previously listed organisms for the long version.

Media

Triple sugar-iron agar slants per designated student group: seven for short version, fourteen for long version.

EQUIPMENT

Bunsen burner and inoculating needle.

PROCEDURE

1. Label each of the TSI agar slants with the name of the test organism to be inoculated and identify each with your initials. The last tube will serve as a control.

2. Using sterile technique, inoculate each experimental organism into its appropriately labeled tube by means of a stab-and-streak inoculation. **Do not fully tighten screw cap.**

3. Incubate for 24 to 48 hours at 37 degrees C.

Observations and Results

1. Examine all agar slant cultures as to the color of both the butt and slant. Based on your observations, determine the type of reaction that has taken place (acid, alkaline, or none) and the carbohydrate that has been fermented (dextrose, lactose and/or sucrose, all, or none) in each culture. Record your observations and results in the chart.

2. Examine all cultures for the presence or absence of blackening within the medium. Based on your observations, determine whether or not each organism was capable of H_2S production. Record your observations and results in the chart.

3. Refer to the color plate illustrations of these reactions.

Bacterial species	Carbohydrate fermentation			H_2S production	
	Butt color and reaction	Slant color and reaction	Carbohydrate fermented	Blackening	H_2S $(+)$ or $(-)$
E. coli					
E. aerogenes					
K. pneumoniae					
S. dysenteriae					
S. typhimurium					
P. vulgaris					
P. aeruginosa					
A. faecalis					
M. luteus					
S. lactis					
S. aureus					
B. cereus					
C. xerosis					
Control					

REVIEW QUESTIONS

1. What is the purpose of the TSI test?

2. Explain why the TSI medium contains a lower concentration of glucose than of lactose and sucrose.

3. Explain the purpose of the phenol red in the medium.

4. Explain the purpose of thiosulfate in the medium.

5. Explain why the test observations must be made between 18 and 24 hours after inoculation.

IMViC Test

Identification of enteric bacilli is of prime importance in controlling intestinal infections by preventing contamination of food and water supplies. The groups of bacteria that can be found in the intestinal tract of humans and lower mammals are classified as members of the family **Enterobacteriaceae.** They are short, gram-negative, non–spore-forming bacilli. Included in this family are:

1. **Pathogens** such as members of the genera *Salmonella* and *Shigella*.

2. **Occasional pathogens** such as members of the genera *Proteus* and *Klebsiella*.

3. **Normal intestinal flora** such as members of the genera *Escherichia* and *Enterobacter*, which are saprophytic inhabitants of the intestinal tract.

Differentiation of the principal groups of Enterobacteriaceae can be accomplished on the basis of their biochemical properties and enzymatic reactions in the presence of specific substrates. The **IMViC** series of tests, **indole, methyl-red, Voges-Proskauer,** and **citrate utilization** can be used.

The following experiments are designed for either a short or long version. The short version uses selected members of the enteric family. The long procedure makes use of bacterial species that do not solely belong to the Enterobacteriaceae. Nonenteric forms are included to acquaint students with the biochemical activities of other organisms grown in these media and also to use these data for further comparisons of both types of bacteria. Selected organisms to be used in the long-version procedures are listed. The enteric organisms are subdivided as lactose fermenters and nonfermenters.

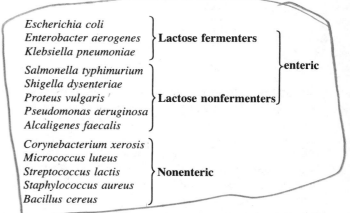

Escherichia coli
Enterobacter aerogenes } **Lactose fermenters**
Klebsiella pneumoniae

Salmonella typhimurium
Shigella dysenteriae
Proteus vulgaris } **Lactose nonfermenters**
Pseudomonas aeruginosa
Alcaligenes faecalis

} **enteric**

Corynebacterium xerosis
Micrococcus luteus
Streptococcus lactis } **Nonenteric**
Staphylococcus aureus
Bacillus cereus

Figure 24.1 on page 148 shows the biochemical reactions that occur during the IMViC tests. It is designed to assist students in the execution and interpretation of each test.

PART A: Indole Production Test
PURPOSE

To determine the ability of microorganisms to degrade the amino acid tryptophan.

PRINCIPLE

Tryptophan is an essential amino acid that can undergo oxidation by way of the enzymatic activities of some bacteria. Conversion of tryptophan into metabolic products is mediated by the enzyme **tryptophanase.** This reaction is illustrated below.

Tryptophan   →Tryptophanase→   **Indole**   **Pyruvic acid**   **Ammonia**

143

Quinoidal red-violet compound

This ability to hydrolyze tryptophan with the production of indole is not a characteristic of all microorganisms and therefore serves as a biochemical marker.

In this experiment, SIM agar, which contains the substrate tryptophan, is used. The presence of indole is detectable by adding Kovac's reagent, which produces a cherry-red reagent layer. This color is produced by the reagent, which is composed of p-dimethylaminobenzaldehyde, butanol and hydrochloric acid. Indole is extracted from the medium into the reagent layer by the acidified butanol component and forms a complex with the p-dimethylaminobenzaldehyde, yielding the cherry red color. This reaction is illustrated at the top of this page.

Cultures producing a red reagent layer following addition of Kovac's reagent are indole-positive. The absence of a red coloration demonstrates that the substrate tryptophan was not hydrolyzed and indicates an indole-negative reaction.

MATERIALS

Cultures

24-hour to 48-hour trypticase soy broth cultures of *Escherichia coli*, *Proteus vulgaris*, and *Enterobacter aerogenes* for the short version.

24-hour to 48-hour trypticase soy broth cultures of the 13 previously listed organisms for the long version.

Media

SIM agar deep tubes per designated student group: four for the short version, fourteen for the long version.

Reagent

Kovac's reagent.

Equipment

Bunsen burner and inoculating needle.

PROCEDURE

1. Label each of the SIM deep tubes with the name of the bacterial organism to be inoculated and identify each with your initials.

2. Using sterile technique, inoculate each experimental organism into its appropriately labeled deep tube by means of a stab inoculation.

3. Incubate tubes for 24 to 48 hours at 37 degrees C.

PART B: Methyl Red Test

PURPOSES

1. To determine the ability of microorganisms to oxidize glucose with the production and stabilization of high concentrations of acid end products.

2. To differentiate between all glucose-oxidizing enteric organisms, particularly *Escherichia coli* and *Enterobacter aerogenes*.

PRINCIPLE

The hexose monosaccharide **glucose** is the major substrate oxidized by all enteric organisms for energy production. The end products of this process will vary depending on the specific enzymatic pathways present in the bacteria. In this test the pH indicator methyl red detects the presence of large concentrations of acid end products. Although all enteric microorganisms ferment glucose with the production of organic acids, this test

$$\text{Glucose} + H_2O \longrightarrow \begin{bmatrix} \text{Lactic acid} \\ \text{Acetic acid} \\ \text{Formic acid} \end{bmatrix} \longrightarrow CO_2 + H_2 \text{ (pH 4.4)}$$

Methyl red indicator red color

is of value in the separation of *E. coli* and *E. aerogenes*.

Both of these organisms initially produce organic acid end products during the early incubation period. The low acidic pH (4) is stabilized and maintained by *E. coli* at the end of incubation. During the later incubation period, *E. aerogenes* enzymatically converts these acids to nonacidic end products such as ethanol and acetonin (acetylmethylcarbinol), resulting in an elevated pH of approximately 6. The glucose fermentation reaction generated by *E. coli* is illustrated at the top of the this page.

As shown, the methyl red indicator in the pH range of 4 will turn red, which is indicative of a positive test. At a pH of 6, still indicating the presence of acid but with a lower hydrogen ion concentration, the indicator turns yellow and is a negative test. Production and detection of the nonacidic end products from glucose fermentation by *E. aerogenes* is amplified in Part C of this exercise, the Voges-Proskauer test, which is performed simultaneously with the methyl red test.

MATERIALS

Cultures

24-hour to 48-hour trypticase soy broth cultures of *Escherichia coli, Enterobacter aerogenes,* and *Klebsiella pneumoniae* for the short version. 24-hour to 48-hour trypticase soy broth cultures of the 13 previously listed organisms for the long version.

Media

MR-VP broth per designated student group: four for the short version, fourteen for the long version.

Reagent

Methyl red indicator.

Equipment

Bunsen burner, inoculating loop, and test tubes.

PROCEDURE

1. Label each of the MR-VP broth media with the name of the bacterial organism to be inoculated and identify each with your initials.

2. Using sterile technique, inoculate each experimental organism into its appropriately labeled tube of medium by means of a loop inoculation.

3. Incubate all cultures for 24 to 48 hours at 37 degrees C.

PART C: Voges-Proskauer Test
PURPOSE

To enable students to differentiate further among enteric organisms such as *Escherichia coli, Enterobacter aerogenes,* and *Klebsiella pneumoniae.*

PRINCIPLE

The Voges-Proskauer test determines the capability of some organisms to produce nonacidic or neutral end products such as acetylmethylcarbinol, from the organic acids that result from glucose metabolism. This glucose fermentation, which is characteristic of *E. aerogenes,* is illustrated below.

$$\text{Glucose} + O_2 \longrightarrow \begin{array}{c} \text{Acetic} \\ \text{acid} \end{array} \longrightarrow \begin{bmatrix} \text{2,3 butanediol} \\ \text{acetylmethylcarbinol} \end{bmatrix} \longrightarrow CO_2 + H_2$$

The reagent used in this test consists of a mixture of alcoholic alpha-naphthol and 40 percent potassium hydroxide solution. Detection of acetylmethylcarbinol requires that this end product be oxidized to a diacetyl compound. This reaction will occur in the presence of the alpha-naphthol catalyst and a guanidine group that is present in the peptone of the MR-VP medium. As a result, a pink complex is formed, imparting a rose color to the medium. The chemistry of this reaction is illustrated at the top of this page.

Development of a deep rose color in the culture 15 minutes following the addition of Barritt's reagent is indicative of the presence of acetylmethylcarbinol and represents a positive result. The absence of rose coloration is a negative result.

MATERIALS

Cultures

24-hour to 48-hour trypticase soy broth cultures of *Escherichia coli, Enterobacter aerogenes,* and *Klebsiella pneumoniae* for the short version.

24-hour to 48-hour trypticase soy broth cultures of the 13 previously listed organisms for the long version.

Aliquots of these experimental cultures must be set aside from the methyl red test.

Reagent

Barritt's reagent.

Equipment

Bunsen burner and inoculating loop.

PROCEDURE

Refer to methyl red test, Part B of this exercise.

PART D: Citrate Utilization Test
PURPOSE

To enable students to differentiate among enteric organisms on the basis of their ability to ferment citrate as a sole carbon source.

PRINCIPLE

In the absence of fermentable glucose or lactose, some microorganisms are capable of using **citrate** as a carbon source for their energy. This ability depends on the presence of a **citrate permease** that facilitates the transport of citrate into the cell. Citrate is the first major intermediate in the Krebs cycle and is produced by the condensation of active acetyl with oxalacetic acid. Citrate is acted on by the enzyme **citrase,** which produces oxalacetic acid and acetate. These products are then enzymatically converted to pyruvic acid and carbon dioxide. During this reaction the medium becomes alkaline, as the carbon dioxide that is generated combines with sodium and water to form sodium carbonate, an alkaline product. The presence of sodium carbonate changes the bromthymol blue indicator incorporated into the medium from green to deep Prussian blue. This reaction is illustrated at the top of page 147.

Following incubation, citrate-positive cultures are identified by the presence of growth on the surface of the slant, which is accompanied by blue coloration. Citrate-negative cultures will show no growth and the medium will remain green.

1.

$$\underset{\text{Citric acid}}{\text{HO}-\underset{\underset{\text{COOH}}{|}}{\overset{\overset{\text{COOH}}{|}}{\underset{\underset{\text{CH}_2}{|}}{\overset{\overset{\text{CH}_2}{|}}{\text{C}}}}}-\text{COOH}} \xrightarrow{\text{Citrase}} \underset{\text{Oxalacetic acid}}{\overset{\overset{\text{COOH}}{|}}{\underset{\underset{\text{COOH}}{|}}{\underset{\text{CH}_2}{\overset{\text{C}=\text{O}}{|}}}}} + \underset{\text{Acetic acid}}{\overset{\text{CH}_3}{\underset{\text{COOH}}{|}}} \longrightarrow \underset{\text{Pyruvic acid}}{\overset{\overset{\text{COOH}}{|}}{\underset{\underset{\text{CH}_3}{|}}{\text{C}=\text{O}}}} + \underset{\substack{\text{Excess}\\\text{carbon}\\\text{dioxide}}}{\text{CO}_2}$$

2. $CO_2 + 2Na^+ + H_2O \longrightarrow Na_2CO_3 \longrightarrow$ Alkaline pH $\longrightarrow$ Color change from green to blue

MATERIALS

Cultures

24-hour to 48-hour trypticase soy broth cultures of *Escherichia coli, Enterobacter aerogenes,* and *Klebsiella pneumoniae* for the short version.

24-hour to 48-hour trypticase soy broth cultures of the 13 previously listed organisms for the long version.

Media

Simmons' citrate agar slants per designated student group: four for the short version, fourteen for the long version.

Equipment

Bunsen burner and inoculating needle.

PROCEDURE

1. Label each of the Simmons' citrate agar slants with the name of the bacterial organism to be inoculated and identify each with your initials.

2. Using sterile technique, inoculate each organism into its appropriately labeled tube by means of a stab-and-streak inoculation.

3. Incubate all cultures for 24 to 48 hours at 37 degrees C.

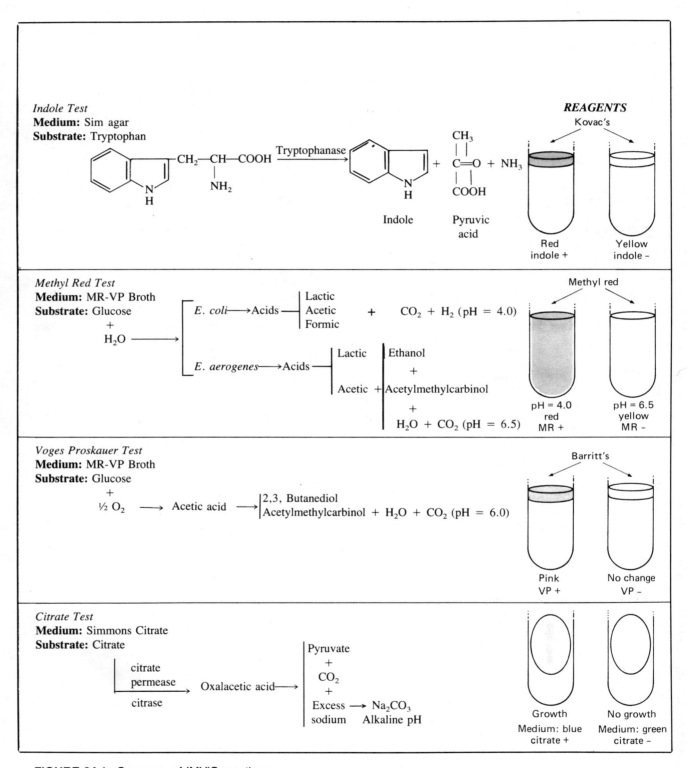

FIGURE 24.1. Summary of IMViC reactions

Hydrogen Sulfide Test

PURPOSE

To enable students to determine the ability of microorganisms to produce hydrogen sulfide from substrates such as the sulfur-containing amino acids or inorganic sulfur compounds.

PRINCIPLE

There are two major fermentative pathways by which some microorganisms are able to produce hydrogen sulfide (H_2S).

PATHWAY 1: Gaseous H_2S may be produced by the reduction (hydrogenation) of organic sulfur present in the amino acid cysteine, which is a component of peptones contained in the medium. These peptones are degraded by microbial enzymes to amino acids, including the sulfur-containing amino acid cysteine. This amino acid in the presence of a **cysteine desulfurase** loses the sulfur atom, and is then reduced by the addition of hydrogen from water to form hydrogen sulfide gas as illustrated:

$$\begin{array}{c} CH_2-SH \\ | \\ CH-NH_2 \\ | \\ COOH \end{array} \xrightarrow[\text{desulfurase}]{\text{Cysteine}} \begin{array}{c} CH_3 \\ | \\ C=O \\ | \\ COOH \end{array} + H_2S\uparrow + NH_3$$

Cysteine   **Pyruvic** **Hydrogen** **Ammonia**
acid **sulfide**
gas

PATHWAY 2: Gaseous H_2S may also be produced by the reduction of inorganic sulfur compounds such as the thiosulfates ($S_2O_3^=$), sulfates ($SO_4^=$), or sulfites ($SO_3^=$). The medium contains sodium thiosulfate, which certain microorganisms are capable of reducing to sulfite with the liberation of hydrogen sulfide. The sulfur atoms act as hydrogen acceptors during oxidation of the inorganic compound as illustrated at the top of the next column.

$$3S_2O_3^= + 4H^+e \xrightarrow[\textbf{reductase}]{\textbf{Thiosulfate}} 2SO_3^= + 2H_2S\uparrow$$

Thiosulfate   **Sulfite** **Hydrogen**
sulfide gas

In this experiment the SIM medium contains peptone and sodium thiosulfate as the sulfur substrates, ferrous ammonium sulfate, $Fe(NH_4)_2SO_4$, which behaves as the H_2S indicator, and sufficient agar to make the medium semisolid and thus enhance anaerobic respiration. Regardless of which pathway is used, the hydrogen sulfide gas is colorless and therefore not visible. Ferrous ammonium sulfate in the medium serves as an indicator by combining with the gas, forming an insoluble black ferrous sulfide precipitate that is seen along the line of the stab inoculation and is indicative of H_2S production. Absence of the precipitate is evidence of a negative reaction. The overall reactions for both pathways and their interpretation are illustrated:

$$\begin{array}{c} CH_2-SH \\ | \\ CH-NH_2 \\ | \\ COOH \end{array} \xrightarrow[\textbf{desulfurase}]{\textbf{Cysteine}} \begin{array}{c} CH_3 \\ | \\ C=O \\ | \\ COOH \end{array} + NH_3 + H_2S\uparrow$$

Cysteine   **Pyruvic** **Ammonia** **H_2S** gas
acid (colorless)

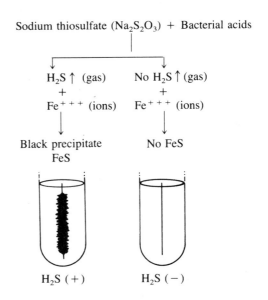

Sodium thiosulfate ($Na_2S_2O_3$) + Bacterial acids

$H_2S\uparrow$ (gas)   No $H_2S\uparrow$ (gas)
+   +
Fe^{+++} (ions) Fe^{+++} (ions)

Black precipitate   No FeS
FeS

H_2S (+) H_2S (−)

SIM agar may also be used to detect motile organisms. Motility is recognized when culture growth (turbidity) is not restricted to the line of inoculation. Growth of nonmotile organisms is confined to the line of inoculation.

MATERIALS

Cultures

24-hour to 48-hour trypticase soy broth cultures of *Enterobacter aerogenes, Proteus vulgaris,* and *Salmonella typhimurium* for the short version. 24-hour to 48-hour trypticase soy broth cultures of the 13 previously listed organisms for the long version.

Media

SIM agar deep tubes per designated student group: four for the short version, fourteen for the long version.

Equipment

Bunsen burner and inoculating needle.

PROCEDURE

1. Label each of the SIM agar deep tubes with the name of the organism to be inoculated and identify each with your initials.

2. Aseptically inoculate each experimental organism into its appropriately labeled tube by means of stab inoculation.

3. Incubate all cultures for 24 to 48 hours at 37 degrees C.

NAME *Chantalle Dezil*

Observations and Results

1. Examine all SIM cultures as to the presence or absence of black coloration along the line of the stab inoculation. Record your results in the chart.

2. Based on your observations, determine and record whether or not each organism was capable of hydrogen sulfide production.

3. Observe all cultures for the presence (+) or absence (−) of motility. Record your results in the chart.

4. Refer to the color plate for illustrations of these reactions.

Bacterial species	*Color of medium*	*H₂S production (+) or (−)*	*Motility (+) or (−)*
E. coli	Red	+	
E. aerogenes	yellow	−	
K. pneumoniae	"	−	
S. dysenteriae			
S. typhimurium		+	
P. vulgaris		+	
P. aeruginosa		−	
A. faecalis		−	
M. luteus		−	
S. lactis		−	
S. aureus		−	
B. cereus		−	
C. xerosis			
Control			

REVIEW QUESTIONS

1. Distinguish between the types of substrates available to cells for H₂S production.

2. Explain how SIM medium is used to detect motility.

3. Explain the function of the ferrous ammonium sulfate in SIM agar.

4. Why is *P. vulgaris* H$_2$S-positive and *E. aerogenes* H$_2$S-negative?

Urease Test

PURPOSE

To enable students to determine the ability of micro-organisms to degrade urea by means of the enzyme urease.

PRINCIPLE

Urease, which is produced by some microorganisms, is an enzyme especially helpful in the identification of *Proteus vulgaris*. Although other organisms may produce urease, their action on the substrate urea tends to be slower than that seen with *Proteus* species. Therefore this test serves rapidly to distinguish members of this genus from other lactose-nonfermenting enteric microorganisms.

Urease is a hydrolytic enzyme that attacks the nitrogen and carbon bond in amide compounds such as urea and forms the alkaline end product ammonia. This chemical reaction is illustrated at the bottom of this page.

The presence of urease is detectable when the organisms are grown in a urea broth medium containing the pH indicator phenol red. As the substrate urea is split into its products, the presence of ammonia creates an alkaline environment that causes the phenol red to turn to a deep pink. This is positive for the presence of urease. Failure of a deep pink color to develop is evidence of a negative reaction.

MATERIALS

Cultures

24-hour to 48-hour trypticase soy broth cultures of *Escherichia coli, Proteus vulgaris, Klebsiella pneumoniae,* and *Salmonella typhimurium* for the short version.

24-hour to 48-hour trypticase soy broth cultures of the 13 previously listed organisms for the long version.

Media

Urea broth per designated student group: five for the short version, fourteen for the long version.

Equipment

Bunsen burner and inoculating loop.

PROCEDURE

1. Label each of the tubes of the urea broth medium with the name of the bacterial organisms to be inoculated and identify each with your initials.

2. Using sterile technique, inoculate each experimental organism into its appropriately labeled tube by means of loop inoculation.

3. Incubate cultures 24 to 48 hours at 37 degrees C.

$$\underset{\textbf{Urea}}{\underset{\displaystyle \underset{O}{\overset{\displaystyle NH_2 \quad NH_2}{\underset{\|}{C}}}}{}} + 2H_2O \xrightarrow{\text{Urease}} \underset{\substack{\textbf{Carbon} \\ \textbf{dioxide}}}{CO_2} + \underset{\textbf{Water}}{H_2O} + \underset{\textbf{Ammonia}}{2NH_3}$$

Litmus Milk Reactions

PURPOSE

To enable students to differentiate among microorganisms that enzymatically transform different milk substrates into varied metabolic end products.

PRINCIPLE

The major milk substrates capable of transformation are the milk sugar lactose and the milk proteins casein, lactoalbumin, and lactoglobulin. To distinguish among the metabolic changes produced in milk, a pH indicator, the oxidation-reduction indicator litmus, is incorporated into the medium. Litmus milk now forms an excellent differential medium in which microorganisms can metabolize milk substrates depending on their enzymatic complement. A variety of different biochemical changes result, as follows:

$$
\text{Litmus milk}\begin{cases}
\text{Lactose fermentation}\\
\text{Gas production}\\
\text{Litmus reduction}\\
\text{Curd formation}\\
\text{Proteolysis}\\
\text{Alkaline reaction}
\end{cases}
$$

Lactose Fermentation

Organisms capable of using **lactose** as a carbon source for energy production, utilize the inducible enzyme **β-galactosidase** and degrade lactose as follows:

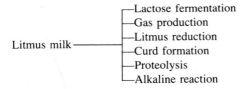

The presence of **lactic acid** is easily detected, as litmus

is purple at a neutral pH and turns pink when the medium is acidified to an approximate pH of 4.

Gas Formation

The end products of the microbial fermentation of lactose may include the **gases $CO_2\uparrow$ + $H_2\uparrow$**. The presence of gas may be seen as separations of the curd, or by the development of tracks or fissures within the curd as gas rises to the surface.

Litmus Reduction

Fermentation is an anaerobic process involving biooxidations that occur in the absence of molecular oxygen. These oxidations may be visualized as the removal of hydrogen (dehydrogenation) from a substrate. Since hydrogen ions cannot exist in the free state, there must be an immediate and concomitant electron acceptor available to bind these hydrogen ions, or else oxidation-reduction reactions are not possible and cells cannot manufacture energy. In the litmus milk test, **litmus** acts as such an acceptor. While in the oxidized state, the litmus is purple; when it accepts hydrogen from a substrate it will become reduced and turn white or milk-colored. This oxidation of lactose, which produces lactic acid, butyric acid, $CO_2\uparrow$, and $H_2\uparrow$, is as follows:

$$
\text{Lactose} \rightarrow \text{Glucose} \rightarrow \text{Pyruvic acid} \rightarrow \begin{cases}\text{Lactic acid}\\ \text{Butyric acid}\\ CO_2 + H_2\end{cases}
$$

The excess hydrogen is now accepted by the hydrogen acceptor litmus, which turns white and is said to be reduced.

Curd Formation

The biochemical activities of different microorganisms grown in litmus milk may result in the production of two distinct types of curds (clots). Curds are designated either as acid or rennet, depending on the biochemical mechanism responsible for their formation.

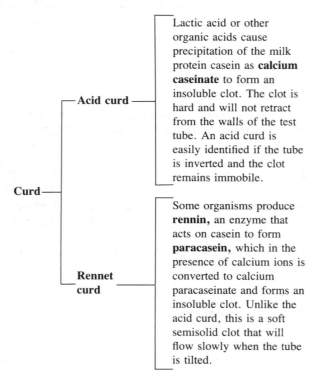

Lactic acid or other organic acids cause precipitation of the milk protein casein as **calcium caseinate** to form an insoluble clot. The clot is hard and will not retract from the walls of the test tube. An acid curd is easily identified if the tube is inverted and the clot remains immobile.

Some organisms produce **rennin,** an enzyme that acts on casein to form **paracasein,** which in the presence of calcium ions is converted to calcium paracaseinate and forms an insoluble clot. Unlike the acid curd, this is a soft semisolid clot that will flow slowly when the tube is tilted.

Proteolysis (Peptonization)

The inability of some microorganisms to obtain their energy by way of lactose fermentation means they must use other nutritional sources such as proteins for this purpose (see Figure 22.2). By means of **proteolytic enzymes,** these organisms hydrolyze the milk proteins, primarily **casein,** to their basic building blocks, namely **amino acids.** This digestion of proteins is accompanied by the evolution of large quantities of ammonia resulting in an alkaline pH in the medium. The litmus turns deep purple in the upper portion of the tube, while the medium begins to lose body and produces a translucent, brown, wheylike appearance as the protein is hydrolyzed to amino acids.

Alkaline Reaction

An alkaline reaction is evident when the color of the medium remains unchanged or changes to a deeper blue. This reaction is indicative of the partial degradation of **casein** into **shorter polypeptide chains,** with the simultaneous release of alkaline end products that are responsible for the observable color change.

Figure 27.1 is a summary of the possible litmus milk reactions and their appearance following appropriate incubation of the cultures.

MATERIALS

Cultures

24-hour to 48-hour trypticase soy broth cultures of *Escherichia coli, Alcaligenes faecalis, Streptococcus lactis,* and *Pseudomonas aeruginosa* for the short version.

24-hour to 48-hour trypticase soy broth cultures of the 13 previously listed organisms for the long version.

Media

Litmus milk broth per designated student group: five for the short version, fourteen for the long version.

Equipment

Bunsen burner and inoculating loop.

PROCEDURE

1. Label each of the tubes of the litmus milk medium with the name of the bacterial organism to be inoculated and identify each with your initials.

2. Using sterile technique, inoculate each experimental organism into its appropriately labeled tube by means of a loop inoculation.

3. Incubate all cultures for 24 to 48 hours at 37 degrees C.

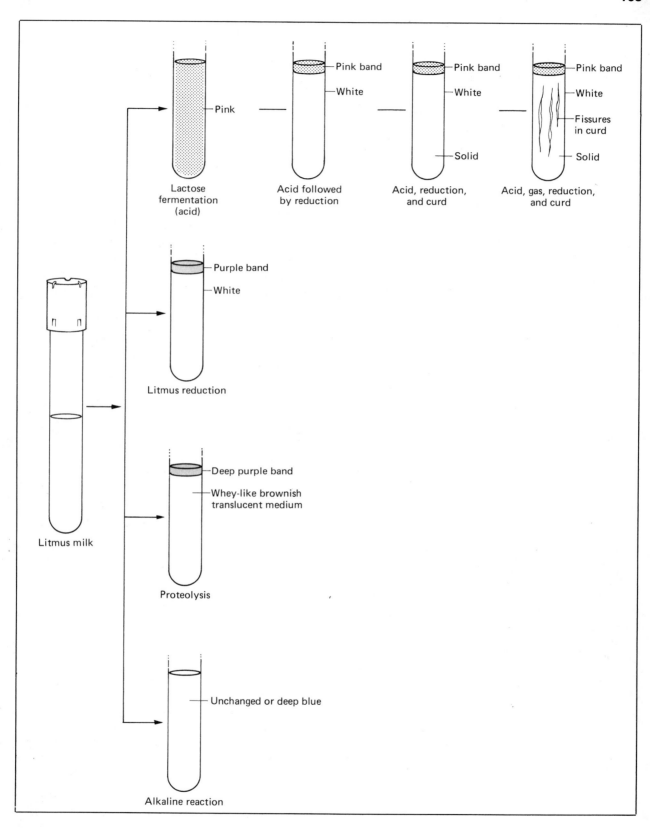

FIGURE 27.1. Litmus milk reactions

Nitrate Reduction Test

PURPOSE

To enable students to determine the ability of some microorganisms to reduce nitrates (NO_3^-) to nitrites (NO_2^-) or beyond the nitrite stage.

PRINCIPLE

The reduction of nitrates by some aerobic and facultative anaerobic microorganisms occurs in the absence of molecular oxygen, an anaerobic process. In these organisms anaerobic respiration is an oxidative process whereby the cell uses inorganic substances such as nitrates (NO_3^-) or sulfates ($SO_4^=$) to supply oxygen that is subsequently utilized as a final hydrogen acceptor during energy formation. The biochemical transformation may be visualized as follows:

$$NO_3^- \ + \ 2H_{e-}^+ \ \xrightarrow[\textbf{Reductase}]{\textbf{Nitrate}} NO_2^- \ + \ H_2O$$

Nitrate + Hydrogen   **Nitrite + Water**
electrons

Some organisms possess the enzymatic capacity to act further on nitrites to reduce them to ammonia (NH_3^+) or molecular nitrogen (N_2). These reactions may be described as follows:

$$NO_2^- \longrightarrow NH_3$$

Nitrite   **Ammonia**

or

$$2NO_3^- \ + \ 10_e^- \ + \ 12H^+ \rightarrow \ N_2 + 6H_2O$$

Nitrate   **Molecular Nitrogen**

Nitrate reduction can be determined by cultivating organisms in a nitrate broth medium. It is basically a nutrient broth supplemented with 0.1 percent potassium nitrate (KNO_3) as the nitrate substrate. In addition, the medium is made into a semisolid by the addition of 0.1 percent agar. The semisolidity impedes the diffusion of oxygen into the medium, thereby favoring the anaerobic requirement necessary for nitrate reduction.

Following incubation of the cultures, an organism's ability to reduce nitrates to nitrites is determined by the addition of two reagents: solution A, which is sulfanilic acid, followed by solution B, which is alphanaphthylamine. Following reduction, the addition of solutions A and B will produce an immediate cherry-red color.

$$NO_3^- \ \xrightarrow[\text{Reductase}]{\text{Nitrate}} \ NO_2^- \ \text{(Red color on addition of solutions A and B)}$$

Cultures not producing a color change suggest one of two possibilities: (1) nitrates were not reduced by the organism or (2) the organism possessed such potent **nitrate reductase** enzymes that nitrates were rapidly reduced beyond nitrites to ammonia or even molecular nitrogen. To determine whether or not nitrates were reduced past the nitrite stage, a small amount of zinc powder is added to the basically colorless cultures already containing solutions A and B. Zinc reduces nitrates to nitrites. The development of red color therefore verifies that nitrates were not reduced to nitrites by the organism. If nitrates were not reduced, a negative nitrate reduction reaction has occurred. If the addition of zinc does not produce a color change, the nitrates in the medium were reduced beyond nitrites to ammonia or nitrogen gas. This is a positive reaction, as shown below.

| Sulfanilic acid (colorless) | α-naphthylamine (colorless) | Nitrous acid | Sulfobenzene azo-α-naphthylamine (red) | |

MATERIALS

Cultures

24-hour to 48-hour trypticase soy broth cultures of *Escherichia coli, Alcaligenes faecalis,* and *Pseudomonas aeruginosa* for the short version. 24-hour to 48-hour trypticase soy broth cultures of the 13 previously listed organisms for the long version.

Media

Trypticase nitrate broth per designated student group: four for the short version, fourteen for the long version.

Reagents

Solution A (sulfanilic acid), solution B (alpha-naphthylamine), and zinc powder.

Equipment

Bunsen burner and inoculating loop.

PROCEDURE

1. Label each of the tubes of the trypticase nitrate broth with the name of the bacterial organism to be inoculated and identify each with your initials.

2. Using sterile technique, inoculate each experimental organism into its appropriately labeled tube by means of a loop inoculation.

3. Incubate all cultures for 24 to 48 hours at 37 degrees C.

$$NO_3 \xrightarrow{\text{nitrate}} NO_2$$

$$NO_3 \xrightarrow{\text{Reduction}} N_2 + H_2O$$

nitrate $\longrightarrow$ nitrite $\bar{c}$ addition of solutions A + B (sulfanilic acid), it will turn red

* from nitrate to nitrite it will turn red
if not turn red, Add Zn to it
if it stays clear, it has passed
nitrate stage from Nitrogen to H$_2$O

nitrate + Zn $\Rightarrow$ Red Color

Study about litmus

Catalase Test

PURPOSE

To enable students to determine the ability of some microorganisms to degrade hydrogen peroxide by producing the enzyme catalase.

PRINCIPLE

During aerobic respiration, microorganisms produce hydrogen peroxide and, in some cases, an extremely toxic superoxide. Accumulation of these substances will result in death of the organism unless they can be enzymatically degraded. These substances are produced when aerobes, facultative anaerobes, and microaerophiles use the aerobic respiratory pathway, in which oxygen is the final electron acceptor, during degradation of carbohydrates for energy production. Organisms capable of producing **catalase** or **peroxidase** rapidly degrade hydrogen peroxide as illustrated:

$$2H_2O_2 \xrightarrow{\textbf{Catalase}} 2H_2O + O_2\uparrow$$

Hydrogen peroxide   **Water**   **Free oxygen**

Superoxide dismutase is the enzyme responsible for degradation of the especially toxic superoxides in catalase-negative aerobic organisms.

The inability of strict anaerobes to synthesize catalase, peroxidase, or superoxide dismutase may explain why oxygen is poisonous to these microorganisms. In the absence of these enzymes, the toxic concentration of H_2O_2 cannot be degraded when these organisms are cultivated in the presence of oxygen.

Catalase production can be determined by adding the substrate H_2O_2 to an appropriately incubated trypticase soy agar slant culture. If catalase is present, the chemical reaction mentioned is indicated by bubbles of free oxygen gas ($O_2\uparrow$). This is a positive catalase test; the absence of bubble formation is a negative catalase test.

MATERIALS

Cultures

24-hour to 48-hour trypticase soy broth cultures of *Staphylococcus aureus, Micrococcus luteus,* and *Streptococcus lactis* for the short version. 24-hour to 48-hour trypticase soy broth cultures of the 13 previously listed organisms for the long version.

Media

Trypticase soy agar slants per designated student group: four for the short version, fourteen for the long version.

Reagent

3 percent hydrogen peroxide.

Equipment

Bunsen burner and inoculating loop.

PROCEDURE

1. Label each of the trypticase soy agar slants with the name of the bacterial organism to be inoculated and identify each with your initials.

2. Using sterile technique, inoculate each experimental organism into its appropriately labeled tube by means of a streak inoculation.

3. Incubate all cultures for 24 to 48 hours at 37 degrees C.

Oxidase Test

PURPOSE

To perform an experimental procedure that will enable students to distinguish between groups of bacteria on the basis of cytochrome oxidase activity.

PRINCIPLE

Oxidase enzymes play a vital role in the operation of the electron transport system during aerobic respiration. **Cytochrome oxidase** catalyzes the oxidation of a reduced cytochrome by molecular oxygen (O_2), resulting in the formation of H_2O or H_2O_2. Aerobic bacteria, as well as some facultative anaerobes and microaerophiles, exhibit oxidase activity. The oxidase test aids in differentiation among members of the genera *Neisseria* and *Pseudomonas,* which are oxidase-positive, and the Enterobacteriaceae, which are oxidase-negative.

The ability of bacteria to produce cytochrome oxidase can be determined by the addition of the test reagent, tetramethyl-p-phenylenediamine dihydrochloride, to colonies grown on a plate medium. This light pink reagent serves as an artificial substrate, donating electrons and thereby becoming oxidized to a blackish compound in the presence of the oxidase and free oxygen. Following the addition of the test reagent, the development of pink, then maroon, and finally black coloration on the surface of the colonies is indicative of cytochrome oxidase production and represents a positive test. No color change or a light pink coloration on the colonies is indicative of the absence of oxidase activity and is a negative test.

MATERIALS

Cultures

24-hour to 48-hour trypticase soy broth cultures of *Escherichia coli, Pseudomonas aeruginosa,* and *Al-*

caligenes faecalis for the short version. 24-hour to 48-hour trypticase soy broth cultures of the 13 previously listed organisms for the long version.

Media

Trypticase soy agar plates per designated student group: one for the short version, four for the long version.

Reagent

Tetramethyl-p-phenylenediamine dihydrochloride.

Equipment

Bunsen burner, inoculating loop, and wax pencil.

PROCEDURE

1. Prepare the trypticase soy agar plate(s) for inoculation as follows:

   a. Short procedure: With a wax pencil, divide the bottom of a Petri dish into three sections and label each section with the name of the test organism to be inoculated and identify each with your initials.

   b. Long procedure: Repeat step 1a, dividing three plates into three sections and one plate into four sections to accommodate the 13 test organisms.

2. Using sterile technique, make a single line streak inoculation of each test organism on the agar surface of its appropriate section of the plate(s).

3. Incubate the plate(s) in an inverted position for 24 to 48 hours at 37 degrees C.

Observations and Results　　　　EXPERIMENT **30**

1. Add two to three drops of the tetramethyl-p-phenylenediamine dihydrochloride to the surface of the growth of each test organism.

2. Observe the growth for the presence or absence of a color change from pink, to maroon, and finally to black. Record the result on the chart below.

3. Based on your observations, determine and record whether or not each organism was capable of producing cytochrome oxidase.

Bacterial species	*Color of colonies*	*Oxidase production* *(+) or (−)*
E. coli		
E. aerogenes		
K. pneumoniae		
S. dysenteriae		
S. typhimurium		
P. vulgaris		
P. aeruginosa		
A. faecalis		
M. luteus		
S. lactis		
S. aureus		
B. cereus		
C. xerosis		
Control		

REVIEW QUESTIONS

1. What is the function of cytochrome oxidase?

2. Why are strict aerobes oxidase-positive?

3. The oxidase test is used to differentiate among which groups of bacteria?

4. What is the function of the test reagent in this procedure?

Identification
of Unknown Bacterial Cultures

PURPOSE

To allow students to use previously studied staining, cultural, and biochemical procedures for the independent identification of an unknown bacterial culture. This may involve a computer-assisted method, a traditional method, or both.

PRINCIPLE

Identification of unknown bacterial cultures is one of the major responsibilities of the microbiologist. Samples of blood, tissue, food, water, and cosmetics are examined daily in laboratories throughout the world for the presence of contaminants. In addition, industrial organizations are constantly screening materials to isolate new antibiotic-producing organisms or organisms that will increase the yield of marketable products such as vitamins, solvents, and enzymes. Once isolated, these unknown organisms must be identified and classified.

The science of classification is called **taxonomy** and deals with the separation of living organisms into interrelated groups. *Bergey's Manual of Determinative Bacteriology,* 8th ed., is the official taxonomic key containing the orders, families, genera, and species of all known, classified bacteria. Students should obtain a copy of this manual from the laboratory instructor or library and become familiar with its use.

At this point students have developed sufficient knowledge in staining methods, isolation techniques, microbial nutrition, biochemical activities, and characteristics of microorganisms to be able to work independently in attempting to identify an unknown culture. Characteristics of the major organisms that have been used in experiments thus far are given in Table 31.1. Students are to use this table for the identification of their unknown cultures. The observations and results obtained following the experimental procedures are the basis of this identification.

These same data may also be analyzed and interpreted with a computer program available to the instructor for the Apple II, IBM PC, or TRS-80. The program is designed to identify the unknown culture on the basis of percentage of similarity between the unknown and the expected results, as indicated in Table 31.1. If the computer option is to be performed, the student will use Table 31.2 to determine the computer code number for the individual test observations and results. These code numbers are to be recorded in the observations and results chart and then entered into the computer for identification of the unknown organism.

MATERIALS

Cultures

Number-coded 24-hour to 48-hour trypticase soy agar slant cultures of the 13 bacterial species used in the long experimental procedures. Each student is to be provided with one unknown pure culture.

Media

Two trypticase soy agar slants, and one each of the following per student: phenol red sucrose broth, phenol red lactose broth, phenol red dextrose broth, SIM agar deep tube, MR-VP broth, tryptic nitrate broth, Simmons' citrate agar slant, urea broth, litmus milk, trypticase soy agar plate, nutrient gelatin deep tube, starch agar plate, and tributyrin agar plate.

Reagents

Crystal violet, Gram's iodine, 95 percent ethyl alcohol, safranin, methyl red, 3 percent hydrogen peroxide, Barritt's reagent, solutions A and B, Kovac's reagent,

TABLE 31.1   Cultural and biochemical characteristics of unknown organisms

Organism	Gram stain	Agar slant cultural characteristics	Litmus milk reaction	Lactose	Dextrose	Sucrose	H₂S production	NO₃ reduction	Indole production	MR reaction	VP reaction	Citrate use	Urease activity	Catalase activity	Oxidase activity	Gelatin liquefaction	Starch hydrolysis	Lipid hydrolysis
				Fermentation														
E. coli	Rod −	White, moist, glistening growth	Acid, curd ±, gas ±, reduction ±	AG	AG	A±	−	+	+	+	−	−	−	+	−	−	−	−
E. aerogenes	Rod −	Abundant, thick, white, glistening growth	Acid	AG	AG	AG±	−	+	−	−	+	+	−	+	−	−	−	−
K. pneumoniae	Rod −	Slimy, white, somewhat-translucent, raised growth	Acid, gas, curd ±	AG	AG	AG	−	+	−	−	±	+	+	+	−	−	−	−
S. dysenteriae	Rod −	Thin, even, grayish growth	Alkaline	−	A	A±	−	+	+	+	−	−	−	+	−	−	−	−
S. typhimurium	Rod −	Thin, even, grayish growth	Alkaline	−	AG±	A±	+	+	−	+	−	+	−	+	−	−	−	−
P. vulgaris	Rod −	Thin, blue-gray, spreading growth	Alkaline	−	AG	AG±	+	+	+	+	−	−	+	+	−	+	−	−
P. aeruginosa	Rod −	Abundant, thin, white growth, with medium turning green	Rapid peptonization	−	−	−	−	+	−	−	−	+	−	+	+	+ Rapid	−	+
A. faecalis	Rod* −	Thin, white, spreading, viscous growth	Alkaline	−	−	−	−	−	−	−	−	−	−	+	+	−		
S. aureus	Coccus +	Abundant, opaque, golden growth	Acid, reduction ±	A	A	A	−	+	−	+	±	−	−	+	−	+	−	+
S. lactis	Coccus +	Thin, even, growth	Acid, rapid reduction with curd	A	A	A	−	−	−	+	−	−	−	−	−	−	−	−
M. luteus	Cocci +	Soft, smooth, yellow growth	Alkaline	−	−	−	±	−	−	−	−	−	+	+	−	+ Slow		
C. xerosis	Rod +	Grayish, granular, limited growth	Alkaline	−	A±	A±	−	−	−	−	−	−	−	+	−	−	−	−
B. cereus	Rod +	Abundant, opaque, white waxy growth	Peptonization	−	A	A	−	+	−	−	±	−	−	+	−	+ Rapid	+	+

Note: AG = Acid and gas
     ± = Variable reaction
     rod* = Coccobacillus

TABLE 31.2 Code numbers for computer-assisted determination of unknown bacterial culture

Gram stain
0—not done
1—blue (Gram +)
2—red (Gram −)

Acid-fast stain
0—not done
1—red (acid-fast)
2—blue (non-acid fast)

Shape (morphology)
0—not done
1—straight rod
2—curved rod
3—spiral
4—spirochete
5—coccus
6—pleomorphic rod
7—coccobacillus

Cultural characteristics
0—not done
1—white, moist, glistening growth
2—abundant, white, thick, glistening growth
3—slimy, transluscent, raised growth
4—thin, even, grayish growth
5—thin, bluish-gray, spreading growth
6—thin, transparent growth with medium turning green
7—thin, white, spreading, viscous growth
8—abundant, opaque, golden growth
9—thin, even, transparent growth
10—soft, smooth, yellow-pigmented growth
11—grayish, granular, united growth

Litmus milk reactions
0—not done
1—acid
2—acid, reduction
3—acid, reduction, curd
4—acid, reduction, curd, gas
5—peptonization (proteolysis)
6—alkaline

Lactose fermentation
Dextrose fermentation
Sucrose fermentation
Mannitol fermentation
0—not done
1—yellow medium (acid)
2—yellow medium + air bubble (acid + gas)
3—red medium (negative)

Hydrogen sulfide
0—not done
1—medium black (positive)
2—medium clear (negative)

Nitrate reduction
0—not done
1—red color with solutions A + B (positive)
2—no color change with solutions A + B + Zn (positive)
3—red color with solutions A + B + Zn (negative)

Indole production
0—not done
1—red reagent layer (positive)
2—yellow reagent layer (negative)

Methyl red test
0—not done
1—red coloration of medium (positive)
2—yellow-orange coloration of medium (negative)

Voges-Proskauer test
0—not done
1—pink coloration of medium (positive)
2—no color change in medium (negative)

Citrate utilization test
0—not done
1—growth, medium blue (positive)
2—no growth, medium green (negative)

Urease activity
0—not done
1—medium deep pink (positive)
2—no color change in medium (negative)

Catalase activity
0—not done
1—evolution of oxygen bubbles (positive)
2—no evolution of oxygen bubbles (negative)

Starch hydrolysis
0—not done
1—clear zone with iodine (positive)
2—no clear zone with iodine (negative)

Lipid hydrolysis
0—not done
1—zone of lipolysis (positive)
2—no zone of lipolysis (negative)

zinc powder, tetramethyl-p-phenylenediamine dihydrochloride.

Equipment

Bunsen burner, inoculating loop and needle, staining tray, immersion oil, lens paper, bibulous paper, microscope, and wax marking pencil.

PROCEDURE

1. Perform a Gram staining procedure of the unknown organism. Determine and record the reaction and the morphology and arrangement of the cells.

2. Using sterile inoculating technique, inoculate two trypticase soy agar slants by means of a streak inoculation. Following incubation, one slant culture will be used to determine the cultural characteristics of the unknown microorganism. The second is to be used as a stock subculture should it be necessary to repeat any of the tests.

3. Exercising care in sterile technique, so as not to contaminate cultures and thereby obtain spurious results, inoculate the media for the following biochemical tests:

Medium	Test
a. Phenol red lactose broth	
b. Phenol red dextrose broth	Carbohydrate fermention
c. Phenol red sucrose broth	
d. Litmus milk	Litmus milk reactions
e. SIM medium	Indole production
	H₂S production
f. Tryptic nitrate broth	Nitrate reduction
g. MR-VP broth	Methyl red test
	Voges-Proskauer test
h. Simmons citrate agar slant	Citrate utilization
i. Urea broth	Urease activity
j. Trypticase soy agar slant	Catalase activity
k. Starch agar plate	Starch hydrolysis
l. Tributyrin agar plate	Lipid hydrolysis
m. Nutrient gelatin deep tube	Gelatin liquefication
n. Trypticase soy agar plate	Oxidase test

4. Incubate all cultures for 24 to 72 hours at 37 degrees C.

VI

The Protozoa

PURPOSES

1. To acquaint students with the distinguishing characteristics of the members of the phylum Protozoa.

2. To perform experimental procedures to identify free-living and parasitic protozoans.

INTRODUCTION

The protozoa are a large and diverse group of unicellular, eukaryotic organisms; most are free living, however, some are parasites. Their major distinguishing characteristics are:

1. The absence of a cell wall; some however, possess a flexible layer, a pellicle, or a rigid shell of inorganic materials outside of the cell membrane.

2. The ability during their entire life cycle or part of it to move by locomotor organelles or by a gliding mechanism.

3. Heterotrophic nutrition whereby the free-living forms ingest particulates such as bacteria, yeast, and algae, while the parasitic forms derive nutrients from the body fluids of their hosts.

4. Primarily asexual means of reproduction, although in some groups sexual modes also occur.

Taxonomically, the classification of protozoa depends on their means of locomotion. The phylum is subdivided into the following four classes (or subphyla by some taxonomists):

1. **Sarcodina:** Motility is due to the streaming of ectoplasm, producing protoplasmic projections called pseudopods (false feet). Prototypic amoebas include the free-living *Amoeba proteus* and its parasitic congener *Entamoeba histolytica*.

2. **Mastigophora:** Locomotion is effected by one or more whiplike, thin structures called flagella. Free-living members include the genera *Cercomonas, Heteronema,* and *Euglena,* photosynthetic protists that may be classified as flagellated algae. The parasitic forms include *Trichomonas vaginalis, Giardia lamblia,* and the *Trypanosoma* species.

3. **Ciliophora:** Locomotion is carried out by means of short hairlike projections called cilia, whose synchronous beating propels the organisms. The characteristic example of a free-living member of this group is *Paramecium caudatum,* and the parasitic member is *Balantidium coli.*

4. **Sporozoa:** Unlike other members of this phylum, sporozoa do not have locomotor organelles in their mature stage; however, immature forms exhibit some type of movement. All the members of this group are parasites. The most significant members belong to the genus *Plasmodium,* the malarial parasites of animals and humans.

Free-Living Protozoa

PURPOSE

To acquaint students with the protozoan flora of pond water.

PRINCIPLE

There are more than 20,000 known species of free-living protozoa. It is not within the scope of this manual to present an in-depth study of this large and diverse population. Therefore in this procedure the student will use Table 32.1 to become familiar with the general structural characteristics of representative protozoa and will identify these in a sample of pond water.

MATERIALS

Cultures

Stagnant pond water.

Reagent

Methyl cellulose.

Equipment

Microscope, glass slides, coverslips, and Pasteur pipettes.

TABLE 32.1.   Structural Characteristics of Free-Living Protozoa

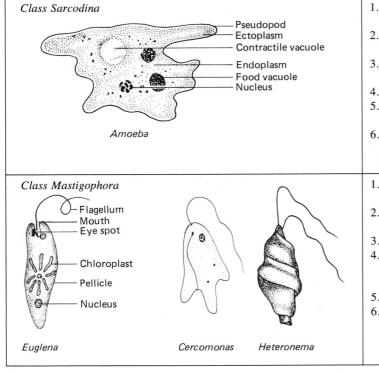

Class Sarcodina — Amoeba	1. **Pseudopods:** Protoplasmic projections that function for locomotion 2. **Ectoplasm:** Outer layer of cytoplasm clear in appearance 3. **Endoplasm:** Inner cytoplasmic region granular in appearance 4. **Nucleus:** One present 5. **Food vacuoles:** Contain engulfed food undergoing digestion 6. **Contractile vacuole:** Large, clear circular structure that regulates internal water pressure
Class Mastigophora — Euglena, Cercomonas, Heteronema	1. **Flagella:** One to several long whiplike structures that function for locomotion 2. **Pellicle:** Elastic layer outside of cell membrane 3. **Mouth:** Present but indistinct 4. **Chloroplast:** Organelles containing chlorophyll present in photosynthetic forms only 5. **Eye spot:** Light-sensitive pigmented spot 6. **Nucleus:** One present

Labels in Amoeba diagram: Pseudopod, Ectoplasm, Contractile vacuole, Endoplasm, Food vacuole, Nucleus

Labels in Euglena diagram: Flagellum, Mouth, Eye spot, Chloroplast, Pellicle, Nucleus

TABLE 32.1   (*continued*)

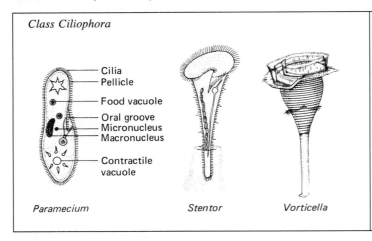

Class Ciliophora	
Cilia Pellicle Food vacuole Oral groove Micronucleus Macronucleus Contractile vacuole *Paramecium* *Stentor* *Vorticella*	1. **Cilia:** Numerous short hairlike structures that function for locomotion 2. **Pellicle:** Outermost flexible layer 3. **Contractile vacuole** with radiating canals: Regulates osmotic pressure 4. **Oral groove:** Indentation that leads to the mouth and gullet 5. **Food vacuoles:** Sites of digestion of ingested food 6. **Macronucleus:** A large nucleus that functions to control the cell's activities; one to several may be present 7. **Micronucleus:** A small nucleus that functions in conjugation, a mode of sexual reproduction

PROCEDURE

1. Obtain a drop of pond water from the bottom of the culture and place it in the center of a clean slide.

2. Add a drop of methyl cellulose to the culture to slow down the movement of the protozoa.

3. Apply a coverslip in the following manner to prevent formation of air bubbles.

   a. Place one edge of the coverslip against the outer edge of the drop of culture.

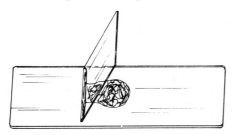

b. After the drop of culture spreads along the inner aspect of the edge of the coverslip, gently lower the coverslip onto the slide.

4. Examine your slide preparation under the low-power and high-power objectives with diminished light.

Observations and Results EXPERIMENT **32**

1. Observe your slide preparation for the different protozoans present.

2. In the space provided, draw a representative sketch of several of the observed protozoans and label their structural components. Identify each organism according to its class based on its mode of locomotion and its genus.

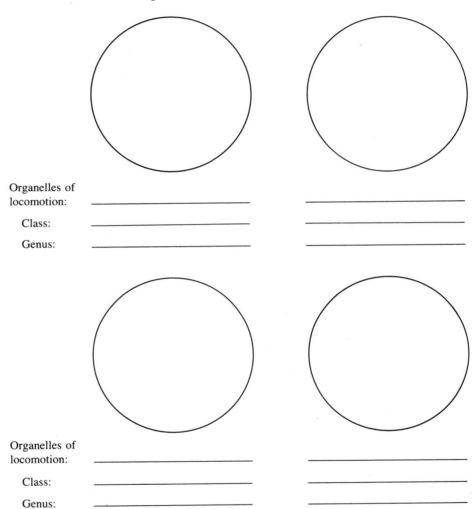

Organelles of
locomotion: _____   _____

Class: _____   _____

Genus: _____   _____

Organelles of
locomotion: _____   _____

Class: _____   _____

Genus: _____   _____

REVIEW QUESTIONS

1. What are the distinguishing characteristics of the free-living members of Sarcodina, Mastigophora, and Ciliophora?

2. Identify and give the function of each of the following:

    a. Pseudopods:

    b. Contractile vacuole:

    c. Eye spot:

    d. Micronucleus:

    e. Pellicle:

    f. Oral groove:

Parasitic Protozoa

PURPOSE

To acquaint students with parasitic protozoan forms.

PRINCIPLE

Unlike the free-living forms, the life cycles of parasitic protozoa vary greatly in complexity. Knowledge of the various developmental stages in these life cycles is essential in the diagnosis, clinical management, and chemotherapy of parasitic infections.

Parasites with the simplest or most direct life cycles not requiring an intermediate host are:

1. *Entamoeba histolytica:* Member of the class Sarcodina that causes amoebic dysentery. Infective, resistant cysts are released from the lumen of the intestine through the feces and are deposited in water, soil, or on vegetation. On ingestion, the mature quadrinucleated cyst wall disintegrates and

the nuclei divide, producing eight active trophozoites that move to the colon where they establish infection.

2. *Balantidium coli:* The ciliated parasitic protozoa exhibits a life cycle similar to *E. histolytica* except that no multiplication occurs within the cyst.

3. *Giardia lamblia:* The intestinal mastigophoric flagellate with a life cycle comparable to those of the above parasites.

The mastigophoric hemoflagellate responsible for various forms of African sleeping sickness has a more complex life cycle. The *Trypanosoma* must have two hosts to complete its cyclic development: A vertebrate, and an invertebrate, blood-sucking, insect host. Humans are the definitive hosts harboring the sexually mature forms; the tsetse fly (*Glossina*) and the Reduvig bug are the invertebrate hosts in which the developmental forms occur.

Table 33.1 illustrates the morphological characteristics of prototypic members of the parasitic protozoa except the Sporozoa.

TABLE 33.1 Characteristics of Representative Parasitic Protozoa

Class, organism, and infection	Structural characteristics	Locomotor organelles	Site of infection	Isolation of parasitic form
Sarcodina ENTAMOEBA HISTOLYTICA INFECTION: AMOEBIC DYSENTERY Peripheral chromatin / Central karyosome / Uniform cytoplasm / Red blood cell / Nuclei / Chromatoid body / Early / Late	**Trophozoite:** Shape: Variable Nucleus: Discrete nuclear membrane with central karyosome and peripheral chromatin granules Cytoplasm: Clear, red blood cells may be present **Cyst:** Shape: Round to oval with thick wall Nuclei: 1–4 present; mature cyst is quadrinucleated Chromatoid bodies: Sausage-shaped with rounded ends present in young cysts only	Pseudopods None	Large intestine by ingestion of mature cysts	Diarrhetic stool Formed stool

TABLE 33.1 (*continued*)

Class, organism, and infection	Structural characteristics	Locomotor organelles	Site of infection	Isolation of parasitic form
Mastigophora *TRYPANOSOMA GAMBIENSE* INFECTION: AFRICAN SLEEPING SICKNESS — Flagellum — Undulating membrane — Nucleus — Volutin granules — Kinetoplast	**Trophozoite:** Shape: Crescent Nucleus: Large, central, and polymorphic Cytoplasm: Granular **Cyst:** None	Single flagellum along undulating membrane	Peripheral blood stream by means of tsetse fly vector	Peripheral blood
GIARDIA LAMBLIA INFECTION: DYSENTERY — Nucleus — Karyosome — Median bodies — Axonemes — Nuclei — Median bodies — Retracted protoplasm — Axonemes	**Trophozoite:** Shape: Pear-shaped with concave sucking disc Nuclei: 2 bilaterally located with central karyosome and no peripheral chromatin Cytoplasm: Uniform and clear **Cyst:** Shape: Oval to ellipsoidal Nuclei: 2–4 present and protoplasm retracted from cyst wall Axostyle Parabasal body	4 pairs of flagella 4 pairs of flagella within cyst	Small intestine through ingestion of cysts	Diarrhetic stool Formed stool
Ciliophora *BALANTIDIUM COLI* INFECTION: DYSENTERY — Cytostome — Cilia — Micronucleus — Macronucleus — Cyst wall — Macronucleus	**Trophozoite:** Shape: Oval Nuclei: Kidney-shaped macronucleus and a micronucleus Cytoplasm: Vacuolated **Cyst:** Shape: Round and thick-walled Nuclei: 1 macronucleus and a micronucleus that is not visible	Cilia Cilia within cyst	Large intestine by the ingestion of cysts	Diarrhetic stool Formed stool

Protozoans demonstrating the greatest degree of cyclic complexity are found in the class Sporozoa. They are composed of exclusively obligate parasitic forms, such as members of the genus *Plasmodium,* and are responsible for malaria in both humans and animals. The life cycle requires two hosts, a human being and the female *Anopheles* mosquito. It is significant to note that in this life cycle the mosquito, and not the human, is the definitive host harboring the sexually mature parasite.

Malaria is initiated when a person is bitten by an infected mosquito, during which time infective sexually mature sporozoites are injected with the insect's saliva. These parasites pass rapidly from the blood into the liver where they infect the parenchymal cells. This is the **pre-erythrocytic stage.** The parasites develop asexually within the liver cells by a process called **schizogony,** producing **merozoites.** This cycle may be repeated or the merozoites released from the ruptured liver cells may now infect red blood cells and initiate the **erythrocytic stage.** During this asexual development the parasite undergoes a series of morphological changes that are of diagnostic value. These forms are designated as **signet rings, trophozoites, schizonts, segmenters, merozoites,** and **gametocytes.** The merozoites are capable of reinfecting other red blood cells or liver cells. Ingestion of the **microgametocytes** (♂) and **macrogametocytes** (♀) by another mosquito during a blood meal initiates the sexual cycle called **sporogamy.** Male and female gametes give rise to a zygote in the insect's gut. The zygote is then transformed into an **ookinite** that burrows through the gut wall to form an **oocyst** in which the sexually mature **sporozoites** develop, thereby completing the life cycle.

In this experiment students will study the parasitic protozoa using prepared slides and the diagnostic characteristics shown in Figure 33.1 and Table 33.1. The purpose of the experiment is to gain an understanding of life cycles of parasitic protozoa.

MATERIALS

Prepared Slides

Entamoeba histolytica trophozoite and cyst, *Giardia lamblia* trophozoite and cyst, *Balantidium coli* trophozoite and cyst, *Trypanosoma gambiense,* and *Plasmodium vivax* in human blood smears.

Equipment

Microscope, immersion oil, and lens paper.

PROCEDURE

Examine all available slides under the oil-immersion objective. Use Table 33.1 and Figure 33.1 to identify the distinguishing microscopic characteristics of each parasite studied.

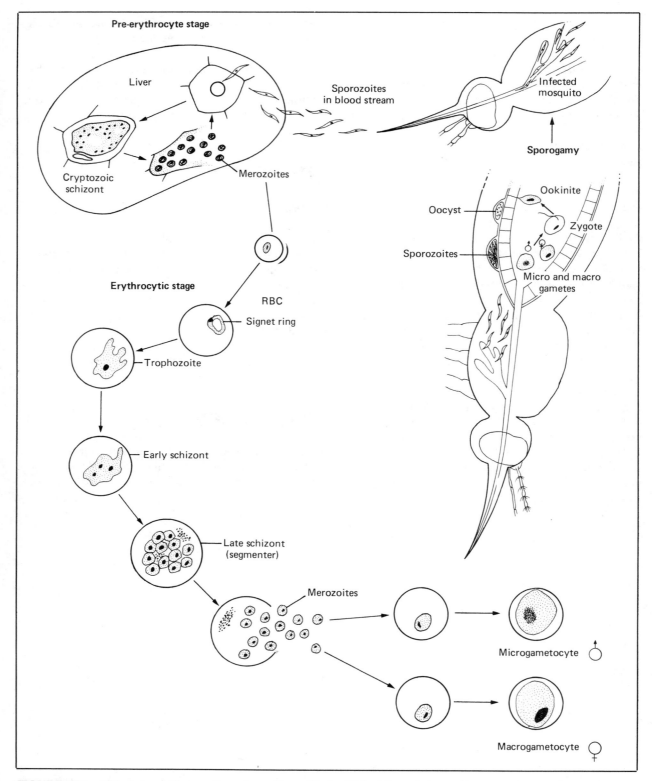

FIGURE 33.1. Life cycle of *Plasmodium vivax*. (From *Basic Clinical Parasitology*, 4th ed., by Harold S. Brown. New York: Appleton-Century-Crofts, 1975. By permission.)

Observations and Results

Draw a representative sketch of the parasitic organisms studied and label the distinguishing structural characteristics you were able to observe.

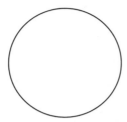 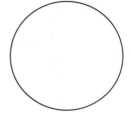

E. histolytica

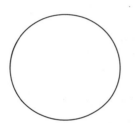 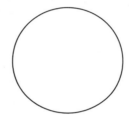 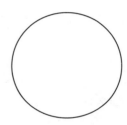

G. lamblia   **T. gambiense**

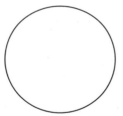 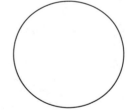

B. coli

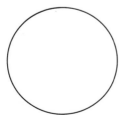  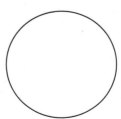 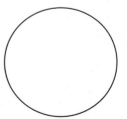

P. vivax: Erythrocytic stages

REVIEW QUESTIONS

1. In malarial infections, the sexually mature parasite is found in which host?
   Is this true for all other protozoan parasitic infections? Explain.

2. Describe the developmental stages of the malarial parasite during sporogamy and schizogony.

3. What role does the invertebrate host serve in the life cycle of the trypanosomes? Explain.

4. Distinguish between the pre-erythrocytic and erythrocytic stages in the life cycle of the malarial parasite.

VII

The Fungi

PURPOSES

1. To acquaint students with the macroscopic and microscopic structure of yeasts and molds.

2. To instruct students with basic mycological culturing and staining procedures.

3. To provide students with the opportunity to identify a common fungal organism.

INTRODUCTION

The branch of microbiology that deals with the study of fungi (yeasts and molds) is called **mycology.** True fungi are separated into the following four classes on the basis of their sexual mode of reproduction:

1. **Phycomycetes:** Water, bread, and terrestrial molds. Reproductive spores are external and uncovered.

2. **Ascomycetes:** Yeasts and molds. Sexual spores, called ascospores, are produced in a saclike structure called an ascus.

3. **Basidiomycetes:** Fleshy fungi, toadstools, mushrooms, puffballs, and bracket fungi. Reproductive spores, basidiospores, separate from specialized stalks called basidia.

4. **Deuteromycetes:** Also called **Fungi Imperfecti** because no sexual reproductive phase has been described.

Nutritionally, the fungi are heterotrophic, eukaryotic microorganisms that are enzymatically capable of metabolizing a wide variety of organic substrates. Fungi can have beneficial or detrimental effects on humans. Those that inhabit the soil play a vital role in decomposing dead plant and animal tissues, thereby maintaining a fertile soil environment. The fermentative fungi are of industrial importance in producing beer and wine, bakery products, cheeses, industrial enzymes, and antibiotics. The detrimental activities of some fungi include spoilage of foods by rots, mildews, and rusts found on fruit, vegetables, and grains. Some species are capable of producing toxins (aflatoxin) and hallucinogens. A few fungal species are of medical significance because of their capacity to produce diseases in humans. Most of the pathogenic fungi are Deuteromycetes and can be divided into two groups based on site of infection. The **superficial mycoses** cause infections of the skin, hair, and nails, for example, ringworm infections. The **systemic mycoses** cause infections of the subcutaneous and deeper tissues such as lungs, genital areas, and nervous system.

197

Cultivation and Morphology of Molds

PURPOSES

1. To acquaint students with mycological culture techniques.

2. To visualize and identify the structural components of molds.

PRINCIPLE

Molds are the major fungal organisms that can be seen by the naked eye. We have all seen them growing on foods such as bread or citrus fruit as a cottony, fuzzy, black, green, or orange growth, depending on the mold. Examination with a simple hand lens shows that these organisms are composed of an intertwining branching mat called **mycelium.** The filaments that make up this mycelial mat are called **hyphae.** Most of the mat grows on or in the surface of the nutrient medium so that it can extract nutrients; the mat is therefore called **vegetative mycelium.** Some of the mycelium mat rises upward from the mat and is referred to as **aerial mycelium.** Specialized hyphae are produced from the aerial mycelium and give rise to spores that are the reproductive elements of the mold. Figure 34.1 shows the reproductive structures of some fungi.

As the structural components of molds are very delicate, special techniques are used to study them. Simple handling with an inoculating loop may result in mechanical disruption of their components. To avoid this, the following microtechnique is used. One needs a deep concave slide containing a suitable nutrient medium with an acidic pH, such as Sabouraud agar, which is then covered by a removable coverslip. Mold spores are deposited on the surface of the agar and incubated in a moist chamber at room temperature. Direct microscopic observation is then possible without fear of disruption or damage to anatomical components. Molds can be identified as to spore type and shape, type of sporangia, and type of mycelium, as shown in Figure 34.1 and Table 36.1.

MATERIALS

Cultures

7-day to 10-day Sabouraud agar cultures of *Penicillium notatum* and *Aspergillus niger*.

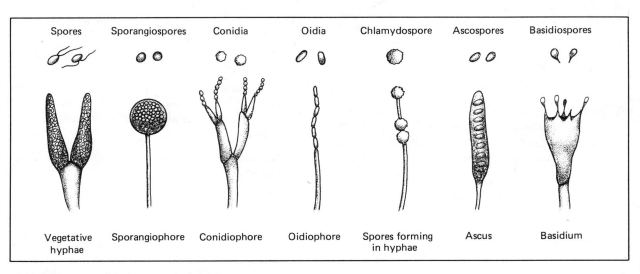

FIGURE 34.1. Spore and sporangia types

Media

One Sabouraud agar deep tube per designated student group.

Equipment

Bunsen burner, water bath, two concave glass slides, two coverslips, petrolatum, sterile Pasteur pipettes, toothpicks, two sterile Petri dishes, forceps, inoculating loop and needle, two sterile U-shaped bent glass rods, thermometer, dissecting microscope, and beaker with 95 percent ethyl alcohol.

PROCEDURE

1. Melt the deep tube of Sabouraud agar in a boiling water bath and cool to 45 degrees C.

2. Place a piece of filter paper in the bottom of each Petri dish, insert a sterile glass bent rod into each dish, and replace the covers as illustrated:

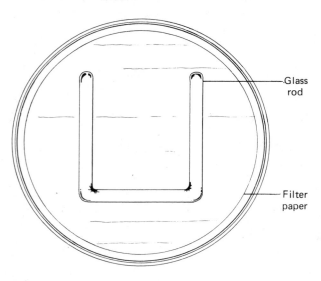

3. Using a forceps dip the concave slides and coverslips in a beaker of 95 percent ethanol, pass through Bunsen burner flame, remove from flame, and hold until all the alcohol has burned off the slides and coverslips.

4. Cool slides and coverslips. Place a slide and coverslip inside each Petri dish on the glass rod concave side up.

5. With a toothpick, add petrolatum to three sides surrounding the concave section of each slide. The fourth side will serve as a vent for air passage.

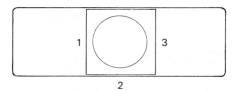

6. With a sterile Pasteur pipette, add one to two drops of cooled Sabouraud agar to the concavity of each slide.

7. Place a coverslip over the concave portion of each slide so that it is completely sealed.

8. With forceps, stand both slides upright inside their respective Petri dishes until the agar solidifies. Following solidification the slide will resemble the following illustration:

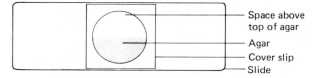

9. When agar is fully hardened, move coverslips downward and with a sterile needle inoculate each prepared slide with the spores from the test cultures.

10. Push the coverslips upward to their original position, thereby sealing off the slide.

11. With a Pasteur pipette, moisten the filter paper with sterile water to provide a moist atmosphere. Remoisten filter paper when necessary during the incubation period.

12. Place slide on U-shaped bent rod, replace Petri dish cover, and label the covers with the name of the organism inoculated and identify each with your initials.

13. Incubate the preparations for seven days at 25 degrees C.

Yeast Morphology

PURPOSE

To acquaint students with the morphological characteristics of yeast cells.

PRINCIPLE

Yeast are unicellular organisms about 5 to 10 times larger than bacteria. Morphologically, they may be ellipsoidal, spherical, and in some cases, cylindrical. Most yeast reproduce asexually by a process called **budding.** A **bud** is an outgrowth from the parent cell that pinches off, producing a daughter cell. Some yeasts may also undergo sexual reproduction when two ascospores conjugate, giving rise to a zygote or diploid cell. The nucleus of this cell divides by meiosis, producing four haploid nuclei called **ascospores** contained within the cell that is now called the **ascus.** When the ascus ruptures, the ascospores are released and conjugate, starting the cycle again.

Yeasts are important for many reasons. Most are beneficial but a few are not. Yeasts such as the *Saccharomyces* are used for brewing beer and for baking; some wild yeasts are important in the fermentation of wine. Also, the high vitamin content of yeasts makes them particularly valuable as a food supplement. Harmful yeasts must be excluded during the manufacture of fruit juices such as grape juice or apple cider to prevent fermentation of the fruit sugars to alcohol. Also, contamination of soft cheeses by some forms of yeast will destroy the product. Finally, some yeasts such as *Candida albicans* are pathogenic and responsible for urinary and vaginal infections known as **moniliasis,** and infections of the mouth called **thrush.**

Yeasts are eucaryotic cells

MATERIALS

Cultures

Seven-day Sabouraud agar culture of *Saccharomyces cerevisiae*.

Reagents

Water-iodine solution, lactophenol-cotton-blue solution.

Equipment

Bunsen burner, inoculating loop and needle, glass slides, coverslips, and microscope.

PROCEDURE

1. Prepare two wet mounts of the yeast culture in the following manner:

    a. Suspend a loopful of yeast culture into a few drops of the water-iodine solution on a microscope slide and cover with a coverslip.

    b. Suspend a loopful of yeast culture into a few drops of lactophenol-cotton-blue solution on a microscope slide and cover with a coverslip.

Identification of Unknown Fungi

PURPOSE

To identify a fungal unknown.

PRINCIPLE

In this experiment, each student will be provided with a number-coded pure culture of a representative fungal organism for cultivation and subsequent identification. Table 36.1 is to be used to aid in identification of the unknown culture.

MATERIALS

Cultures

Number-coded, seven-day Sabouraud broth spore suspensions of *Aspergillus, Mucor, Penicillium, Alternaria, Rhizopus, Cladosporium, Cephalosporum,* and *Fusarium.*

Media

One Sabouraud agar plate per student.

TABLE 36.1. Identification of Fungi

Diagram	*Colonial morphology*	*Microscopic appearance*
Molds — Sporangium — Columella — Sporangiophore — Mycelium — Rhizoid *RHIZOPUS:* BLACK BREAD MOLD; COMMON LABORATORY CONTAMINANT	Rapidly growing white-colored fungus swarms over entire plate; aerial mycelium cottony and fuzzy	Spores are oval, colorless or brown; nonseptate mycelium gives rise to straight sporangiophores that terminate with black sporangium containing a columella; rootlike hyphae (rhizoids) penetrate the medium
 — Sporangium — Columella — Sporangiophore — Mycelium *MUCOR:* FOOD CONTAMINANT	Resembles the colonies of *Rhizopus*	Spores are oval; nonseptate mycelium gives rise to single sporangiophores with globular sporangium containing a columella; there are no rhizoids

207

TABLE 36.1.   (continued)

Diagram	Colonial morphology	Microscopic appearance
Molds — Conidia — Conidiophore — Mycelium *ALTERNARIA:* NORMALLY FOUND ON PLANT MATERIAL	Grayish-green or black colonies with gray edges rapidly swarming over entire plate; aerial mycelium not very dense, appears grayish to white	Multicelled spores (conidia) are pear-shaped and attached to single conidiophores arising from a septate mycelium
— Conidia — Conidiophore *FUSARIUM:* FOUND IN SOIL	Woolly, white, fuzzy colonies changing color to pink, purple, or yellow	Multicelled spores (conidia) are oval or crescent-shaped and attached to conidiophores arising from a septate mycelium
— Conidia — Sterigma — Vesicle — Conidiophore — Mycelium *ASPERGILLUS:* PLANT AND ANIMAL PATHOGENS; SOME SPECIES USED INDUSTRIALLY	White colonies become greenish-blue, black, or brown as culture matures	Single-celled spores (conidia) in chains developing at the end of the sterigma arising from the terminal bulb of the conidiophore, the vesicle; long conidiophores arise from a septate mycelium
— Conidia — Sterigma — Metula — Conidiophore — Mycelium *PENICILLIUM:* ANTIBIOTIC-PRODUCING CITRUS FRUIT CONTAMINANT; SOIL INHABITANT	Mature cultures usually greenish or blue-green	Single-celled spores (conidia) in chains develop at the end of the sterigma arising from the metula of the conidiophore; branching conidiophores arise from a septate mycelium

TABLE 36.1. (*continued*)

Diagram	*Colonial morphology*	*Microscopic appearance*
 CLADOSPORIUM: DEAD AND DECAYING PLANTS	Small, heaped colonies are greenish-black and powdery	Spores (conidia) develop at the end of complex conidiophores arising from a septate mycelium that is usually brownish
 CEPHALOSPORUM: ANTIBIOTIC PRODUCTION	Rapidly growing compact and moist colonies becoming cottony with aerial hyphae that are gray or rose-colored	Single-celled conical or elliptical spores (conidia) held together in clusters at the tips of the conidiophores by a mucoid substance; erect, unbranched conidiophores arise from a septate mycelium
Yeast *TORULA:* CHEESE AND FOOD CONTAMINANT	Colonies are pink, moist, with unbroken, even edges	Cells are oval, colorless, and reproduce by budding
 CANDIDA: HUMAN PATHOGEN	Colonies are small, round, moist, and colorless, with unbroken, even edges	Yeastlike fungus produces pseudomycelium

Reagent

Lactophenol-cotton-blue solution.

Equipment

Bunsen burner, dissecting microscope, hand lens, sterile cotton swabs, glass slides, and coverslips.

PROCEDURE

1. Label the Sabouraud agar plate with the culture number and identify with your initials.

2. With a sterile cotton swab, inoculate the Sabouraud agar plate with the provided culture.

3. Incubate the plates in an inverted position for one week at 25 degrees C in a moist incubator.

Observations and Results   EXPERIMENT **36**

1. Observe mold cultures with a hand lens, noting and recording their colonial morphology.

2. Prepare a wet mount by suspending some of the culture in a few drops of lactophenol-cotton-blue solution. This should be done gently to avoid damaging the fungal structures.

3. Examine the preparation under high-power and low-power magnifications with the aid of a dissecting microscope, and record your observations.

4. Draw a representative microscopic field of your culture in the chart.

5. Using Table 36.1, identify your unknown fungal organism.

   Description of microscopic structure _____

   _____

   _____

   _____

   Description of cultural morphology _____

   _____

   _____

   _____

Diagram of microscopic appearance

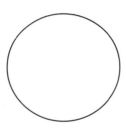

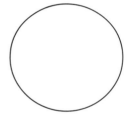

Low-power   High-power

Genus of fungal unknown _____

VIII

The Viruses

PURPOSES

1. To instruct students in the chemical structure, morphology, and replicative activities of bacterial viruses.

2. To perform a phage dilution procedure for the cultivation and enumeration of bacterial viruses.

3. To isolate bacteriophages from sewage.

4. To cultivate animal viruses in embryonated chick eggs.

INTRODUCTION

Viruses are noncellular biological entities composed solely of a nucleic acid surrounded by a protein envelope called the capsid. Because of their limited and simplistic structure, viruses can be chemically defined as **nucleoproteins.** They are devoid of the sophisticated enzymatic and biosynthetic machinery essential for independent activities of cellular life. This lack of metabolic machinery mandates that they must exist as parasites, and they cannot be cultivated outside of a susceptible living cell. Viruses are differentiated from cellular forms of life on the following bases:

1. They are ultramicroscopic and can only be visualized with the electron microscope.

2. They are filterable and able to pass through bacterial-tight filters.

3. They do not increase in size.

4. They must replicate within a susceptible cell.

5. Replication occurs because the viral nucleic acid subverts the synthetic machinery of the host cell for the purpose of producing new viral components.

6. Viruses are designated either RNA or DNA viruses, as they contain one of the nucleic acids but never both.

Much of our knowledge of the mechanism of animal viral infection and replication has been based on our understanding of infection in bacteria by bacterial viruses, called the **bacteriophages** or **phages.** The bacteriophages were first described in 1915 almost simultaneously by Twort and

213

d'Herelle. The name bacteriophage, which in Greek means *to eat bacteria,* was coined by d'Herelle because of the destruction through lysis of the infected cell. Bacteriophages exhibit notable
variability in their size, shape, and complexity of structure. The T-even (T_2, T_4, and T_6) phages
illustrated here demonstrate the greatest morphological complexity.

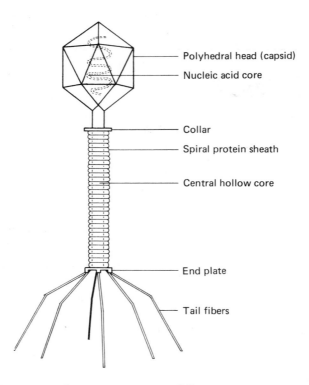

The functions of these structural components are as follows:

Component	Function
Capsid (protein coat)	Protection of nucleic acid from destruction by DNAses
Nucleic acid core	Phage genome carrying genetic information necessary for replication of new phage particles
Spiral protein sheath	Retracts so that nucleic acid can pass from capsid into host cell's cytoplasm
End plate and **tail fibers**	Attachment of phage to specific receptor sites on a susceptible host's cell wall

Phage replication depends on the ability of the phage particle to infect a suitable bacterial
host cell. Infection consists of the following sequential events:

1. **Adsorption:** Tail portion of the phage particle binds to receptor sites on host's cell wall.

2. **Penetration** (infection): Spiral protein sheath retracts, and an enzyme, early muramidase,
   perforates the bacterial cell wall enabling the phage nucleic acid to pass through the hollow

core into the host cell's cytoplasm. The empty protein shell remains attached to the cell wall and is called the protein ghost.

3. **Replication:** The phage genome subverts the synthetic machinery, which is then used for the production of new phage components.

4. **Maturation:** The period during which the new phage components are assembled and form complete, mature virulent phage particles.

5. **Release:** Late muramidase lyses the cell wall, liberating infectious phage particles that are now capable of infecting new susceptible host cells, thereby starting the cycle over again.

These five events are illustrated in the lytic portion of the diagram below.

The lytic and lysogenic life cycles of a bacteriophage

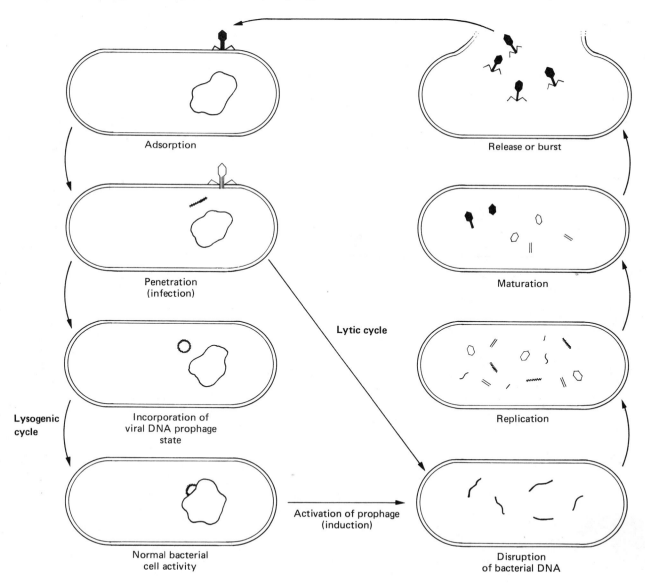

Adsorption

Release or burst

Penetration
(infection)

Maturation

Lytic cycle

Incorporation of
viral DNA prophage
state

Replication

**Lysogenic
cycle**

Activation of prophage
(induction)

Normal bacterial
cell activity

Disruption
of bacterial DNA

   Virulent phage particles that infect susceptible host cells always initiate the **lytic cycle** as described. Other phage particles, called **temperate phages,** incorporate their nucleic acid into the host's chromosome. Lysis of the host cell does not occur until it is induced by exogenous physical agents such as ultraviolet, ionizing radiation or chemical mutagenic agents. Bacterial cells containing the incorporated phage nucleic acid, the **prophage,** are called **lysogenic cells.** Lysogenic cells appear and function as normal cells, and reproduce by fission. When induced by physical or chemical agents, these cells will release a virulent prophage from the host's genome, which then initiates the lytic cycle.

Cultivation and Enumeration of Bacteriophages

PURPOSE

To enable students to develop techniques for cultivating and enumerating bacteriophages.

PRINCIPLE

This exercise demonstrates the ability of viruses to replicate inside a susceptible host cell. For this purpose, the students will be provided with a virulent phage and a susceptible host cell culture. This technique also enables students to enumerate phage particles on the basis of plaque formation in a solid agar medium. Plaques will be observed as clear areas in an agar medium previously seeded with a diluted phage sample and a host cell culture. Each clear zone represents the lysis of a phage-infected bacterial cell.

The procedure requires the use of a double-layered cultural technique in which the hard agar serves as a base layer and a mixture of phage and host cells in a soft agar forms the upper overlay. Susceptible *Escherichia coli* cells multiply rapidly and produce a lawn of confluent growth on the medium. When one phage particle adsorbs to a susceptible cell, penetrates the cell, replicates, and goes on to lyse the host cell, the destroyed cell produces a circular clear area in the bacterial lawn. This clear zone is a **plaque-forming unit (PFU)**, which is illustrated in Figure 37.1.

The number of phage particles contained in the original stock phage culture is determined by counting the number of plaques formed on the seeded agar plate and multiplying this by the dilution factor. To determine a valid phage count, the number of plaques per plate should not exceed 300 or be less than 30.

Example: 200 PFUs are counted in a 10^{-6} dilution. $(200) \times (10^6) = 200 \times 10^6$ or 2×10^8 PFU per ml of stock phage culture.

Plates showing greater than 300 PFUs are **too numerous to count (TNTC);** plates showing fewer than 30 PFUs are **too few to count (TFTC).**

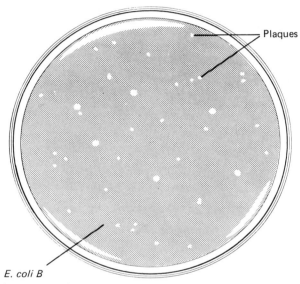

E. coli B
(Lawn of confluent growth)

FIGURE 37.1. Plaque-forming units (PFU)

MATERIALS

Cultures

24-hour nutrient broth cultures of *Escherichia coli B* and T_2 coliphage.

Media

Five each of the following per designated student group: tryptone agar plates and tryptone soft agar, 2 ml per tube, and 10 tryptone broth, 9 ml per tube.

Equipment

Bunsen burner, water bath, thermometer, 1-ml sterile pipettes, sterile Pasteur pipettes, and wax pencil.

PROCEDURE

1. Label all dilution tubes and media as follows:

   a. 5 tryptone soft agar tubes: 10^{-5}, 10^{-6}, 10^{-7}, 10^{-8}, 10^{-9}.

   b. 5 tryptone hard agar plates: 10^{-5}, 10^{-6}, 10^{-7}, 10^{-8}, 10^{-9}.

   c. 10 tryptone broth tubes: 10^{-1} through 10^{-10}.

2. Place the five labeled soft tryptone agar tubes into a water bath. Water should be of a depth just slightly above that of the agar in the tubes. Bring the water bath to 100 degrees C to melt the agar. Cool and maintain the melted agar at 45 degrees C.

3. With 1-ml pipettes aseptically perform a 10-fold serial dilution of the provided phage culture using the 10 9-ml tubes of tryptone.

4. To the tryptone soft agar tube labeled 10^{-5}, aseptically add two drops of the *E. coli B* culture with a Pasteur pipette and 1 ml of the 10^{-5} tryptone broth phage dilution. Rapidly mix by rotating the tube between the palms of your hands and pour the contents over the hard tryptone agar plate labeled 10^{-5}, thereby forming a double-layered plate culture preparation. Swirl the plate gently and allow to harden.

5. Using separate Pasteur pipettes and 1-ml sterile pipettes, repeat step 4 for the tryptone broth phage dilution tubes labeled 10^{-6} through 10^{-9}.

6. Incubate all plate cultures in an inverted position for 24 hours at 37 degrees C.

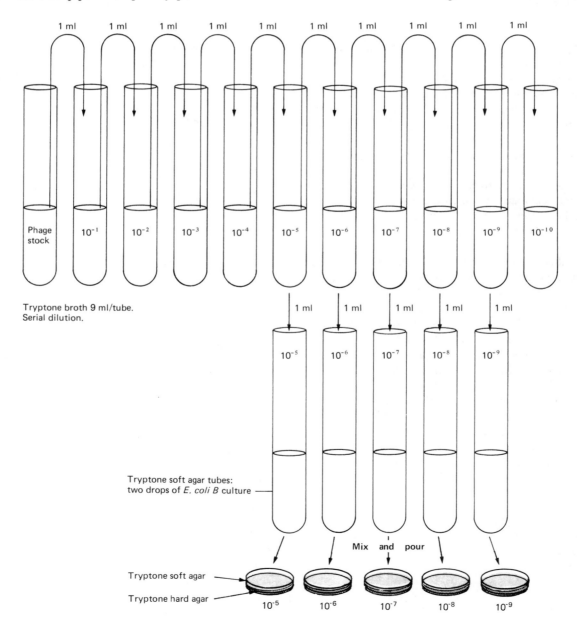

Observations and Results   EXPERIMENT **37**

1. Observe all plates for presence of plaque-forming units that develop on the bacterial lawn.

2. Count the number of PFUs on each plate in the range of 30 to 300.

3. Calculate the number of phage particles per ml of the stock phage culture based on your PFU count.

4. Record your results in the chart.

Phage dilution	Number of PFUs	Calculation: PFU × dilution	PFU/ml of stock phage culture
10^{-5}			
10^{-6}			
10^{-7}			
10^{-8}			
10^{-9}			

REVIEW QUESTIONS

1. Discuss the effects of lytic and lysogenic infections on the life cycle of the host cell.

2. Discuss the factors responsible for the transformation of a lysogenic infection to one that is lytic.

3. Distinguish between the replicative and maturation stages of a lytic phage infection.

4. In this experimental procedure, why is it important to use both a hard and a soft agar overlay technique to demonstrate plaque formation?

5. Explain what is meant by plaque-forming units.

6. Determine the number of PFUs per ml in a 10^{-9} phage culture that shows 204 PFUs in the agar lawn.

Isolation of Coliphages from Raw Sewage

PURPOSE

To isolate virulent coliphages from raw sewage.

PRINCIPLE

Isolates of bacterial viruses (bacteriophages) can be obtained from a variety of natural sources, including soil, intestinal contents, raw sewage, and some insects such as cockroaches and flies. Their isolation from these environments is not an easy task, as the phage particles are usually present in low concentrations. Therefore isolation requires a series of steps:

1. Collection of the phage-containing sample at its source.

2. Addition of an enriched susceptible host cell culture to the sample to increase the number of phage particles for subsequent isolation.

3. Following incubation, centrifugation of the enriched sample for the removal of gross particles.

4. Filtration of the supernatant liquid through a bacteria-retaining membrane filter.

5. Inoculation of the bacteria-free filtrate onto a lawn of susceptible host cells grown on a soft agar plate medium.

6. Incubation and observation of the culture for the presence of phage particles, which is indicated by plaque formation in the bacterial lawn.

In the following experiment, students will use the previously outlined procedure, illustrated in Figure 38.1 on page 222, for the isolation of *E. coli* phage particles from raw sewage. Bacteriophages that infect *E. coli* (coliphages) are designated by the letter T, indicating types. Seven types have been identified and are labeled T_1 through T_7. The T-even phages (T_2, T_4, and T_6) differ from the T-odd phages in that the former vary in size, form, and chemical compositions. All of the T phages are capable of infecting the non-motile susceptible *E. coli B* host cell.

MATERIALS

Cultures

Part 1: 5-ml 24-hour broth cultures of *Escherichia coli B* and 45-ml samples of fresh sewage collected in screw-capped bottles.
Part 2: 10-ml 24-hour broth culture of *Escherichia coli B*.

Media

Media per designated student group:
Part 1: One 5-ml tube of bacteriophage broth ($10 \times$).
Part 2: Five tryptone agar plates and five 3-ml-tubes of tryptone soft agar.

Equipment

Part 1: Sterile 250-ml Erlenmeyer flask and stopper.
Part 2: Sterile membrane filter apparatus, sterile 125-ml Erlenmeyer flask and stopper, 125-ml flask, 1000-ml beaker, centrifuge, Bunsen burner, forceps, 1-ml sterile disposable pipettes, and sterile Pasteur pipette.

PROCEDURE

Part 1: Enrichment of sewage sample

1. Aseptically add 5 ml of bacteriophage broth, 5 ml of the *E. coli B* broth culture, and 45 ml of the raw sewage sample to an appropriately labeled sterile 250-ml Erlenmeyer flask. **Handle the raw sewage with caution.**

2. Incubate the culture for 24 hours at 37 degrees C.

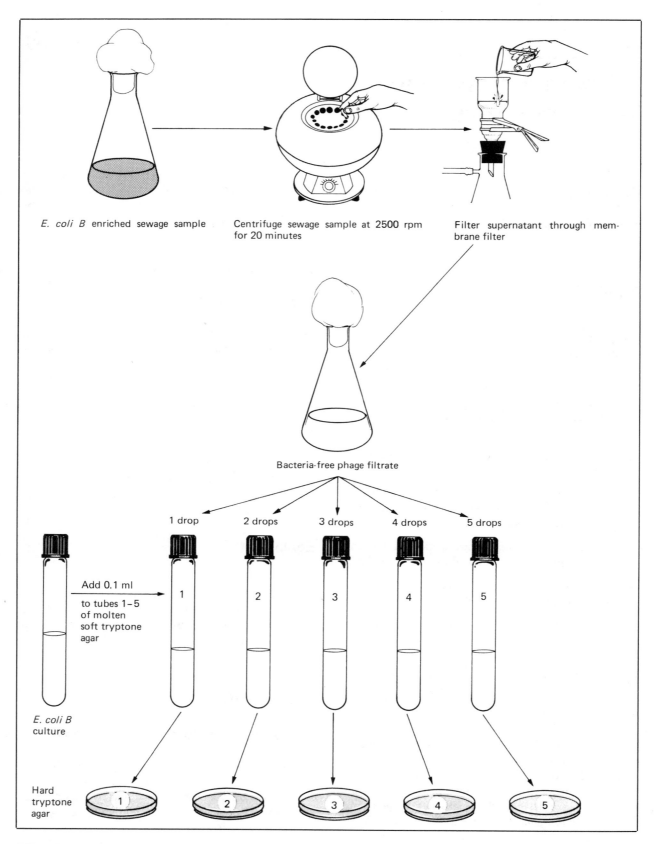

FIGURE 38.1. Procedure for isolation of coliphages from raw sewage

Part 2: Filtration and seeding

1. Following incubation, pour the phage-infected culture into 100-ml centrifuge bottle or several centrifuge tubes and centrifuge at 2500 rpm for 20 minutes.

2. Remove the centrifuge bottle or tubes, being careful not to stir up the sediment, and carefully decant the supernatant into a 125-ml flask.

3. Pour the supernatant solution through a sterile membrane filter apparatus to collect the bacteria-free, phage-containing filtrate in the vacuum flask below. Refer to Experiment 50 for the procedure in assembling the filter membrane apparatus.

4. Melt the soft tryptone agar by placing the five tubes in a boiling water bath and cool to 45 degrees C.

5. Label the five tryptone agar plates and the five tryptone agar tubes 1, 2, 3, 4, and 5, respectively.

6. Using a sterile 1-ml pipette, aseptically add 0.1 ml of the *E. coli B* culture to all the molten soft agar tubes.

7. Using a sterile Pasteur pipette, aseptically add 1, 2, 3, 4, and 5 drops of the filtrate to the respectively labeled molten soft agar tubes. Mix and pour each tube of soft agar into its appropriately labeled agar plate.

8. Allow agar to harden.

9. Incubate all the plates in an inverted position for 24 hours at 37 degrees C.

Observations and Results   EXPERIMENT **38**

1. Examine all the culture plates for plaque formation, which is indicative of the presence of coliphages in the culture.

2. Indicate the presence (+) or absence (−) of plaques in each of the cultures in the chart.

	Plaque formation				
Drops of phage filtrate	*1*	*2*	*3*	*4*	*5*

3. Based on your observations, what is the relationship between the number of plaques observed and the number of drops of filtrate in each culture?

REVIEW QUESTIONS

1. Why is enrichment of the sewage sample necessary for the isolation of phage?

2. How is enrichment of the sewage sample accomplished?

3. How are sterile phage particles obtained?

4. Why is it necessary to exercise caution when handling raw sewage samples?

5. What are the differences between T-even and T-odd coliphages?

Animal Virus Cultivation in Embryonated Chick Eggs

PURPOSES

1. To acquaint students with *in vitro* cultivation of animal viruses.

2. To perform an experiment using the chick embryo to demonstrate *in vitro* viral replication and its effect on the chick embryo.

PRINCIPLE

In vitro cultivation of animal viruses requires the use of susceptible living host cell systems. The three major sources of these systems are **susceptible animals, chick embryos,** and **tissue cultures.** Susceptible animals no longer serve as major host systems, having been replaced by the following more recent, efficient, and economical techniques.

The **embryonated egg,** although an *in vivo* system, functions essentially as an *in vitro* laboratory system. The egg containing the embryo is comparable to a test tube; and the embryo itself, with its cavities, yolk material, and membranes, provides an excellent culture medium for viral growth. The choice of a specific inoculation site depends on the purpose for which the virus is to be used. The possible sites are outlined as follows:

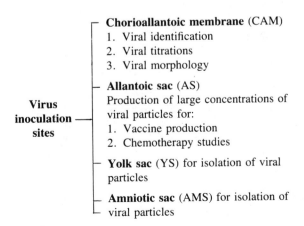

Virus inoculation sites ─

- **Chorioallantoic membrane** (CAM)
  1. Viral identification
  2. Viral titrations
  3. Viral morphology
- **Allantoic sac** (AS)
  Production of large concentrations of viral particles for:
  1. Vaccine production
  2. Chemotherapy studies
- **Yolk sac** (YS) for isolation of viral particles
- **Amniotic sac** (AMS) for isolation of viral particles

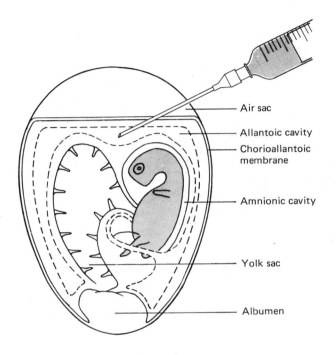

Labels: Air sac, Allantoic cavity, Chorioallantoic membrane, Amnionic cavity, Yolk sac, Albumen

FIGURE 39.1. Embryonated egg

Since the early 1950s animal cells grown in **tissue cultures** have become the major vehicle for the cultivation of animal viruses. This highly sophisticated technique requires placing the cells to be cultivated in a glass or plastic vessel containing an extremely rich nutritional medium. The cells attach to the surface of the vessel and continue dividing until the entire surface is covered with a confluent monolayer of cells. Viruses inoculated into a susceptible tissue culture infect the cells, subvert their metabolic machinery, and replicate rapidly.

This experiment is designed to familiarize students with the technique used to infect a 12-day-old chick embryo and to observe the cytopathogenic effect produced when the embryo is infected with Newcastle disease virus (NDV). This virus causes a respiratory disease that is responsible for the destruction of large flocks of commercial chickens. In humans, NDV is responsible for a mild form of **conjunctivitis.**

MATERIALS

Cultures

1:2 saline dilution of Newcastle virus.

Media

Two 10-day or 12-day embryonated chick eggs per designated student group.

Reagents

Tincture of iodine and 70 percent alcohol.

Equipment

Candling device, small electric or hand drill, sterile 1-ml tuberculin syringe with a 27-gauge 1¼-inch needle, vaspar, absorbent cotton, small beaker with 70 percent alcohol, Bunsen burner, and 1000-ml beaker containing household bleach.

PROCEDURE

1. Using the candling device, candle the two embryonated eggs to determine the viability of the embryo. This is verified if the embryo shows movement in response to heat from the light. Also determine the position of the air sac and large blood vessels, and mark their location on the egg shell.

2. Disinfect the shell over the air sac with tincture of iodine, allow to dry, and swab with absorbent cotton saturated in 70 percent alcohol.

3. Using a small drill with a bit that has been sterilized by dipping it in 70 percent alcohol and flaming, make a small hole in the shell over the air sac in a region away from the blood vessels. **Do not puncture the air sac membrane.**

4. Using a sterile 1-ml tuberculin syringe with a 27-gauge 1¼-inch needle, inject 0.2 ml of the 1:2 dilution of Newcastle virus into the allantoic cavity. This is done by holding the egg in a vertical position, inserting the needle through the hole in the shell to its hilt at a 45-degree angle, piercing the air sac membrane, and penetrating into the allantoic cavity as shown in Figure 39.1.

5. Following inoculation, withdraw the needle and seal the hole in the shell with sterile hot vaspar.

6. Repeat steps 2 through 5 to inoculate the second embryonated egg with 0.2 ml of sterile saline, which will serve as a control.

7. Incubate the eggs at 37 degrees C with properly supplied humidity.

Observations and Results   EXPERIMENT **39**

1. Candle the eggs each day during the incubation period and observe for embryonic death as evidenced by cessation of movement, which usually occurs three to four days following inoculation of virus.

2. Once death has been determined, dislodge the embryo from its shell, pour contents into Petri dish, and observe for necrotic lesions and evidence of hemorrhage. **Discard all contaminated material in a beaker containing household bleach.**

3. Examine the control egg for evidence of cytopathogenicity.

4. Record your results in the chart.

	Days post-incubation							
	NDV-infected egg				*Saline control*			
	1	*2*	*3*	*4*	*1*	*2*	*3*	*4*
Movement (+) or (−)								
Viability (+) or (−)								
Presence of necrotic lesions (+) or (−)								
Hemorrhages (+) or (−)								

REVIEW QUESTIONS

1. Discuss the advantages of using tissue culture techniques instead of the embryonated egg for viral cultivation.

2. Explain the tissue culture procedure for cultivation of animal viruses.

3. Are the pathogenic effects produced by a specific virus the same in both humans and lower animals? _____ Explain.

4. Why are susceptible animals used less frequently than before for the cultivation of viruses?

5. Cite the major reasons for the addition of antibiotics to tissue culture media.

6. Explain why it is necessary to disinfect the egg shell at the site of viral inoculation.

IX

Physical and Chemical Agents for the Control of Microbial Growth

---

PURPOSES

1. To acquaint students with the basic methods for inhibiting microbial growth and their modes of antimicrobial action.

2. To demonstrate the effects of physical agents, moist heat, osmotic pressure, and ultra-violet radiation on selected microbial populations.

3. To demonstrate effects on selected microbial populations of chemical agents used as disinfectants, antiseptics, and antibiotics.

INTRODUCTION

Control of microorganisms is essential in the home, industry, and medical fields to prevent and treat diseases, and to inhibit the spoilage of foods and other industrial products. Common methods of control involve chemical and physical agents that adversely affect microbial structures and functions, thereby producing a microbicidal or microbistatic effect. A **microbicidal effect** is one that kills the microbes immediately; a **microbistatic effect** inhibits the reproductive capacities of the cells and maintains the microbial population at a constant size.

CHEMICAL METHODS FOR CONTROL OF MICROBIAL GROWTH

1. **Antiseptics:** Chemical substances used on living tissue that inhibit the growth of microorganisms.

2. **Disinfectants:** Chemical substances that inhibit the growth of vegetative microbial forms on nonliving materials.

3. **Chemotherapeutic agents:** Chemical substances that destroy or inhibit the growth of microorganisms in living tissues.

PHYSICAL METHODS FOR CONTROL OF MICROBIAL GROWTH

The following illustrates major acceptable physical methods used for the control of microbial growth:

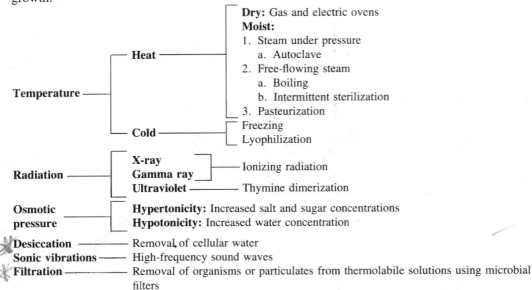

The modes of action of the different chemical and physical agents of control vary, although they all produce damaging effects to one or more essential cellular structures or molecules to cause cell death or inhibition of growth. Sites of damage that can result in malfunction are the cell wall, cell membrane, cytoplasm, enzymes, or nucleic acids. The adverse effects manifest themselves in the following ways:

1. **Cell-wall injury:** This can result in one of two ways. First, lysis of the cell wall will leave the protoplast susceptible to osmotic damage, and a hypotonic environment may cause lysis of the vulnerable protoplast. Second, certain agents inhibit cell wall synthesis, which is essential during microbial cell reproduction. Again, failure to synthesize a missing segment of the cell wall results in an unprotected protoplast.

2. **Cell-membrane damage:** This may be the result of lysis of the membrane, which will cause immediate cell death. Also, the selective nature of the membrane may be affected without causing its complete disruption. As a result, there may be a loss of essential cellular molecules or interference with the uptake of nutrients. In both cases, metabolic processes will be adversely affected.

3. **Alteration of the colloidal state of cytoplasm:** If this occurs, it will produce irreversible cell damage, as certain agents cause denaturing of cytoplasmic proteins. Uncoiling of these macromolecules leads to coagulation of the proteins, which in their clumped state are biologically inactive.

4. **Inactivation of enzymes:** When this results from the activity of certain physical and chemical agents, the cell cannot perform essential life functions. Inactivation of enzymes by these agents occurs through the mechanisms of competitive or noncompetitive inhibition, which will be discussed in a later exercise.

5. **Interference with the structure and function of the DNA molecule:** The DNA molecule is the control center of the cell and may also represent a cellular target area for destruction or inhibition. Some agents have an affinity for DNA and cause breakage or distortion of the molecule, thereby interfering with its replication and role in protein synthesis.

Awareness of the mode of action of the physical and chemical agents is absolutely essential for their proper selection and application in microbial control. The exercises in this section are designed to acquaint the students more fully with several commonly employed agents and their uses.

Physical Agents of Control:
Moist Heat

PURPOSE

To acquaint students with the susceptibility of microbial species to destruction by the application of moist heat.

PRINCIPLE

Temperature has an effect on cellular enzyme systems and therefore a marked influence on the rate of chemical reactions and thus the life and death of microorganisms. Despite the diversity among microorganisms as to the temperature sensitivity of their enzymes and their requirements for growth, extremes in temperature can be used in microbial growth control. Sufficiently low temperatures will inactivate enzymes and produce a static effect. High temperatures destroy cellular enzymes, which become irreversibly denatured.

The application of heat is a common means of destroying microorganisms. Both dry and moist heat are effective, however, because of its greater penetrating capacity that causes coagulation of proteins, moist heat destroys them more rapidly and at lower temperatures than dry heat. **Sterilization,** the destruction of all forms of life, is accomplished in 15 minutes at 121 degrees C with moist heat; dry heat requires a temperature of 160 degrees C to 180 degrees C for one-and-one-half to three hours.

Microbes exhibit differences in their resistance to moist heat. As a general rule, bacterial spores require temperatures above 100 degrees C for destruction, whereas most bacterial vegetative cells are killed at temperatures of 60 degrees C to 70 degrees C in 10 minutes. Fungi can be killed at 50 degrees C to 60 degrees C and fungal spores require 70 degrees C to 80 degrees C for 10 minutes for destruction. Because of this variability, moist heat can either sterilize or disinfect. Common applications include free-flowing steam under pressure (autoclaving), free-flowing steam at 100 degrees C (tyndallization), and the use of lower temperatures to destroy only pathogenic microorganisms (pasteurization).

Free-flowing steam under pressure requires the use of an autoclave, a double-walled metal vessel that allows steam to be pressurized in the outer jacket (see Figure 40.1 on page 234). At a designated pressure, the saturated steam is released into the inner chamber from which all the air has been evacuated. The steam under pressure in the vacuumed inner chamber is now capable of achieving temperatures in excess of 100 degrees C. The temperature is determined by the pounds of pressure applied per square inch, as illustrated:

Pressure (pounds/inch^2)	Temperature (degrees C)
0 (free-flowing steam)	100
10	115
15	121
20	126
25	130

A pressure of 15 pounds per inch^2 achieves a temperature of 121 degrees C, and sterilizes in 15 minutes. This is the usual procedure; however, depending on the heat sensitivity of the material to be sterilized, the operating pressure and time conditions can be adjusted.

Application of **free-flowing steam** requires exposure of the contaminated substance to temperatures of 100 degrees C, achieved by boiling water. Exposures to boiling water for 30 minutes will result in disinfection only, as all vegetative cells will be killed, but not necessarily the more heat-resistant spores. Another procedure is **Tyndallization,** also referred to as intermittent or fractional sterilization. This procedure requires exposure of the material to free-flowing steam at 100 degrees C for 20 minutes on three consecutive days with intermittent incubation at 37 degrees C. The steaming kills all vegetative cells. Any spores that may be present germinate during the period of incubation and are destroyed during subsequent exposure to a temperature of 100 degrees C. Repeating this procedure for three days ensures germination of all spores and their destruction in the vegetative form. Because Tyndallization requires so much time, it is used only for sterilization of materials that are composed of thermolabile chemicals and that might be subject to decomposition at higher temperatures.

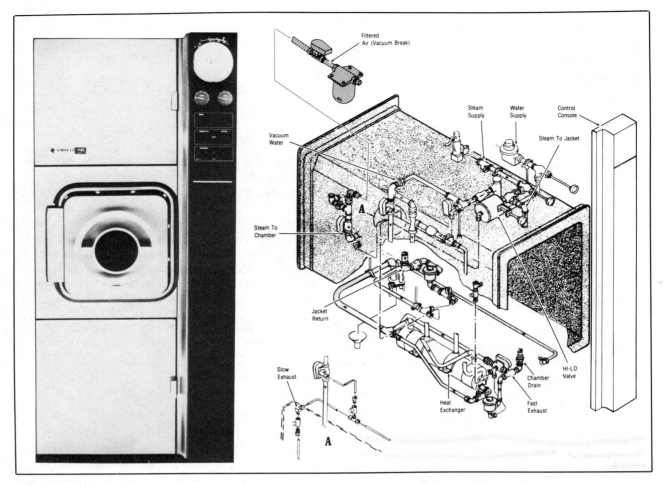

FIGURE 40.1. The Autoclave (Courtesy AMSCO/American Sterilizer Company, Erie, PA 16512.)

Pasteurization exposes fairly thermolabile products such as milk, wine, and beer for a given period of time to a temperature that is high enough to destroy pathogens that may be present, without necessarily destroying all vegetative cells. There are two types of pasteurization: The high-temperature, short-time (flash) procedure requires a temperature of 71 degrees C for 15 seconds; and the low-temperature, long-time method requires 63 degrees C for 30 minutes.

MATERIALS

Cultures

48-hour to 72-hour nutrient broth cultures of *Staphylococcus aureus* and *Bacillus cereus;* 72-hour to 96-hour Sabouraud broth cultures of *Aspergillus niger* and *Saccharomyces cerevisiae.*

Media

Five nutrient agar plates, five Sabouraud agar plates, and one 10-ml tube of nutrient broth per designated student group.

Equipment

Bunsen burner, 800-ml beaker (water bath), tripod and asbestos screen, thermometer, wax marking pencil, and inoculating loop.

PROCEDURE

1. Label the covers of each of the nutrient agar and Sabouraud agar plates, indicating the experimental heat temperatures to be used: 25 degrees C (control), 40, 60, 80, and 100 degrees C.

2. Score the underside of all plates with a wax marking pencil into two sections. On the nutrient agar plates, label one section *S. aureus* and the other *B. cereus*. On the Sabouraud agar plates, label one section *A. niger* and the second *S. cerevisiae*.

3. Using sterile technique, inoculate the nutrient agar and Sabouraud agar plates labeled 25 degrees C by means of a loop, making a single line of inoculation of each test organism in its respective section of the plate.

4. Set up the water bath as illustrated.

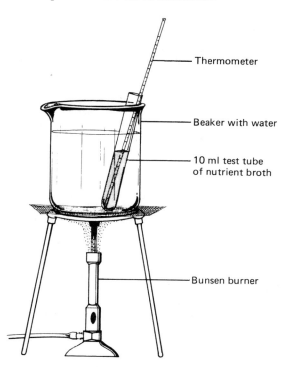

Thermometer

Beaker with water

10 ml test tube
of nutrient broth

Bunsen burner

5. Slowly heat the water to 40 degrees C; check the thermometer frequently so as not to exceed the desired temperature. Place the four cultures of the experimental organisms into the beaker and maintain the temperature at 40 degrees C for 10 minutes. Remove the cultures and aseptically inoculate each organism in its appropriate section on the two plates labeled 40 degrees C.

6. Raise the water bath temperature to 60 degrees C and repeat step 5 for the inoculation of the two plates labeled 60 degrees C.

7. Raise the water bath temperature to 80 degrees C and repeat step 5 for the inoculation of the two plates labeled 80 degrees C.

8. Raise the water bath temperature to 100 degrees C and repeat step 5 for the inoculation of the two plates labeled 100 degrees C.

9. Incubate the nutrient agar plate cultures in an inverted position for 24 to 48 hours at 37 degrees C and the Sabouraud agar plate cultures for 4 to 5 days at 25 degrees C in a moist chamber.

Observations and Results

1. Observe all plates as to the amount of growth of the test organisms at each of the temperatures.

2. Record your results as 0 = none, 1+ = slight, 2+ = moderate, or 3+ = abundant, in the chart.

Microbial species	*Amount of growth (degrees C)*				
	25	*40*	*60*	*80*	*100*
B. cereus					
S. aureus					
A. niger					
S. cerevisiae					

3. List the microbial organisms in order of their increasing heat resistance:

_____
_____
_____
_____

REVIEW QUESTIONS

1. Account for the microbistatic effect produced by low temperatures as compared to the microbicidal effect produced by high temperatures.

2. Cite the advantages of each of the modes of sterilization, Tyndallization and autoclaving.

3. Discuss the detrimental effects of control agents on:

   a. Cytoplasm:

   b. Cell wall:

   c. Nucleic acids:

   d. Cell membrane:

4. Explain why milk is subjected to Pasteurization rather than sterilization.

Physical Agents of Control: Environmental Osmotic Pressure

PURPOSE

To acquaint students with the possible effects of osmotic pressure environments on microorganisms.

PRINCIPLE

Osmosis is the net movement of water molecules (solvent) across a semipermeable membrane from a solution of their higher concentration to a solution of their lower concentration. The determining factor in the relative water concentrations of the two solutions is their respective solute concentration. In comparing two solutions, the **hypertonic solution** possesses a higher osmotic pressure and has a higher solute concentration, and therefore a lower water concentration; it tends to draw in water. The **hypotonic solution** possesses a lower osmotic pressure and has a lower solute concentration, and therefore a higher water concentration; it tends to lose water. If two solutions separated by a semipermeable membrane have equal concentrations of solutes and therefore water concentrations, there is no osmosis and the solutions are **isotonic.**

The cell and its environment represent two solutions separated by the semipermeable cell membrane. The cell's cytoplasm contains colloidal and solute particles dispersed in water, as does the cell's environment. Water is essential for the transport of materials across the cell membrane. As such the **osmotic pressure of the environment** in relation to that of the cytoplasm of the cell plays a vital role in the life and death of a cell. The ideal environment of animal cells, which are bounded only by the fragile cell membrane, is completely or closely isotonic so that the cells are not susceptible to damage due to osmotic pressure. In a hypertonic, high-pressure environment, all cells lose water by osmosis and become shriveled. This effect is called **plasmolysis;** it inhibits cell reproduction. As water is necessary for the occurrence of many chemical reactions, water loss adversely affects cell metabolism and

reproduction. In a hypotonic, low-pressure environment, cells take in water and become swollen. This phenomenon is **plasmoptysis.** In an environment with sufficiently low osmotic pressure, animal cells undergo **lysis,** which causes their death. Microorganisms and other organisms whose cells possess a rigid cell wall are not usually susceptible to lysis in hypotonic environments. Instead, these organisms usually prefer a slightly hypotonic environment to maintain themselves in a turgid state. The effects of environmental osmotic pressure on cells unprotected by a cell wall are illustrated in Figure 41.1 on page 240.

Osmotic pressure can be used as an effective antimicrobial agent. Microorganisms are not usually adversely affected by low environmental osmotic pressure because of their small size and the presence of a rigid cell wall that permits rapid adjustment to the hypotonicity. However, hypertonicity is a commonly used method of inhibiting microbial growth.

Because of their varied habitats, microorganisms are generally well adapted to exist in all types of osmotic pressure environments, such as soil, air, fresh waters, and waters of varying salinity. Different groups of microorganisms require different degrees of salinity for growth, and as a rule can adjust to salt concentrations of 0.5 percent to 3 percent. Concentrations of 10 percent to 15 percent are inhibitory to the growth of most microbes, except for **halophiles,** which are adapted to life in waters of high salinity and require high concentrations for growth. This sensitivity is the basis of food preservation by the process of salting.

MATERIALS

Cultures

24-hour to 48-hour nutrient broth cultures of *Staphylococcus aureus* and *Escherichia coli,* 48-hour *Halobacterium* salt broth culture of *Halobacterium salinarium.*

239

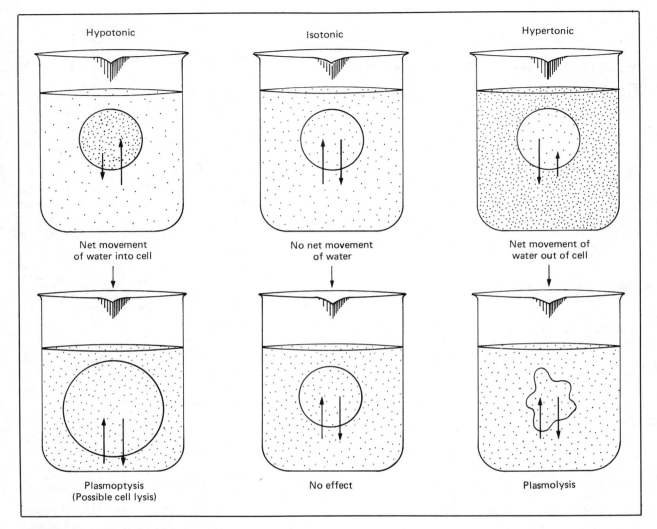

FIGURE 41.1. Osmotic environments

Media

One nutrient agar plate of each of the following sodium chloride concentrations per designated student group: 0.85 percent, 5 percent, 10 percent, 15 percent, and 25 percent.

Equipment

Bunsen burner, inoculating loop, and wax marking pencil.

PROCEDURE

1. Score the underside of each of the five nutrient agar plates into three sections with wax marking pencil.

2. Label each of the three sections on each plate with the name of the organism to be inoculated. Identify each with your initials.

3. Using sterile technique, inoculate each of the agar plates with the three experimental organisms by making a single-line loop inoculation of each organism in its appropriately labeled section.

4. Incubate all plates in an inverted position for four to five days at 25 degrees C.

▲ **1. Eosin-methylene blue agar plate** with *E. coli* exhibiting a green metallic sheen.

▲ **6. Starch agar plate**
Starch hydrolysis on left; no starch hydrolysis on right.

▲ **2. MacConkey agar plate**
Lactose nonfermenter on left; lactose fermenter on right.

▼ **3. Mannitol salt agar plate**
Mannitol nonfermenting staphylococci on left; mannitol fermenting staphylococci (*S. aureus*) on right.

▼ **5. Blood agar plate**
Alpha-hemolysis.

▼ **4. Blood agar plate**
Beta-hemolysis with stab inoculation to show intensified hemolytic activity at the top.

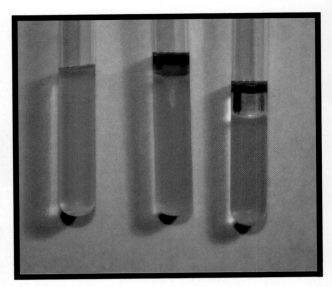

▶ *7. Indole production test*
From left to right—(1) uninoculated; (2) positive;
(3) negative.

▼ *8. Methyl red test*
From left to right—(1) uninoculated; (2) negative;
(3) positive.

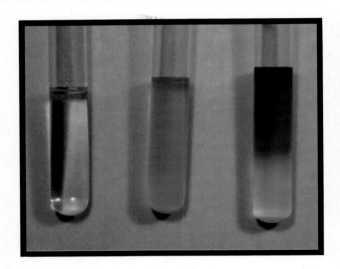

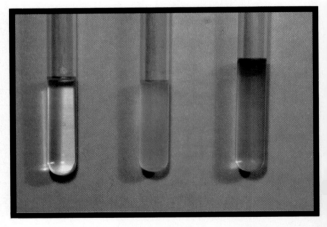

▲ *9. Voges-Proskauer test*
From left to right—(1) uninoculated; (2) negative;
(3) positive.

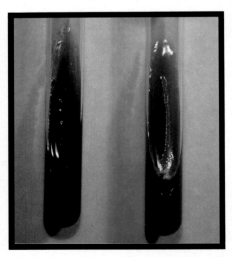

▲ *10. Citrate utilization test*
Negative on left; positive on right.

▼ 11. *Urease test*
Negative on left; positive on right.

▼ 12. *H₂S production*
From left to right—(1) negative; (2) positive;
(3) positive.

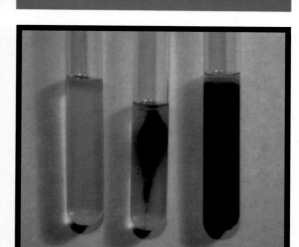

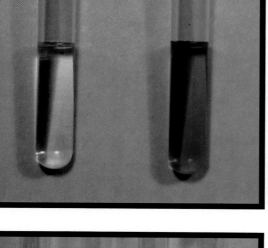

◄ 13. *Triple sugar-iron agar test*
From left to right—(1) uninoculated; (2) acid slant/acid
butt; (3) alkaline slant/acid butt, H₂S, gas; (4) alkaline
slant/acid butt; (5) alkaline slant/alkaline butt.

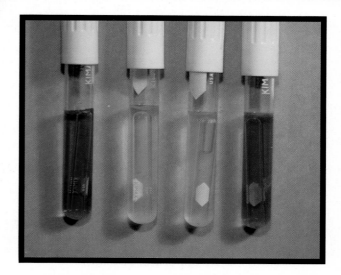

▲ 14. *Carbohydrate fermentation*
From left to right—(1) uninoculated; (2) acid; (3) acid and
gas; (4) negative.

▼ 15. Litmus milk reactions

From left to right—(1) uninoculated; (2) acid; (3) acid, reduction, curd; (4) alkaline, no change; (5) rapid proteolysis; (6) slow proteolysis.

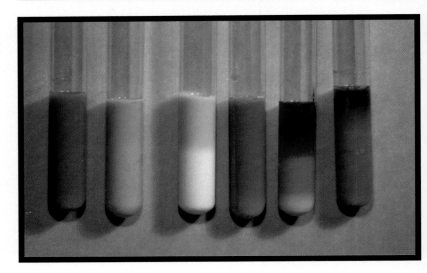

▶ 16. Nitrate reduction test

From left to right—(1) uninoculated; (2) positive with Solutions *A* + *B*; (3) positive with Solutions *A* + *B* + zinc; (4) negative with Solutions *A* + *B* + zinc.

◀ 17. Catalase test

Positive as evidenced by the evolution of O₂ bubbles.

Observations and Results EXPERIMENT **41**

1. Observe each of the nutrient agar plate cultures as to the amount of growth of each of the experimental species.

2. Record your results in the chart as 0 = no growth; + = scant growth; 2+ = moderate growth; 3+ = abundant growth.

Microbial species	Salt concentration of medium (%)				
	0.85	5	10	15	25
S. aureus					
E. coli					
H. salinarium					

3. For each of the experimental species, indicate the salt concentration range in which growth will occur and the optimal salt concentration.

Microbial species	Range for growth	Optimal NaCl concentration
S. aureus		
E. coli		
H. salinarium		

REVIEW QUESTIONS

1. Compare hypertonic, hypotonic, and isotonic solutions and their effects on cells.

2. Explain how hypertonicity can be used as a means of controlling microbial growth.

3. Explain the difference between the effect produced by a hypotonic environment on animal cells and the effect produced on bacterial cells.

4. What is a halophilic organism?

Physical Agents of Control: Electromagnetic Radiations

PURPOSE

To acquaint students with the microbicidal effect of ultraviolet radiation on microorganisms.

PRINCIPLE

Certain forms of electromagnetic radiation are capable of producing a lethal effect on cells and therefore can be used for microbial control. Electromagnetic radiations that possess sufficient energy to be microbicidal are the short-wavelength radiations, below 300 nm. These include ultraviolet, gamma rays, and x-rays. The high-wavelength radiations, those above 300 nm, have insufficient energy to destroy cells. The electromagnetic spectrum and its effect on molecules is illustrated in Figure 42.1.

Gamma radiation, which originates from unstable atomic nuclei, and **x-radiation,** originating from outside of the atomic nucleus, are representative of **ionizing forms of radiation.** Both transfer their energy through quanta (photons) to the matter through which they pass, causing excitation and the loss of electrons from molecules in their path. This injurious effect is nonspecific in that any molecule in the path of the radiation will undergo ionization. Essential cell mole-

cules can be directly affected through loss of their chemical structure and activity brought about by the ionization. Also, water, the most abundant chemical constituent of cells, commonly undergoes radiation breakdown, with the ultimate production of highly reactive H^+, OH^-, and in the presence of oxygen, HO_2 free radicals. These may combine with each other, frequently forming hydrogen peroxide (H_2O_2), which is highly toxic to cells lacking catalase or other peroxidases; or the highly reactive free radicals may combine with any cellular constituents, again resulting in biochemical cell damage.

Because of their high energy content and therefore ability to penetrate matter, x-ray and gamma radiations can be used as means of sterilization, particularly of thermolabile materials. They are not commonly used, however, because of the expense of the equipment and the special facilities necessary for their use.

Ultraviolet light, which has a lower energy content than ionizing radiations, is capable of producing a lethal effect in cells exposed to the low penetrating wavelengths in the range of 210 nm to 300 nm. Cellular components capable of absorbing ultraviolet light are the nucleic acids, with the DNA acting as the primary site of damage. As the pyrimidines especially absorb ultraviolet wavelengths, the major effect of this form of radiation is **thymine dimerization,** which is the

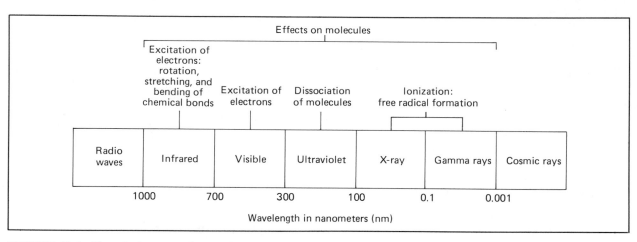

FIGURE 42.1. The electromagnetic spectrum

covalent bonding of two adjacent thymine molecules on one nucleic acid strand in the DNA molecule. This dimer formation distorts the configuration of the DNA molecule and the distortion interferes with DNA replication and transcription during protein synthesis.

Some cell types, including some microorganisms, possess enzyme systems for the repair of radiation-induced DNA damage. Two different systems may be operational: (1) The **dark repair system** that functions in the absence of light; and (2) the **light repair system,** which is made operational by exposure of the irradiated cells to visible light in the wavelength range of 420 nm to 540 nm. The visible light serves to activate an enzyme that splits the dimers and reverses the damage.

Ultraviolet radiation, because of its low penetrability, cannot be used as a means of sterilization and its practical application is only for surface or air disinfection.

MATERIALS

Cultures

24-hour to 48-hour nutrient broth cultures of *Serratia marcescens* and *Bacillus cereus;* sterile saline spore suspension of *Aspergillus niger.*

Media

Six nutrient agar plates per designated student group.

Equipment

Bunsen burner, inoculating loop, ultraviolet radiation source (254 nm), and wax marking pencil.

PROCEDURE

1. Divide all nutrient agar plates into three sections by scoring the underside of each plate with a marking pencil.

2. Label each of the three sections on each plate with the name of the organism to be inoculated, and identify each with your initials.

3. Using sterile technique, inoculate all the plates by means of a streak inoculation as illustrated:

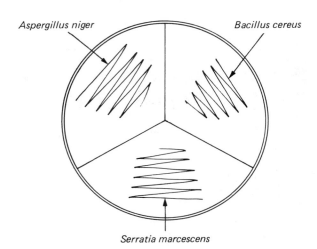

4. Label the cover of each inoculated plate with the exposure time to ultraviolet radiation as 0 (control), 1 minute, 3 minutes, 5 minutes. Label two plates as 7 minutes, one of which will serve as the irradiated, covered control.

5. Irradiate all inoculated plates for the designated period of time by placing them 12 inches below the ultraviolet light source, as illustrated below. Make sure first to remove all Petri dish covers except the 7-minute irradiated control plate. **Do not expose your eyes to the ultraviolet light.**

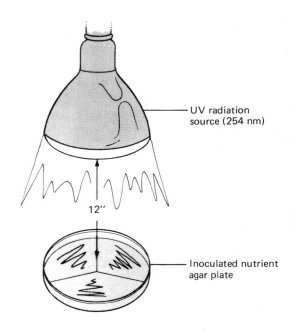

6. Incubate all plates in an inverted position for four to five days at 25 degrees C.

Observations and Results

1. Observe all nutrient agar plate cultures for the amount of growth of each of the microbial species.

2. Record your observations in the chart as 0 = no growth; + = slight growth; 2+ = moderate growth; 3+ = abundant growth.

	Time of irradiation (minutes)					
Microbial species	*0*	*1*	*3*	*5*	*7*	*7**
B. cereus						
S. marcescens						
A. niger						

*Irradiated, covered plate

REVIEW QUESTIONS

1. Discuss the effects of ionizing radiation on cellular constituents.

2. Explain why x-rays can be used for sterilization, whereas ultraviolet rays can be used only for surface disinfection of materials.

3. Explain the mechanism of action of ultraviolet radiation on cells.

4. Account for the greater susceptibility of *S. marcescens* than that of *B. cereus* to the effects of ultraviolet radiation.

5. Why is it not essential to shield your hands from ultraviolet light, whereas great care must be exercised to shield eyes from this type of radiation?

Chemical Agents of Control: Chemotherapeutic Agents

Chemotherapeutic agents are chemical substances used in the treatment of infectious diseases. Their mode of action is to interfere with microbial metabolism without producing a like effect in host cells, thereby producing a static or cidal effect on the microorganisms. These drugs can be separated into two categories:

1. **Antibiotics** are synthesized and secreted by certain bacteria, actinomycetes, and fungi that destroy or inhibit the growth of other microorganisms. Today, some antibiotics are laboratory synthesized or

modified; however, their origins are living cells.

2. **Synthetic drugs** are synthesized in the laboratory.

To determine a therapeutic drug of choice one must know its mode of action, possible adverse side effects in the host, and the scope of its antimicrobial activity. The specific mechanism of action varies among different drugs, and the short-term or long-term use of many drugs can produce systemic side effects in the host. These vary in severity from mild and temporary upsets to permanent tissue damage (Table 43.1).

TABLE 43.1. Prototypic antibiotics

Antibiotic	*Mode of action*	*Possible side effects*
Penicillin	Prevents incorporation of muramic acid into the mucocomplex component of the cell wall, thereby inhibiting cell-wall synthesis	Penicillin resistance or sensitivity (allergic reaction)
Streptomycin	Has an affinity for bacterial ribosomes, causing misreading of codons on mRNA, thereby interfering with protein synthesis	May produce damage to auditory nerve, causing deafness
Chloramphenicol	Has an affinity for bacterial ribosomes, preventing peptide bond formation between amino acids during protein synthesis	May cause aplastic anemia, which is fatal due to destruction of RBC-forming and WBC-forming tissues
Tetracyclines	Have an affinity for bacterial ribosomes; prevent hydrogen bonding between the anticodon on the tRNA-amino acid complex and the codon on mRNA during protein synthesis	Permanent discoloration of teeth in young children
Bacitracin	Inhibits cell-wall synthesis	Toxic if taken internally; used for topical application only
Polymyxin	Destruction of cell membrane	Toxic if taken internally; used for topical application only

Synthetic Agents

Sulfadiazine (sulfonamide) produces a static effect on a wide range of microorganisms by a mechanism of action called **competitive inhibition.** The active component of the drug is sulfanilamide; it acts as an **antimetabolite** and replaces the **essential metabolite,** *p*-aminobenzoic acid (PABA), as an integral part of the folic acid molecule during its synthesis in the microbial cell. Folic acid is an essential cellular coenzyme involved in the synthesis of amino acids and purines. Many microorganisms possess enzymatic pathways for folic acid synthesis and can be adversely affected by sulfonamides. Human cells lack these enzymes, and the essential folic acid enters the cells in a preformed state. Therefore these drugs have a limited effect on human cells. The similarity between the chemical structure of the antimetabolite sulfanilamide and the structure of the essential metabolite PABA is illustrated below.

PART A: The Kirby-Bauer Antimicrobial Sensitivity Test Procedure

PURPOSE

To acquaint students with the Kirby-Bauer procedure for the evaluation of the antimicrobial activity of chemotherapeutic agents.

PRINCIPLE

The available chemotherapeutic agents vary in their scope of antimicrobial activity. Some have a limited spectrum of activity, being effective against only one group of microorganisms. Others exhibit broad-spectrum activity against a range of microorganisms. The drug sensitivities of many pathogenic microorganisms are known, but it is sometimes necessary to test several agents to determine the drug of choice.

A standardized filter-paper disc-agar diffusion procedure, known as the Kirby-Bauer method, is frequently used to determine the drug sensitivity of microorganisms isolated from infectious processes (Figure 43.1). This method allows for the rapid determination of the efficacy of a drug by measuring the diameter of the zone of inhibition that results from diffusion of the agent into the medium surrounding the disc. In this procedure, filter-paper discs of uniform size are impregnated with specified concentrations of different antibiotics and then placed on the surface of an agar plate that has been seeded with the organism to be tested. The medium of choice is Mueller-Hinton agar, with a pH of 7.2 to 7.4, which is poured into plates to a uniform depth of 5 mm and refrigerated on solidification. Prior to use, the plates are transferred to an incubator at 37 degrees C for 10 to 20 minutes to dry off the moisture that develops on the agar surface. The plates are then heavily inoculated with a standardized inoculum by means of a cotton swab to ensure the confluent growth of the organism. The discs are aseptically applied to the surface of the agar plate at well-spaced intervals. Once applied, each disc is gently touched with a sterile applicator stick to ensure its firm contact with the agar surface.

Following incubation the plates are examined for the presence of growth inhibition, a clear zone surrounding each disc. The susceptibility of an organism to a drug is determined by the size of this zone, which itself is dependent on variables such as:

1. The ability and rate of diffusion of the antibiotic into the medium and its interaction with the test organism.

Sulfadiazine (sulfonamide)

Sulfanilamide component (antimetabolite) / Pyrimidine component

p-Aminobenzoic acid

PABA (essential metabolite)

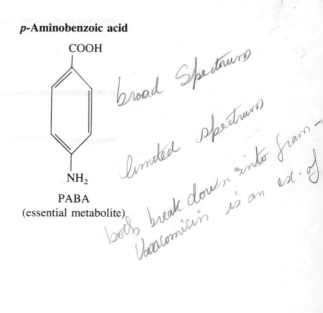

FIGURE 43.1. Kirby-Bauer antibiotic sensitivity procedure

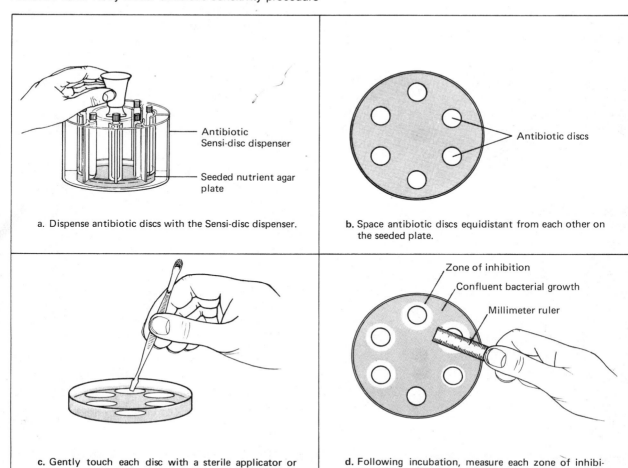

a. Dispense antibiotic discs with the Sensi-disc dispenser.

b. Space antibiotic discs equidistant from each other on the seeded plate.

c. Gently touch each disc with a sterile applicator or forceps.

d. Following incubation, measure each zone of inhibition with a millimeter ruler.

TABLE 43.2.   Interpretation of size of zone of inhibition

Antibiotic	Disc concentration	Diameter of zone inhibition (mm)		
		Resistant	Intermediate	Sensitive
Ampicillin[1]	10 μg	11 or less	12–13	14 or more
Ampicillin or Penicillin G[2]	10 μg	20 or less	21–28	29 or more
Bacitracin	10 u	8 or less	9–12	13 or more
Cephalothin	30 μg	14 or less	15–17	18 or more
Chloramphenicol	30 μg	12 or less	13–17	18 or more
Erythromycin	15 μg	13 or less	14–17	18 or more
Gentamicin	10 μg	12 or less	13–14	15 or more
Novobiocin	30 μg	17 or less	18–21	22 or more
Vancomycin	30 ug	9 or less	10–11	12 or more
Streptomycin	10 μg	11 or less	12–14	15 or more
Sulfanilamide	300 u	12 or less	13–16	17 or more
Tetracycline	30 μg	14 or less	15–18	19 or more

[1]When testing gram-negative enterics or enterococci.
[2]When testing staphylococci.
Source: *The Federal Register, 37*:191, 20527–20529 (1972).

2. The number of organisms inoculated.

3. The growth rate of the organism.

4. The degree of sensitivity of the organism to the antibiotic.

A measurement of the diameter of the zone of inhibition in millimeters is made and its size compared to that contained in a standardized chart, which is shown in Table 43.2. Based on this comparison, the test organism is determined to be resistant, intermediate, or sensitive to the antibiotic.

MATERIALS

Cultures

0.85 percent saline suspensions of *Escherichia coli, Staphylococcus aureus, Streptococcus lactis, Pseudomonas aeruginosa, Proteus vulgaris, Mycobacterium smegmatis,* and *Corynebacterium xerosis* adjusted to an O.D. of 0.1 at 600 mμ.

Media

Seven Mueller-Hinton agar plates per designated student group.

Antimicrobial Sensitivity Discs

Penicillin G, 10 μg; streptomycin, 10 μg; tetracycline, 30 μg; chloramphenicol, 30 μg; gentamicin, 10 μg; vancomycin, 30 μg; and sulfanilamide, 300 u.

Equipment

Sensi-disc dispensers or forceps, Bunsen burner, sterile cotton swabs, wax marking pencil, and millimeter ruler.

PROCEDURE

1. Place agar plates right side up in an incubator heated to 37 degrees C for 10 to 20 minutes with the covers adjusted so that the plates are slightly opened.

2. Label the covers of each of the plates with the name of the test organism to be inoculated and identify each with your initials.

3. Using sterile technique, inoculate all agar plates with their respective test organisms as follows:

a. Dip a sterile cotton swab into a well-mixed saline test culture and remove excess inoculum by pressing the saturated swab against the inner wall of the culture tube.

b. Using the swab, streak the entire agar surface of the plate first in a horizontal direction and then vertically to ensure a heavy growth over the entire surface.

4. Allow all culture plates to dry for about five minutes.

5. Using the sensi-disc dispenser, apply the antibiotic discs by placing the dispenser over the agar surface and pressing the plunger, depositing the discs simultaneously onto the agar surface. If dispensers are not available, the individual discs are distributed at equal distances with forceps dipped in alcohol and flamed.

6. Gently press each disc down with the wooden end of a cotton swab or sterile forceps to ensure that the discs adhere to the surface of the agar. **Do not press the discs into the agar.**

7. Incubate all plate cultures in an inverted position for 24 to 48 hours at 37 degrees C.

PART B: Synergistic Effect of Drug Combinations
PURPOSE

To acquaint students with the disc-agar diffusion technique for determination of synergistic combinations of chemotherapeutic agents.

PRINCIPLE

Combination chemotherapy, the use of two or more antimicrobial or antineoplastic agents, is being employed in medical practice with ever-increasing frequency. The rationale for using drug combinations is the expectation that effective combinations might lower the incidence of bacterial resistance, reduce host toxicity of the antimicrobial agents (because of decreased dosage requirements), or enhance the agents' bacteriocidal activity. Enhanced bacteriocidal activity is known as **synergism.** Synergistic activity is evident when the sum of the effects of the chemotherapeutic agents used in combination is significantly greater than the sum of their effects when used individually. This result is readily differentiated from an **additive (indifferent) effect,** which is evident when the interaction of two drugs produces a combined effect that is no greater than the sum of their separately measured individual effects.

A variety of *in vitro* methods are available to demonstrate synergystic activity. In this experiment, a disc-agar diffusion technique will be performed to demonstrate this phenomenon. This technique uses the Kirby-Bauer antibiotic susceptibility test procedure, as described in part A of this experiment, and requires both Mueller-Hinton agar plates previously seeded with the test organisms and commercially prepared, antimicrobial-impregnated discs. The two discs, representing the drug combination, are placed on the inoculated agar plate and separated by a distance (measured in mm) that is equal to or slightly greater than one-half the sum of their individual zones of inhibition when obtained separately. Following the incubation period, an additive effect is exhibited by the presence of two distinctly separate circles of inhibition. If the drug combination is synergistic, the two inhibitory zones merge to form a "bridge" at their juncture, as illustrated below.

The drug combinations to be used in this experimental procedure are:

1. **Sulfisoxazole, 150 μg, and trimethoprim, 5 μg.** Both antimicrobial agents are enzyme inhibitors that act sequentially in the metabolic pathway leading to folic acid synthesis as shown at the top of the next column.

The antimicrobial effect of each drug is enhanced when used in combination. The pathway thus exemplifies synergism.

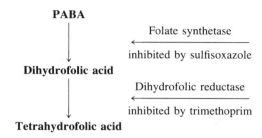

2. **Trimethoprim, 5 μg, and tetracycline, 30 μg.** The mode of antimicrobial activity of these two chemotherapeutic agents differs, as tetracycline acts to interfere with protein synthesis at the ribosomes. Thus, when used in combination, these drugs produce an additive effect.

MATERIALS

Cultures

0.85 percent saline suspensions of *Escherichia coli* and *Staphylococcus aureus* adjusted to an O.D. of 0.1 at 600 mμ.

Media

Four Mueller-Hinton agar plates per designated student group.

Antimicrobial Sensitivity Discs

Tetracycline, 30 μg; trimethoprim, 5 μg; and sulfisoxazole, 150 μg.

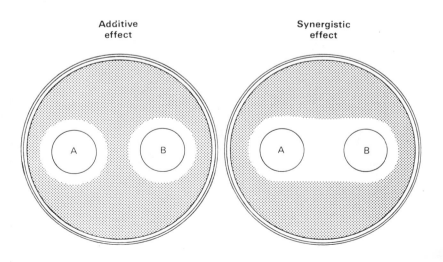

Equipment

Bunsen burner, forceps, sterile cotton swabs, millimeter ruler, and wax pencil.

PROCEDURE

1. Inoculation of Mueller-Hinton agar plates: follow steps 1 through 4 as described under the procedure in Part A of this experiment on page 250.

2. Using the millimeter ruler, determine the center of the underside of each plate and mark with wax pencil.

3. Using the wax pencil, mark the underside of each agar plate culture at both sides from the center mark at the distances specified below:

   a. *E. coli*-seeded plate for trimethoprim and sulfisoxazole combination sensitivity: 12.5 mm on each side of center mark.

   b. *S. aureus*-seeded plate for trimethoprim and sulfisoxazole combination sensitivity: 14.5 mm on each side of center mark.

   c. *E. coli*- and *S. aureus*-seeded plates for trimethoprim and tetracycline combination sensitivity: 14.0 mm on each side of center mark.

4. Using sterile forceps, place the antimicrobial discs, in the combinations specified in step 3, onto the surface of each agar plate culture at the previously marked positions. Gently press each disc down with the sterile forceps to ensure that it adheres to the agar surface.

5. Incubate all plate cultures in an inverted position for 24 to 48 hours at 37 degrees C.

Observations and Results

PART A: Kirby-Bauer Antimicrobial Sensitivity Test Procedure

1. Examine all plate cultures for the presence or absence of a zone of inhibition surrounding each disc.

2. Using a ruler graduated in millimeters, carefully measure each zone of inhibition to the nearest millimeter and record your result in the chart.

3. Compare your results with Table 43.2 and indicate in the chart the susceptibility of each test organism to the chemotherapeutic agent as resistant (R), intermediate (I), or sensitive (S).

Chemothera-peutic agent	Gram-negative							Acid-fast	
	E. coli		P. aeruginosa		P. vulgaris		M. smegmatis		
	Zone size	Susceptibility	Zone size	Susceptibility	Zone size	Susceptibility	Zone size	Susceptibility	
Penicillin	P	resistant		resistant					
Streptomycin	S	Sensitive		intermediate					
Tetracycline	Te	Sensitive		resistant					
Chloram-phenicol		sensitive							
Gentamicin	Gm	Sensitive		sensitive					
Vancomycin	VA	resistant		resistant					
Sulfanilamide	SD	Sensitive		resistant					

Chemothera-peutic agent	Gram-positive					
	S. aureus		S. lactis		C. xerosis	
	Zone size	Susceptibility	Zone size	Susceptibility	Zone size	Susceptiblity
Penicillin		resistant		R Sensitive		R
Streptomycin		resistant		R resistant		R
Tetracycline		resistant		R Sensitive		R
Chloram-phenicol		resistant		R. Sensitive		R
Gentamicin		resistant		R resistant		R
Vancomycin		Sensitive		R Sensitive		R
Sulfanilamide		resistant		R resistant		R

4. For each of the chemotherapeutic agents indicate:

   a. The spectrum of its activity as broad or limited.

   b. The type or types of organisms it is effective against as gram-positive, gram-negative, or acid-fast.

Chemotherapeutic agent	Spectrum of activity	Type(s) of microorganisms
Penicillin		
Streptomycin		
Tetracycline		
Chloramphenicol		
Gentamicin		
Vancomycin		
Sulfanilamide		

PART B: Synergistic Effect of Drug Combinations

Examine all agar plate cultures to determine the zone of inhibition patterns exhibited. Distinctly separate zones of inhibition are indicative of an additive effect, whereas a merging of the inhibitory zones is indicative of synergism. Record your observations and results in the chart.

Cultures	Appearance of Zone of Inhibition	Synergistic or Additive Effect
E. coli: trimethoprim and sulfisoxazole	_____	_____
trimethoprim and tetracycline	_____	_____
S. aureus: trimethoprim and sulfisoxazole	_____	_____
trimethoprim and tetracycline	_____	_____

Chemical Agents of Control: Disinfectants and Antiseptics

Antiseptics and disinfectants are chemical substances used to prevent contamination and infection. Many are available commercially for disinfection and asepsis.

Table 44.1 shows the major groups of antimicrobial agents, their modes and ranges of action, and their practical uses.

TABLE 44.1. Chemical agents—disinfectants and antiseptics

Agent	Mechanism of action	Use
Phenolic compounds Phenol	1. Germicidal effect due to alteration of protein structure resulting in protein denaturation. 2. Surface-active agent (surfactant) precipitates cellular proteins and disrupts cell membranes. (Phenol has been replaced by better disinfectants that are less irritating, less toxic to tissues, and better inhibitors of microorganisms.)	1. 89% solutions: Cauterization of minor wounds. 2. 5% solutions: Disinfection. 3. 0.5% to 1% solutions: antiseptic effect and relief of itching as it exerts a local anesthetic effect on sensory nerve endings.
Cresols (ortho) (meta) (para)	1. Similar to phenol. 2. Poisonous and must be used externally. 3. 50% solution of cresols in vegetable oil known as Lysol.	2% to 5% Lysol solutions used as disinfectants.
Hexachlorophene	Germicidal activity similar to phenol. (This agent is to be used with care, especially in infants, because following absorption it may cause neurotoxic effects.)	1. Reduction of pathogenic organisms on skin; added to detergents, soaps, lotions, and creams. 2. Effective against gram-positive organisms. 3. An antiseptic used topically.

TABLE 44.1.   *(continued)*

Agent	*Mechanism of action*	*Use*
Resorcinol (OH, OH structure)	1. Germicidal activity similar to phenol. 2. Acts by precipitating cell protein.	1. Antiseptic. 2. Keratolytic agent for softening or dissolving keratin in epidermis.
Hexylresorcinol (OH, OH, $CH_2(CH_2)_3CH_2CH_3$ structure)	Germicidal activity similar to phenol.	1. Treatment of worm infections. 2. Urinary antiseptic.
Thymol (OH, C_3H_7, CH_3 structure)	1. Related to the cresols. 2. More effective than phenol.	1. Antifungal activity. 2. Treatment of hookworm infections. 3. Mouthwashes and gargle solutions.

Alcohols
 Ethyl
   CH_3CH_2OH
 Isopropyl
   $(CH_3)_2CHOH$
	1. Lipid solvent.
	 2. Denaturation and coagulation of proteins.
	 3. Wetting agent used in tinctures to increase the wetting ability of other chemicals.
	 4. Germicidal activity increases with increasing molecular weight.
		Skin antiseptics:
		 Ethyl—50% to 70%.
		 Isopropyl—75%.

Halogens
 Chlorine compounds:
   Sodium hypochlorite (Dakin's fluid):
     NaOCl
   Chloramine $CH_3C_6H_4SO_2NNaCl$
	1. Germicidal effect due to rapid combination with proteins.
	 2. Chlorine reacts with water to form hypochlorous acid, which is bactericidal.
	 3. Oxidizing agent.
	 4. Noncompetitively inhibits enzymes, especially those dealing with glucose metabolism, by reacting with SH and NH_2 groups on the enzyme molecule.
		1. Water purification.
		 2. Sanitation of utensils in dairy and restaurant industries.
		 3. Chloramine, 0.1% to 2% solutions, for wound irrigation and dressings.
		 4. Microbicidal.

Iodine compounds:
   Tincture of iodine
   Povidone-iodine solution (Betadine)
	1. Mechanism of action is not entirely known but it is believed that it precipitates proteins.
	 2. Surface-active agent.
		1. Tinctures of iodine are used for skin antisepsis.
		 2. Treatment of goiter.
		 3. Effective against spores, fungi, and viruses.

TABLE 44.1.   *(continued)*

Agent	*Mechanism of action*	*Use*
Heavy metals Mercury compounds: Inorganic: Mercury bichloride Mercurial ointments	1. Mercuric ion brings about precipitation of cellular proteins. 2. Noncompetitive inhibition of specific enzymes caused by reaction with sulfhydryl group (SH) on enzymes of bacterial cells.	1. Inorganic mercurials are irritating to tissues, toxic systemically, adversely affected by organic matter, and they have no action on spores. 2. Mercury compounds are mainly used as disinfectants of laboratory materials.
Organic mercurials: Mercurochrome (merbromin) Merthiolate (thimerosal) Metaphen (nitromersol) Merbak (acetomeroctol)	1. Similar to inorganic mercurials, but in proper concentrations are useful antiseptics. 2. Much less irritating than inorganic mercurials.	1. Less toxic, less irritating, used mainly for skin asepsis. 2. Do not kill spores.
Silver compounds: Silver nitrate	1. Precipitate cellular proteins. 2. Interfere with metabolic activities of microbial cells. 3. Inorganic salts are germicidal.	Asepsis of mucous membrane of throat and eyes.
Surface-active agents Wetting agents: Emulsifiers, soaps, and detergents	1. Lower surface tension and aid in mechanical removal of bacteria and soil. 2. If active portion of the agent carries a negative electric charge it is called an anionic surface-active agent. If active portion of the agent carries a positive electric charge it is called a cationic surface-active agent. 3. Exert bactericidal activity by interfering with or by depressing metabolic activities of microorganisms. 4. Disrupt cell membranes. 5. Alter cell permeability.	Weak action against fungi, acid-fast microorganisms, spores, and viruses.
Cationic agents: Quaternary ammonium compounds Benzalkonium chloride	1. Lower surface tension because of kerotolytic, detergent, and emulsifying properties. 2. Its germicidal activity reduced by soaps.	1. Bactericidal, fungicidal, inactive against spores and viruses. 2. Asepsis of intact skin. 3. Disinfectant for operating-room equipment. 4. Dairy and restaurant sanitization.

TABLE 44.1. *(continued)*

Agent	*Mechanism of action*	*Use*
Anionic agents:	1. Neutral or alkaline salts of high-molecular-weight acids. Common soaps included in this group.	
	2. Exert their maximum activity in an acid medium and are most effective against gram-positive cells.	
Tincture of green soap	3. Same as all surface-active agents.	1. Cleansing agent.
Sodium tetradecyl sulfate	4. Same as all surface-active agents.	2. Sclerosing agent in treatment of varicose veins and internal hemorrhoids.
Acids (H^+) **Alkali** (OH^-)	1. Destruction of cell wall and cell membrane. 2. Coagulation of proteins.	Disinfection, however, of little practical value.
Formaldehyde (liquid or gas)	Alkylating agent causes reduction of enzymes.	1. Room disinfection. 2. Alcoholic solution for instrument disinfection. 3. Specimen preservation.
Ethylene oxide $H_2C \underset{\displaystyle O}{\overset{\displaystyle \diagdown \diagup}{——}} CH_2$	Alkylating agent causes reduction of enzymes.	Sterilization of heat-labile material.
Beta-propiolactone (liquid or gas) $H_2C —— CH_2$ with C=O below	Alkylating agent causes reduction of enzymes.	1. Sterilization of tissue for grafting. 2. Destroys hepatitis virus. 3. Room disinfection.
Basic dyes Crystal violet	Affinity for nucleic acids, interfere with reproduction in gram-positive organisms.	1. Skin antiseptic. 2. Laboratory isolation of gram-negative bacteria.

The efficiency of all these disinfectants and antiseptics is influenced by several factors. Some that warrant consideration are:

1. **Concentration:** The concentration of a chemical substance markedly influences the effect produced on microorganisms, with higher concentrations producing a more rapid death. Concentration cannot be arbitrarily determined; the toxicity of the chemical to the material or tissue being treated must also be considered.

2. **Length of exposure:** Microbes are not all destroyed within the same exposure time. Sensitive forms are destroyed more rapidly than other resistant ones. The longer the exposure to the agent, the greater its antimicrobial activity. The toxicity of the chemical and environmental conditions must be considered in assessing the length of time necessary for disinfection or asepsis.

3. **Type of microbial population to be destroyed:** Microorganisms vary in their susceptibility to destruction by chemicals. Bacterial spores are the most resistant forms. Further variations are exemplified in that capsulated bacteria are more resistant than noncapsulated forms; acid-fast bacteria are more resistant than nonacid-fast; and older, metabolically less active cells are more resistant than younger cells. Awareness of the types of microorganisms that may be present will influence the choice of agent.

4. **Environmental conditions:** Conditions under which the disinfectant or antiseptic being used affects the chemical agent:

   a. **Temperature:** Cells are killed as the result of a chemical reaction between the agent and a cellular component. As increasing temperatures increase the rate of chemical reactions, application of heat during disinfection markedly increases the rate at which the microbial population is destroyed.

   b. **pH:** The pH conditions during disinfection may produce an effect not only on the microorganisms but on the compound. Extremes in pH are harmful to many microorganisms and may enhance the antimicrobial action of a chemical. Deviation from a neutral pH may cause ionization of the disinfectant; depending on the chemical agent, this may serve to increase or decrease the chemical's microbicidal action.

   c. **Type of material on which the microorganisms exist:** The destructive power of the compound on cells is due to its combination with organic cellular molecules. If the material on which the microorganisms are found is primarily organic, such as blood, pus, or tissue fluids, the agent will combine with these extracellular organic molecules and its antimicrobial activity will be reduced.

Numerous laboratory procedures are available for evaluating the antimicrobial efficiency of disinfectants or antiseptics. They provide a general rather than an absolute measure of the effectiveness of any agent, as test conditions frequently differ considerably from those seen during practical use. Two commonly employed procedures are presented.

PART A: Phenol Coefficient

PURPOSE

To enable students to compare the effectiveness of disinfectants.

PRINCIPLE

The **phenol coefficient test** compares the antimicrobial activity of a chemical compound to that of phenol under standardized experimental conditions. Equal quantities of a series of dilutions of the chemical being tested and of pure phenol are placed into sterile test tubes. A standardized quantity of a pure culture of the test microorganisms, such as *Staphylococcus aureus* or *Salmonella typhi* is added to each of the tubes. Subcultures of the test microorganism are made from each dilution of the test chemicals into sterile broth media at intervals of 5, 10, and 15 minutes after introduction of the organisms. All the subcultures are incubated at 37 degrees C for 48 hours and examined for the presence or absence of growth.

To determine the phenol coefficient, divide the highest dilution of the chemical being tested that destroyed the microorganisms in 10 minutes but not in 5 by the highest dilution of phenol that destroyed the microorganisms in 10 minutes, but not in 5 minutes. A phenol coefficient no greater than 1 indicates that this agent is equal to or less effective than phenol. A

Illustration of phenol coefficient determination

Chemical agent and dilution		Presence of growth in subcultures (minutes)		
		5	10	15
Phenol	1:80	−	−	−
	1:90	+	−	−
	1:100	+	+	−
Test chemical	1:400	−	−	−
	1:450	+	−	−
	1:500	+	+	−

Test chemical dilution of 1:450 showed no growth at 10 minutes, but growth at 5 minutes.
Phenol dilution of 1:90 showed no growth at 10 minutes, but growth at 5 minutes.

$$\text{Phenol coefficient of test chemical} = \frac{1/450}{1/90} = 5.$$

phenol coefficient greater than 1 suggests that the chemical is more effective than phenol when employed under test conditions.

A phenol coefficient of 5 indicates that the chemical agent under evaluation is five times as effective as phenol.

MATERIALS

Cultures

24-hour nutrient broth cultures of *Staphylococcus aureus* dispensed in sterile dropper-bottles.

Media

Nineteen nutrient broth tubes per designated student group.

Disinfectants

Phenol dilutions: 1:80, 1:90, 1:100 and Lysol dilutions: 1:400, 1:450, 1:500 per designated student group.

Equipment

Bunsen burner and inoculating loop.

PROCEDURE

1. Label 18 nutrient broth tubes with the name and dilution of the disinfectant, the time interval of subculturing (e.g., phenol 1:80, 5 minutes), and your initials.

2. In a test tube rack, place one test tube of each of the different phenol and Lysol dilutions.

3. Rapidly introduce one drop of the *S. aureus* culture into each of the test tubes of disinfectant. Note the time when you first start introducing the microorganisms into the disinfectant.

4. Agitate all the test tubes to ensure contact of the disinfectant and the microbes.

5. Using sterile technique, at intervals of 5, 10, and 15 minutes transfer one full loopful from each of the test tubes containing the disinfectant and microorganisms into the appropriately labeled sterile tubes of nutrient broth.

6. Incubate all nutrient broth cultures for 48 hours at 37 degrees C.

PART B: Agar Plate-Sensitivity Method

PURPOSE

To enable students to evaluate the effectiveness of antiseptic agents against selected test organisms.

PRINCIPLE

This procedure requires the heavy inoculation of an agar plate with the test organism. Sterile, color-coded filter-paper discs are impregnated with a different antiseptic and equally spaced on the inoculated agar plate. Following incubation the agar plate is examined for zones of inhibition, areas of no microbial growth, surrounding the discs. A zone of inhibition is indicative of microbicidal activity against the organism. Absence of a zone of inhibition indicates that the chemical was ineffective against the test organism. It should be noted that the size of the zone of inhibition is not indicative of the degree of effectiveness of the chemical agent.

MATERIALS

Cultures

24-hour to 48-hour trypticase soy broth cultures of *Escherichia coli, Bacillus cereus, Staphylococcus aureus,* and *Mycobacterium smegmatis.*

Media

Four trypticase soy agar plates per designated student group.

Antiseptics

10 ml of each of the following dispensed in 25-ml beakers per designated student group: Tincture of iodine, 3 percent hydrogen peroxide, 70 percent isopropyl alcohol, Mercurochrome, and Listerine.

Equipment

Five different-colored, sterile Sensi-discs, forceps, sterile cotton-tipped swabs, Bunsen burner, and wax marking pencil.

PROCEDURE

1. Label each of the trypticase soy agar plates with the name of the test organism to be inoculated and identify each with your initials.

2. Aseptically inoculate the agar plates with their respective test organisms by streaking each plate in horizontal and vertical directions with a sterile swab.

3. Color-code the sensi-discs as to the chemical agents to be used (i.e., red = Listerine).

4. Using forceps dipped in alcohol and flamed, expose five same-colored discs by placing them into the solution of one of the chemical agents. Drain the saturated discs on absorbent paper immediately prior to placing one on each of the inoculated agar plates. Place each disc approximately 2 cm in from the edge of the plate. Gently press the discs down with the forceps so that they adhere to the surface of the agar.

5. Impregnate the remaining discs as described in step 4. Place one of each of the four remaining colored discs on the surface of each of the five inoculated agar plates equidistant from each other around the periphery of the plate.

6. Incubate all plate cultures in an inverted position for 24 to 48 hours at 37 degrees C.

Observations and Results

Part A: Phenol Coefficient

1. Observe all nutrient broth cultures for the presence of growth. Use the sterile test tube of nutrient broth as a control for the determination of the presence or absence of growth.

2. Record your observations in chart as growth (+) or no growth (−).

		Growth in subcultures (minutes)		
Disinfectant	**Dilution**	5	10	15
Phenol	1:80			
	1:90			
	1:100			
Lysol	1:400			
	1:450			
	1:500			

3. Calculate the phenol coefficient of Lysol.

Part B: Agar Plate-Sensitivity Method

1. Observe all the plates for the presence of a zone of inhibition surrounding each of the impregnated discs.

2. Record your observations in chart as absence of a zone of inhibition (0), presence of a zone of inhibition (+).

	Antimicrobial agent				
Bacterial species	**Tincture of iodine**	**3% hydrogen peroxide**	**70% isopropyl alcohol**	**Listerine**	**Mercurochrome**
E. coli gram-negative					
S. aureus gram-positive					
M. smegmatis acid-fast					
B. cereus spore-former					

3. Indicate which of the antiseptics exhibited microbicidal activity against each of the following groups of microorganisms.

Bacterial group	Listerine	Tincture of iodine	3% hydrogen peroxide	Mercuro- chrome	70% isopropyl alcohol
Gram-negative					
Gram-positive					
Acid-fast					
Spore-former					

4. Which of the experimental chemical compounds appears to have the broadest range of microbicidal activity? _____

The narrowest range of microbicidal activity? _____

REVIEW QUESTIONS

1. Evaluate the effectiveness of a disinfectant with a phenol coefficient of 40.

2. Explain how environmental factors influence the effectiveness of the chemical compound under consideration.

3. Can the disinfection period (exposure time) be arbitrarily increased? _____ Explain.

X

Microbiology of Food

PURPOSES

To familiarize students with:

1. The endogenous and exogenous organisms that may be found in food products.

2. The analysis of foods as a means of determining their quality from the public health point of view.

3. The microbiological production of wine and sauerkraut.

INTRODUCTION

Microbiologists have always been aware of the fact that foods, especially milk, have served as important inanimate vectors in the transmission of disease. These foods contain the organic nutrients that provide an excellent medium to support the growth and multiplication of microorganisms under suitable temperatures.

Food and dairy products may be contaminated in a variety of ways and from a variety of sources:

1. **Soil and water:** Food-borne organisms that may be found in soil and water and may contaminate food are members of the genera *Alcaligenes, Bacillus, Citrobacter, Clostridium, Pseudomonas, Serratia, Proteus, Enterobacter,* and *Micrococcus.* The common soil and water molds include *Rhizopus, Penicillium, Botrytis, Fusarium,* and *Trichothecium.*

2. **Food utensils:** The type of microorganisms found on utensils depends on the type of food that was handled.

3. **Enteric microorganisms of humans and animals:** The major members of this group are *Bacteroides, Lactobacillus, Clostridium, Escherichia, Salmonella, Proteus, Shigella, Staphylococcus,* and *Streptococcus.* These organisms find their way into the soil and water from which they contaminate plants and are carried by wind currents onto utensils or prepared and exposed foods.

4. **Food handlers:** People who handle food are especially likely to contaminate foods, as microorganisms on hands and clothing are easily transmitted. A major offending organism is *Staphylococcus,* generally found on hands, skin, and in the upper respiratory tract. Food handlers with poor personal hygiene and unsanitary habits are most likely to contaminate foods with enteric organisms.

5. **Animal hides and feeds:** Microorganisms found in water, soil, feed, dust, and fecal debris can be found on animal hides. Infected hides may serve as a source of infection for workers or the microorganisms may migrate into the musculature of the animal and remain viable following its slaughter.

By enumerating microorganisms in milk and foods the quality of a particular sample can be determined. Although the microorganisms cannot be identified, the presence of a high number suggests a good possibility that pathogens may be present. Even if a sample contains a low microbial count, it can still transmit infection.

In the laboratory procedures that follow, students will have an opportunity directly and indirectly to enumerate the number of microorganisms present in milk and other food products and thereby determine the quality of the samples.

Methylene Blue Reductase Test

PURPOSE

To familiarize students with an enzymatic test to determine the quality of a milk sample.

PRINCIPLE

A milk sample that contains a large population of actively metabolizing microorganisms will contain a markedly decreased concentration of dissolved oxygen because of the vigorous growth of the organisms. In other words, the oxidation-reduction potential of the sample is greatly lowered. The dye methylene blue (MB), a redox indicator, loses its color in an anaerobic environment and is said to be reduced.

The methylene blue reductase test is designed to screen the quality of raw milk that contains large populations of enteric organisms and *Streptococcus lactis,* which are potent reducers of the dye. The speed at which reduction occurs following addition of MB to a sample of milk indicates the milk's quality. This determination is made as follows:

1. Reduction within 30 minutes is indicative of very poor quality.

2. Reduction occurring between one-half hour and two hours is indicative of poor quality.

3. Reduction occurring between two and six hours is indicative of fair quality.

4. Reduction occurring between six and eight hours is indicative of good quality.

MATERIALS

Cultures

Two raw milk samples of different quality.

Reagent

Methylene blue solution (1:250,000).

Equipment

Sterile screw-cap test tubes, sterile 10-ml and 1-ml pipettes, 37 degrees C water bath, and Bunsen burner.

PROCEDURE

1. Label the test tubes 1 and 2, and identify each with your initials.

2. Using a different 10-ml pipette both times, transfer 10 ml of each type of milk into its test tube.

3. Add 1 ml of methylene blue dye to each test tube.

4. Stopper, invert the test tubes gently about four times, and place in water bath. Record the time of incubation.

5. Allow the tubes to stabilize for five minutes, remove them from the water bath, invert them gently once and replace them in the water bath.

Observations and Results   EXPERIMENT **45**

1. Observe the milk samples for methylene blue reduction every 30 minutes for 3 to 6 hours. Reduction is demonstrated by a change in the color of the sample to white.

2. Record in the chart the time required for reduction in both milk samples.

3. Based on your observations, determine and record in the chart the quality of each sample as very poor, poor, fair, or good.

	Tube 1	*Tube 2*
Reduction time		
Quality of milk sample		

REVIEW QUESTIONS

1. If milk is a sterile body fluid, explain how it may become contaminated during the milking process.

2. Explain the function of the methylene blue dye in this experiment.

3. If a milk sample is judged to be of good quality, can it still serve as a source of human infection? _____ Explain.

4. Explain why milk sours in the absence of refrigeration.

Microbiological Analysis of Food Products: Bacterial Count

PURPOSES

1. To determine the total number of microorganisms present in food products.

2. To determine the presence of coliform bacteria in the selected food products.

PRINCIPLE

The presence of microorganisms in food may be considered harmful in some cases, while in others it is definitely beneficial. Certain microorganisms are necessary in preparation of foods such as cheese, pickles, sauerkraut, yogurt, and sausage. Whereas the presence of other microorganisms is responsible for serious and sometimes fatal food poisoning and toxicity as well as spoilage. As with milk or water, the presence and number of coliform bacteria and other enteric organisms in food is indicative of fecal contamination and may suggest the presence of pathogens.

Figure 46.1 on page 272 illustrates the procedure to enumerate microorganisms in foods and to determine the presence of coliform bacteria.

MATERIALS

Cultures

Samples of thawed frozen vegetables, ground beef, and dried fruit.

Media

Per designated student group: nine brain-heart infusion agar deep tubes, three eosin methylene blue agar plates, three 99-ml sterile water blanks, and three 180-ml sterile water blanks.

Equipment

Bunsen burner, water bath, Quebec or electronic colony counter, balance, sterile glassine weighing paper, Waring blender with three sterile jars, sterile Petri dishes, 1-ml pipettes, inoculating loop, and wax marking pencil.

PROCEDURE

1. Label three sets of three Petri dishes for each of the food samples to be tested and their dilution (10^{-2}, 10^{-3}, 10^{-4}), and identify each with your initials. Label the three eosin methylene blue agar plates with the name of the food and identify each with your initials.

2. Melt the brain-heart infusion agar deep tubes in a water bath, cool, and maintain at 45 degrees C.

3. Place 20 gm of each food sample, weighed on sterile glassine paper, into its labeled Waring blender jar. Add 180 ml of sterile water to each of the blender jars and blend each mixture for five minutes. A 1:10 (10^{-1}) dilution of each food sample has been made.

4. Transfer 1 ml of the 10^{-1} ground beef suspension into its labeled 99-ml sterile water blank, thereby effecting a 10^{-3} dilution, and 0.1 ml to the appropriately labeled 10^{-2} Petri dish. Shake the 10^{-3} sample dilution, and using a different pipette, transfer 1 ml to the plate labeled 10^{-3} and 0.1 ml to the plate labeled 10^{-4}. Add a 15-ml aliquot of the molten and cooled agar to each of the three plates. Swirl the plates gently to obtain a uniform distribution and allow the plates to solidify.

5. Repeat step 4 for the remaining two 10^{-1} test food sample dilutions.

6. Aseptically prepare a four-way streak plate as described in Experiment 11 of each 10^{-1} food sam-

ple dilution on its appropriately labeled eosin methylene blue agar plate.

7. Incubate all plates in an inverted position for 24 to 48 hours at 37 degrees C.

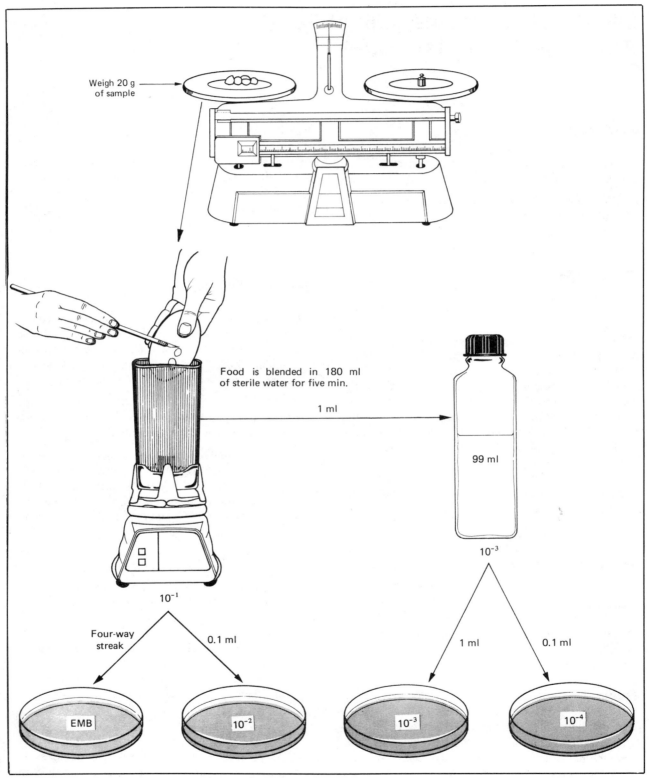

Weigh 20 g of sample

Food is blended in 180 ml of sterile water for five min.

1 ml

99 ml

10^{-3}

10^{-1}

Four-way streak

0.1 ml

1 ml

0.1 ml

EMB

10^{-2}

10^{-3}

10^{-4}

FIGURE 46.1. Preparation of food sample for analysis

Observations and Results

1. Using either the Quebec or electronic colony counter, count the number of colonies on each plate. Count only statistically valid plates that contain between 30 and 300 colonies. Designate plates with fewer than 30 colonies as **too few to count (TFTC)** and plates with more than 300 colonies as **too numerous to count (TNTC).**

2. Determine the number of organisms per milliliter of each food sample by multiplying the number of colonies counted by the dilution factor.

3. Record in the chart the number of colonies per plate and the number of organisms per milliliter of each food sample.

Type of food	Dilution	Number of colonies per plate	Number of organisms per ml
Ground beef	10^{-2} 10^{-3} 10^{-4}		
Frozen vegetables	10^{-2} 10^{-3} 10^{-4}		
Dried fruit	10^{-2} 10^{-3} 10^{-4}		

4. Examine the eosin methylene blue agar plate cultures for colonies presenting a metallic green sheen on their surface, which is indicative of *E. coli*. Record your results as to the presence or absence of *E. coli* growth and the possibility of fecal contamination of the food.

Sample	E. coli (+) or (−)	Fecal contamination (+) or (−)
Ground beef		
Frozen vegetables		
Dried fruit		

REVIEW QUESTIONS

1. Explain the purpose of the eosin methylene blue agar plates in this procedure.

2. Explain why it is not advisable to thaw and then refreeze food products.

3. Indicate some possible ways in which foods may become contaminated with enteric organisms.

Wine Production

PURPOSE

To acquaint students with wine production by the fermentative activities of yeast cells.

PRINCIPLE

Wine is a product of the natural fermentation of the juice of grapes and other fruits such as peaches, pears, plums, and apples by the action of yeast cells. This biochemical conversion of juice to wine occurs when the yeast cells enzymatically degrade the fruit sugars fructose and glucose first to acetaldehyde and then to alcohol, as illustrated at the bottom of this page.

Grapes containing 20 percent to 30 percent sugar concentration will yield wines with an alcohol content of approximately 10 percent to 15 percent. Also present in grapes are acids and minerals whose concentrations are increased in the finished product and are responsible for the characteristic tastes and bouquets of different wines. To produce red wine the crushed grapes must be fermented with their skins to allow extraction of their color into the juice. White wine is produced from the juice of white grapes.

The commercial production of wine is a long and exacting process. First, the grapes are crushed or pressed to express the juice, which is called **must.** Potassium metabisulfite is added to the must to retard the growth of acetic acid bacteria, molds, and wild yeast that are endogenous to grapes in the vineyard. A wine-producing strain of yeast, *Saccharomyces cerevisiae* var. *ellipsoideus,* is used to inoculate the must, which is then incubated for three to five days under aerobic conditions at 21 degrees to 32 degrees C. This is followed by an anaerobic incubation period. The wine is then aged for a period of one to five years in aging tanks or wooden barrels. During this time the wine is clarified of any turbidity, thereby producing volatile esters that are responsible for characteristic flavors. The clarified product is then filtered, pasteurized at 60 degrees C for 30 minutes, and bottled.

This experiment is a modified method in which white wine is produced from white grape juice. Students will examine the fermenting wine at one-week intervals during the incubation period for:

1. Total acidity (expressed as % tartaric acid): To a 10-ml aliquot of the fermenting wine add 10 ml of distilled water and 5 drops of 1 percent phenolphthalein solution. Mix and titrate to the first persistent pink color with 0.1N sodium hydroxide. Calculate total acidity using the following formula:

$$\% \; tartaric \; acid = \frac{ml \; alkali \times normality \; of \; alkali \times 7.5}{weight \; of \; sample \; in \; gm*}$$

*1 ml = 1 gm.

2. Volatile acidity (expressed as % acetic acid): Following titration, calculate volatile acidity using the following formula:

$$\% \ acetic \ acid = \frac{ml \ alkali \ \times \ normality \ of \ alkali \ \times \ 6.0}{weight \ of \ sample \ in \ gm^*}$$

3. Alcohol (expressed as volume %): Optional, can be determined by means of an ebulliometer.

4. Taste: Bitter, sour, sweet, salty.

5. Odor: Fruity, yeast-like, sweet, none.

6. Clarity: Clear, turbid.

MATERIALS

Cultures

50 ml white grape juice broth culture of *Saccharomyces cerevisiae* var. *ellipsoideus* incubated for 48 hours at 25 degrees C.

Media

500 ml of pasteurized Welch's commercial white grape juice per designated student group.

*1 ml = 1 gm.

Reagents

1 percent phenolphthalein solution, 0.1N sodium hydroxide, and sucrose.

Equipment

One-liter Erlenmeyer flask, one-holed rubber stopper containing a 2-inch glass tube plugged with cotton, pan balance, spatula, glassine paper, 10-ml graduated cylinder, ebulliometer (optional), and burette or pipette for titration.

PROCEDURE

1. Pour 500 ml of the white grape juice into the 1-liter Erlenmeyer flask. Add 20 gm of sucrose and the 50 ml *S. cerevisiae* grape juice broth culture (10 percent starter culture). Close the flask with the stopper containing a cotton plugged air vent, and identify with your initials.

2. After two days and four days of incubation, add 20 gm of sucrose to the fermenting wine.

3. Incubate the fermenting wine for 21 days at 25 degrees C.

Observations and Results   EXPERIMENT 47

1. Using uninoculated white grape juice:

    a. Perform a titration to determine total acidity and volatile acidity.

    b. Note taste, aroma, and clarity.

    c. Determine volume % alcohol (optional).

2. Record your results in the chart.

3. At seven-day intervals, using samples of the fermenting wine, repeat steps 1a through 1c and record your results in the chart.

	Grape juice	*Fermenting wine*		
		7 days	*14 days*	*21 days*
% Tartaric acid				
% Acetic acid				
Volume % alcohol				
Taste				
Aroma				
Clarity				

REVIEW QUESTIONS

1. What is the purpose of adding sulfite to the must?

2. Explain what occurs during the aging process in the commercial preparation of wine.

3. What are the chemical end products of fermentation?

4. How are white and red wines produced?

5. Why is wine pasteurized?

Sauerkraut Production

PURPOSE

To acquaint students with the microbiological production of sauerkraut.

PRINCIPLE

Sauerkraut is a classic example of a food of plant origin produced by microbial fermentation. Its preparation requires the fermentative activities of a mixed microbial flora including *Leuconostoc mesenteroides, Lactobacillus plantarum, Lactobacillus brevis,* and *Streptococcus faecalis.*

In the production of sauerkraut, shredded cabbage is treated with sodium chloride, which creates an osmotic environment in which plasmolysis occurs, extracting the juice from the cabbage tissue. The resultant brine solution favors the growth of lactic acid-producing microorganisms and inhibits the growth of other microorganisms. The lactic acid is responsible for the characteristic kraut flavor and also acts as a preservative by inhibiting the growth of microorganisms that cause food spoilage.

Production of the lactic acid is initiated by *Leuconostoc mesenteroides,* which are cocci, and sustained by *Lactobacillus plantarum,* which are bacilli. When the acid concentration reaches a level of 0.7 percent to 1 percent the fermentative activities of *Leuconostoc mesenteroides* cease and the final stages of the process are carried out by *Lactobacillus plantarum, Lactobacillus brevis,* and *Streptococcus faecalis.* The finished product contains a total acidity of 1.5 percent to 2 percent, of which lactic acid represents 1 percent to 1.5 percent.

In this experiment the students will prepare two samples of sauerkraut, one for sampling and testing of the final product and the other for testing at specific intervals during incubation for:

1. Odor: Acid, earthy, spicy, or putrid.

2. Color: Brown, pink, straw yellow, pale yellow, or colorless.

3. Taste: Sour (acid), sweet, bitter, or salty.

4. Texture:

   a. Soft: Fermentation initiated by *Lactobacillus plantarum* rather than *Leuconostoc mesenteroides.*

   b. Slimy: Rapid growth of *Lactobacillus cucumeris* at elevated temperatures.

   c. Rotted: Spoilage by bacteria, yeast, or molds.

5. pH: The pH of the finished product should be in the range of 3.1 to 3.7.

6. Total acidity expressed as % lactic acid, determination made as follows:

   a. Place 10 ml of the fermentation juice and 10 ml distilled water into an Erlenmeyer flask. Boil to drive off the CO_2.

   b. Cool and add five drops of 1 percent phenolphthalein to the diluted juice.

   c. Titrate to the first persistent pink color with 0.1N NaOH.

   d. Calculate % lactic acid as follows:

$$\% \text{ lactic acid} = \frac{ml \text{ of alkali} \times \text{ normality of alkali} \times 9}{\text{weight of sample in gm*}}$$

7. Microscopic appearance of the microbial flora.

MATERIALS

Media

Two heads of cabbage per designated student group.

*1 ml = 1 gm.

Reagents

1 percent phenolphthalein, 0.1N NaOH, methylene blue, and uniodized table salt.

Equipment

Two wide-mouthed jars with covers, two wooden boards to fit into jars, two heavy weights, cheesecloth, pH meter or indicator paper, pan balance, microscope, Bunsen burner, inoculating loop, glass slides, coverslips, 10-ml disposable pipettes, knife, and Erlenmeyer flask.

PROCEDURE

1. Remove the outer leaves and all bruised tissues from each of the cabbage heads.

2. Halve, core, and wash the heads in tap water.

3. Shred the cabbage.

4. Weigh the shredded cabbage on a pan balance and separate into two equal portions.

5. Weigh out the table salt in amounts equal to 3 percent of the weight of each of the portions of shredded cabbage.

6. Place the shredded cabbage and salt in alternating layers in the two wide-mouthed jars, identified with your initials.

7. Place a wooden board over each of the mixtures and press gently to squeeze out a layer of juice from the cabbage.

8. Place a weight on each of the boards and cover the jars with cheesecloth.

9. Incubate the jars for 14 days at 30 degrees C.

Observations and Results   EXPERIMENT **48**

Examine the sauerkraut preparation on days 2, 7, 14, and 21 of incubation as follows:

1. Observe the fermenting cabbage as to its aroma, taste, texture, and color. Record your results in the chart.

2. With a 10-ml pipette, remove 10 ml of the fermentation juice.

   a. Using methylene blue, prepare a stained slide preparation for microscopic examination.

   b. Using the pH meter or indicator paper, determine the pH of the juice.

   c. Perform a titration of the juice to determine the % lactic acid present.

Record your results in the chart.

Result	Sauerkraut preparation			
	2 days	7 days	14 days	21 days
Odor				
Color				
Taste				
Texture				
% lactic acid				
pH				
Drawing of microbial flora				

3. Based on your observations, was there any indication of microbial spoilage of your sauerkraut? _____ Explain.

REVIEW QUESTIONS

1. Discuss the importance of the specific sequential activity of the microbial flora responsible for sauerkraut production.

2. What is the function of the salt in preparation of sauerkraut?

3. How does the process of fermentation aid in food preservation?

4. Why is uniodized salt used in this procedure?

5. Explain the production of slimy or rotten kraut.

XI

Microbiology of Water

PURPOSES

1. To acquaint students with the types of microorganisms present in water.

2. To determine the potability of water using standard qualitative and quantitative procedures.

INTRODUCTION

The importance of potable (drinking) water supplies cannot be overemphasized. With increasing industrialization in the past 50 years, water sources available for consumption and recreation have been adulterated with industrial as well as animal and human wastes. As a result, water has become a formidable factor in disease transmission. Polluted waters contain vast amounts of organic matter that serve as an excellent nutritional source for the growth and multiplication of microorganisms. The presence of nonpathogenic organisms is not of major concern, but intestinal contaminants of fecal origin are important. These pathogens are responsible for intestinal infections such as **bacillary dysentery, typhoid fever, cholera,** and **paratyphoid fever.**

Analysis of water samples on a routine basis would not be possible if each individual pathogen required detection. Therefore water is examined to detect *Escherichia coli,* the bacterium that indicates fecal pollution. Since *Escherichia coli* is always present in the human intestine, its presence in water alerts public health officials to the possible presence of other human or animal intestinal pathogens. Both qualitative and quantitative methods are used to determine the sanitary condition of water.

Standard Qualitative Analysis of Water

The three basic tests to detect coliform bacteria in water are presumptive, confirmed, and completed (Figure 49.1 on page 286). The tests are performed sequentially on each sample under analysis. They detect the presence of coliform bacteria (indicators of fecal contamination) the gram-negative, nonspore-forming bacilli that ferment lactose with the production of acid and gas that is detectable following a 24-hour incubation period at 37 degrees C.

PART A: Presumptive Test: Determination of the Most Probable Number

PURPOSES

1. To determine the presence of coliform bacteria in a water sample.

2. To obtain some index as to the possible number of organisms present in the sample under analysis.

PRINCIPLE

The **presumptive test** is specific for detection of coliform bacteria. Measured aliquots of the water to be tested are added to a lactose fermentation broth containing an inverted gas vial. Because these bacteria are capable of using lactose as a carbon source while the other enteric organisms are not, their detection is facilitated by use of this medium. In addition to lactose, the medium also contains a surface tension depressant, bile salt used to suppress the growth of organisms other than coliform bacteria.

Tubes of this lactose medium are inoculated with 10-ml, 1.0-ml, and 0.1-ml aliquots of the water sample. The series consists of at least three groups, each composed of three tubes of the specified medium. The tubes in each group are then inoculated with the designated volume of the water sample as described under Procedure. The greater the number of tubes per group, the greater the sensitivity of the test. Development of gas in any one of the tubes is *presumptive* evidence of the presence of coliform bacteria in the sample. The

presumptive test also enables the microbiologist to obtain some idea of the number of organisms present by means of the **most probable number test (MPN).** The MPN is estimated by determining the number of tubes in each group that show gas following the incubation period (Table 49.1 on page 287).

MATERIALS

Cultures

Water samples from sewage plant, pond, and tap.

Media

Per designated student group: nine double-strength lactose fermentation broth (LB2X) and eighteen single-strength lactose fermentation broth (LB1X).

Equipment

Bunsen burner, sterile 10-ml pipettes, sterile 1-ml pipettes, and sterile 0.1-ml pipettes.

PROCEDURE

1. Set up three separate series consisting of three groups, a total of nine tubes per series, in a test-tube rack; label tubes as to the water source and volume of sample inoculated as illustrated, and identify with your initials.

Series 1: Sewage water	3 tubes of LB2X-10 ml 3 tubes of LB1X-1 ml 3 tubes of LB1X-0.1 ml
Series 2: Pond water	3 tubes of LB2X-10 ml 3 tubes of LB1X-1 ml 3 tubes of LB1X-0.1 ml
Series 3: Tap water	3 tubes of LB2X-10 ml 3 tubes of LB1X-1 ml 3 tubes of LB1X-0.1 ml

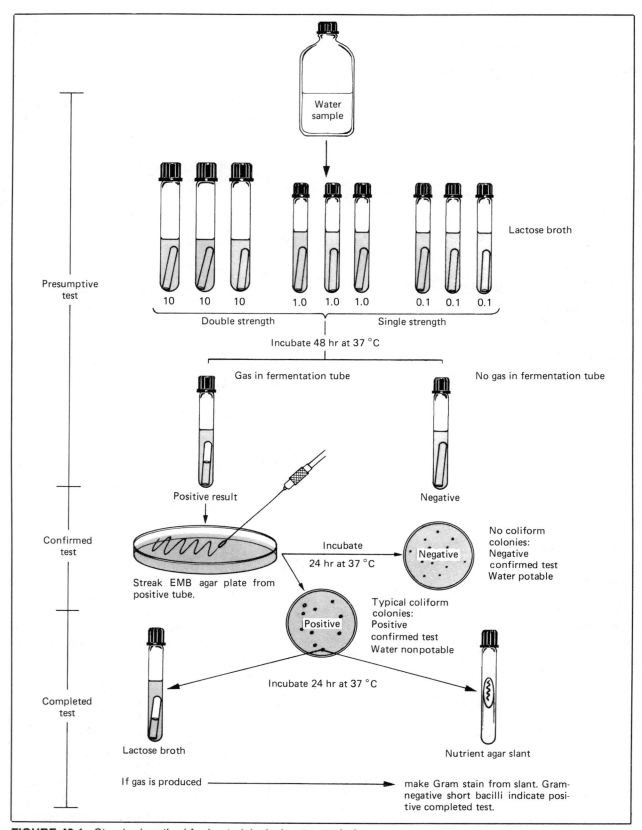

FIGURE 49.1. Standard method for bacteriological water analysis

TABLE 49.1. MPN Determination from Multiple Tube Test

Number of tubes giving positive reaction out of			MPN index per 100 ml	95% Confidence limits	
3 of 10 ml each	3 of 1 ml each	3 of 0.1 ml each		Lower	Upper
0	0	1	3	<0.5	9
0	1	0	3	<0.5	13
1	0	0	4	<0.5	20
1	0	1	7	1	21
1	1	0	7	1	23
1	1	1	11	3	36
1	2	0	11	3	36
2	0	0	9	1	36
2	0	1	14	3	37
2	1	0	15	3	44
2	1	1	20	7	89
2	2	0	21	4	47
2	2	1	28	10	150
3	0	0	23	4	120
3	0	1	39	7	130
3	0	2	64	15	380
3	1	0	43	7	210
3	1	1	75	14	230
3	1	2	120	30	380
3	2	0	93	15	380
3	2	1	150	30	440
3	2	2	210	35	470
3	3	0	240	36	1,300
3	3	1	460	71	2,400
3	3	2	1,100	150	4,800

From: **Standard Methods for the Examination of Water and Wastewater,** 14th edition. American Public Health Association, American Water Works Association, Water Pollution Control Federation, Washington, D.C., 1975.

2. Mix sewage plant water sample by shaking thoroughly. **Exercise care in handling sewage waste water sample, as enteric pathogens may be present.**

3. Flame bottle, and using a 10-ml pipette transfer 10-ml aliquots to the three tubes labeled LB2X-10 ml.

4. Flame bottle, and using a 1-ml pipette transfer 1 ml of water to the three tubes labeled LB1X-1 ml.

5. Flame bottle, and using a 0.1-ml pipette transfer 0.1 ml of water to the three tubes labeled LB1X-0.1 ml.

6. Repeat steps 2 through 5 for the tap and pond water samples.

7. Incubate all tubes for 48 hours at 37 degrees C.

PART B: Confirmed Test

PURPOSE

To confirm the presence of coliform bacteria in a water sample showing a positive presumptive test.

PRINCIPLE

The presence of a positive or doubtful presumptive test immediately suggests that the water sample is nonpotable. Confirmation of these results is necessary, since positive presumptive tests may be the result of organisms of noncoliform origin that are not recognized as indicators of fecal pollution.

The **confirmed test** requires that selective and differential media such as eosin methylene blue (EMB) or endo agar be streaked from a positive lactose broth tube obtained from the presumptive test. The nature of these differential and selective media was discussed in Experiment 15, but is reviewed briefly here. Eosin methylene blue contains the dye methylene blue, which inhibits the growth of gram-positive organisms. In the presence of an acid environment, EMB forms a complex that precipitates out onto the coliform colonies, producing dark centers and a green metallic sheen. This reaction is characteristic for *Escherichia coli,* the major indicator of fecal pollution. Endo agar is a nutrient medium containing the dye fuchsin, which is present in the decolorized state. In the presence of acid produced by the coliform bacteria, fuchsin forms a dark pink complex that turns the *E. coli* colonies and the surrounding medium pink.

MATERIALS

Cultures

One 24-hour-old positive lactose broth culture from each of the three series from the presumptive test.

Media

Three each per designated student group: Eosin methylene blue agar plates and endo agar plates.

Equipment

Bunsen burner, wax pencil, and inoculating loop.

PROCEDURE

1. Label the covers of the three EMB plates and three endo agar plates as to the source of the water sample (sewage, pond, and tap) and identify each with your initials.

2. Using a positive 24-hour lactose broth culture from the sewage water series from the presumptive test, streak the surface of one EMB and one endo agar plate, as described in Experiment 11 to obtain discrete colonies.

3. Repeat step 2 using the positive lactose broth cultures from the pond and tap water series to inoculate the remaining plates.

4. Incubate all plate cultures in an inverted position for 24 hours at 37 degrees C.

PART C: Completed Test

PURPOSE

To confirm the presence of coliform bacteria in a water sample, or if necessary, to confirm a suspicious but doubtful result of the previous test.

PRINCIPLE

The **completed test** is the final analysis of the water sample. It examines the coliform colonies that appeared on the EMB or endo agar plates used in the confirmed test. An isolated colony is picked from the confirmatory test plate and inoculated into a tube of lactose broth and streaked on a nutrient agar slant to perform a Gram stain. Following inoculation and incubation, tubes showing acid and gas in the lactose broth and the presence of gram-negative bacilli on microscopic examination are further confirmation of the presence of *E. coli,* and indicative of a positive completed test.

MATERIALS

Cultures

One 24-hour coliform-positive EMB or endo agar culture from each of the three series of the confirmed test.

Media

Three each per designated student group: nutrient agar slants and lactose fermentation broths.

Reagents

Crystal violet, Gram's iodine, 95 percent ethyl alcohol, and safranin.

Equipment

Bunsen burner, staining tray, inoculating loop, lens paper, bibulous paper, and microscope.

PROCEDURE

1. Label all tubes as to the source of the water sample and identify each with your initials.

2. Inoculate one lactose broth and one nutrient agar slant from the same isolated *E. coli* colony obtained from an EMB or an endo agar plate from each of the experimental water samples.

3. Incubate all tubes for 24 hours at 37 degrees C.

Observations and Results

EXPERIMENT **49**

PART A: PRESUMPTIVE TEST

1. Examine all tubes after 24 and 48 hours of incubation. Record your results in the chart as:

   a. Positive: 10 percent or more of gas appears in a tube in 24 hours.

   b. Doubtful: Gas develops in a tube after 48 hours.

   c. Negative: There is no gas in the tube in the series in 48 hours.

2. Determine and record the MPN using Table 49.1.

   Example: If gas appeared in all three tubes labeled LB2X-10, in two of the tubes labeled LB1X-1, and in one labeled LB1X-0.1, the series would be read as 3-2-1. From the MPN table, such a reading would indicate that there would be approximately 150 microorganisms per 100 ml of water, with a 95 percent probability that there are between 30 and 440 organisms present.

Water sample	*Acid and gas*									*Reading*	*MPN*	*Range 95% probability*
	LB2X-10			*LB1X-1*			*LB1X-0.1*					
Tube	*1*	*2*	*3*	*4*	*5*	*6*	*7*	*8*	*9*			
Sewage												
Pond												
Tap												

PART B: CONFIRMED TEST

1. Examine all the plates as to the presence or absence of *E. coli* colonies. Record your result in the chart.

2. Based on your results, determine and record whether each of the samples is potable or nonpotable. The presence of *E. coli* is a positive confirmed test, indicating that the water is nonpotable. The absence of *E. coli* is a negative test, indicating that the water is not contaminated with fecal wastes and is therefore potable.

Water sample	*Coliforms*		*Potable*	*Nonpotable*
	EMB plate	*Endo agar plate*		
Sewage				
Pond				
Tap				

PART C: COMPLETED TEST

1. Examine all lactose fermentation broth cultures as to the presence or absence of acid and gas. Record your results in the chart.

2. Prepare a Gram stain, using the nutrient agar slant cultures of the organisms that showed a positive result in the lactose fermentation broth (refer to Experiment 6 for the staining procedure).

3. Examine the slides microscopically for the presence of gram-negative, short bacilli, which are indicative of *E. coli* and thus nonpotable water. Record your results as to Gram stain reaction and morphology of the cells in the chart.

Water source	Lactose broth (A/G) (+) or (−)	Gram stain Reaction morphology	Potability Potable	Nonpotable
Sewage				
Pond				
Tap				

REVIEW QUESTIONS

1. What is the rationale for selecting *E. coli* as the indicator of water potability?

2. Why is this procedure qualitative rather than quantitative?

3. Explain why it is of prime importance to analyze water supplies that serve industrialized communities.

4. Is it necessary to perform a completed test on all water samples under analysis? _____ Explain.

Quantitative Analysis of Water: Membrane Filter Method

PURPOSE

To determine the quality of water samples using the membrane filter method.

PRINCIPLE

Bacteria-tight **membrane filters** capable of retaining microorganisms larger than 0.45 micrometers, are frequently used for analysis of water. These filters offer several advantages over the conventional, multiple-tube method of water analysis: (1) Results are available in a shorter period of time, (2) larger volumes of sample can be processed, and (3) because of the high accuracy of this method, the results are readily reproducible. A disadvantage involves the processing of turbid specimens that contain large quantities of suspended materials; particulate matter clogs the pores and inhibits passage of the specified volume of water.

A water sample is passed through a sterile membrane filter that is housed in a special filter apparatus contained in a suction flask. Following filtration, the filter disc that contains the trapped microorganisms is aseptically transferred to a sterile Petri dish containing an absorbent pad saturated with a selective, differential liquid medium. Following incubation, the number of colonies present on the filter is counted with the aid of a microscope.

This experiment analyzes a series of dilutions of water samples collected upstream and downstream from an outlet of a sewage treatment plant. A total count of coliform bacteria determines the potability of the water sources. Also, the types of fecal pollution, if any, are established by means of a fecal coliform count, indicative of human pollution; and a fecal streptococcal count, indicative of pollution from other animal origins. The ratio of the fecal coliforms to fecal streptococci per milliliter of sample is interpreted as follows: Between 2 and 4 indicates human and animal pollution; >4 indicates human pollution; <0.7 indicates poultry and livestock pollution.

MATERIALS

Cultures

Water samples collected upstream (labeled U) and downstream (labeled D) from an outlet of a sewage treatment plant.

Media

Per designated student group for analysis of one water sample: one 20-ml tube of M-endo broth, one 20-ml tube of M-FC broth, one 20-ml tube of K-F broth, four 90-ml sterile water blanks, and one 300-ml flask of sterile water.

Equipment

Sterile Millipore membrane apparatus (base, funnel, and clamp), 1-liter suction flask, 15 sterile Millipore membrane filters and absorbent pads, 15 sterile 50-mm Petri dishes, small beaker of 95 percent alcohol, forceps, and dissecting microscope.

PROCEDURE

The following instructions are for analysis of one of the provided water samples. Different samples may be assigned to individual groups.

1. Label the four 90-ml water blanks with the source of the water sample and dilution (10^{-1}, 10^{-2}, 10^{-3}, and 10^{-4}).

2. Using 10-ml pipettes, aseptically perform a 10-fold serial dilution of the assigned undiluted water sample, using the four 90-ml water blanks to effect the 10^{-1}, 10^{-2}, 10^{-3}, and 10^{-4} dilutions.

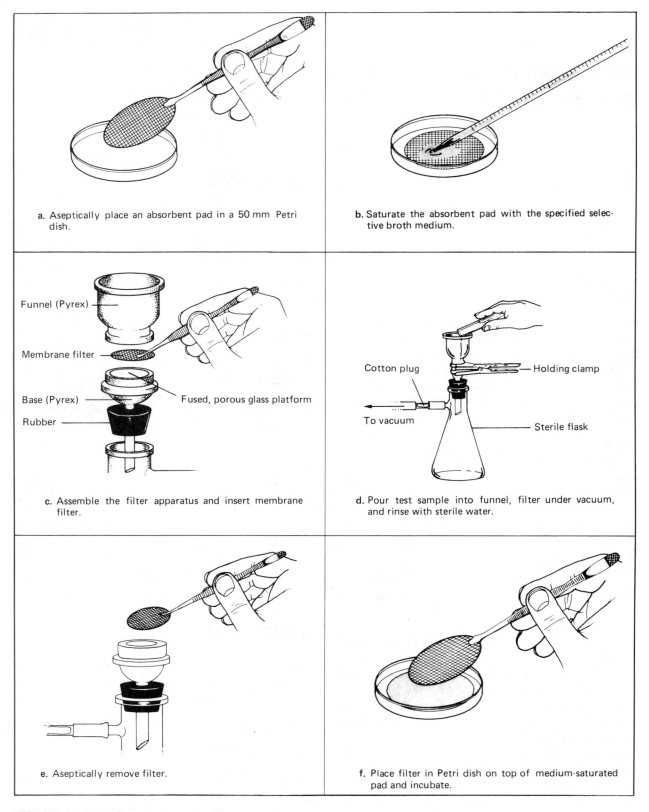

a. Aseptically place an absorbent pad in a 50 mm Petri dish.

b. Saturate the absorbent pad with the specified selective broth medium.

c. Assemble the filter apparatus and insert membrane filter.

d. Pour test sample into funnel, filter under vacuum, and rinse with sterile water.

e. Aseptically remove filter.

f. Place filter in Petri dish on top of medium-saturated pad and incubate.

FIGURE 50.1. Membrane filter technique

3. Arrange the 15 Petri dishes into 3 sets of 5 plates. Label each set as follows:

   a. For total coliform count (TCC) and dilutions (undiluted, 10^{-1}, 10^{-2}, 10^{-3}, and 10^{-4}).

   b. For fecal coliform count (FCC) and dilutions as in step 3a.

   c. For fecal streptococcal count (FSC) and dilutions as in step 3a.

   Identify each plate with your initials.

4. Using a sterile forceps dipped in 95 percent alcohol and flamed, add a sterile absorbent pad to all Petri dishes.

5. With sterile 10-ml pipettes, aseptically add:

   a. Two ml of M-endo broth to each pad in the plates labeled TCC.

   b. Two ml of M-FC broth to each pad in the plates labeled FCC.

   c. Two ml of K-F broth to each pad in the plates labeled FSC.

6. Aseptically assemble the sterile paper-wrapped membrane filter unit as follows:

   a. Unwrap and insert the sintered glass filter base into the neck of a 1-liter side-arm suction flask.

   b. With sterile forceps, place a sterile membrane filter disc, grid side up, on the sintered glass platform.

   c. Unwrap and carefully place the funnel section of the apparatus on top of the filter disc. Using the filter clamp, secure the funnel to the filter base.

   d. Attach a rubber hose from the side-arm on the vacuum flask to a vacuum source.

7. Using the highest sample dilution (10^{-4}), pipette 20 ml of the dilution into the funnel and start the vacuum.

8. When the entire sample has been filtered, wash the inner surface of the funnel with 20 ml of sterile water.

9. Disconnect the vacuum, unclamp the filter assembly, and with sterile forceps remove the membrane filter and place it on the medium saturated pad in the Petri dish labeled TCC, 10^{-4}.

10. Aseptically place a new membrane on the platform, reassemble the filtration apparatus, and repeat steps 7 through 9 twice, adding the filter discs to the 10^{-4} dilution plates labeled FCC and FSC.

11. Repeat steps 8 through 10, using 20 ml of the 10^{-3}, 10^{-2}, and 10^{-1} dilutions, and the undiluted samples.

12. Incubate the plates in an inverted position as follows:

    a. TCC and FSC plates for 24 hours at 37 degrees C.

    b. FCC plates sealed with waterproof tape and placed in a weighted water-tight plastic bag, which is then submerged in 44.5 degrees C water bath for 24 hours.

Observations and Results

1. Remove the filter discs from the Petri dishes and allow to dry on absorbent paper for one hour.

2. Examine all filter discs under a dissecting microscope and perform colony counts on each set of discs as follows:

   a. TCC: Count colonies on M-endo agar that present a golden metallic sheen (performed on a disc showing 20 to 80 of these colonies).

   b. FCC: Count colonies on M-FC agar that are blue (performed on a disc showing 20 to 60 of these colonies).

   c. FSC: Count colonies on K-F agar that are pink to red (performed on a disc showing 20 to 100 of these colonies).

   Dilution samples that show fewer colonies than indicated are designated as TFTC, and those showing a greater number of colonies are designated as TNTC.

3. For each of the three counts, determine the number of fecal organisms present in 100 ml of the water sample, using the following formula:

$$\frac{colony\ count\ \times\ dilution\ factor}{ml\ of\ sample\ used} \times 100$$

4. Record your results in the chart.

Dilution	Upstream water			Downstream water		
	TCC	FCC	FSC	TCC	FCC	FSC
Undiluted						
10^{-1}						
10^{-2}						
10^{-3}						
10^{-4}						

5. Determine the FC:FS ratio and record your results in the chart.

Dilution	Upstream water			Downstream water		
	Cells/ml*		FC:FS ratio	Cells/ml*		FC:FS ratio
	FCC	FSC		FCC	FSC	
Undiluted						
10^{-1}						
10^{-2}						
10^{-3}						
10^{-4}						

*Cells/ml $= \dfrac{Cells/100\ ml}{100}$.

6. Based on your FC:FS ratio, indicate the type of fecal pollution, if any:

Upstream water sample:

Downstream water sample:

REVIEW QUESTIONS

1. What are the advantages of the membrane filter method in the analysis of water samples?

Disadvantages?

2. What is the purpose of determining the FC:FS ratio?

3. Cite some other microbiological applications of the membrane filter technique in environmental studies.

XII

Microbiology of Soil

PURPOSES

1. To acquaint students with the characteristics and activities of soil microorganisms.

2. To demonstrate the role of soil microorganisms in the biotransformation of nitrogenous compounds via the nitrogen cycle.

3. To enumerate soil microorganisms.

4. To demonstrate the ability of some soil microorganisms to produce antibiotics.

INTRODUCTION

Life on this planet could not be sustained in the absence of microorganisms that inhabit the soil. These flora are essential for degradation of organic matter deposited in the soil as animal wastes and dead plant and animal tissues. Hydrolysis of these macromolecules by microbial enzymes supplies and replenishes the soil with basic elemental nutrients such as carbon dioxide, nitrates, nitrites, and recyclable gaseous nitrogen. By means of enzymatic transformations, plants assimilate these nutrients into the organic macromolecules essential for their growth and reproduction. In turn, these plants become a source of nutrition for animals.

These diverse microbial flora play a vital role in elemental cycles such as the carbon, nitrogen, phosphorus, and sulfur cycles. Although each one is of significant importance to all forms of life, it is not in the scope of this manual to deal with them individually. Instead, experiments are included that illustrate the biochemical reactions of the nitrogen cycle.

The **nitrogen cycle** is concerned with the enzymatic conversion of nitrogenous compounds found in the soil and gaseous nitrogen from the atmosphere into inorganic nitrogen compounds that are used by plants for synthesis of essential macromolecules such as nucleic acids and proteins. The four distinct phases of this cycle are as follows:

1. **Ammonification:** Sequential degradation of nitrogenous organic compounds with the release of ammonium.

2. **Nitrification:** Step-by-step oxidation of ammonia to nitrite (NO_2^-) and then to nitrates (NO_3^-), a nutritional form of nitrogen that is assimilated by plants.

3. **Denitrification:** Reduction of nitrates not used by plants to gaseous nitrogen ($N_2 \uparrow$).

4. **Nitrogen fixation:** Chemical combination of free nitrogen ($N_2 \uparrow$) with other elements to form fixed nitrogen (nitrogen-containing compounds).

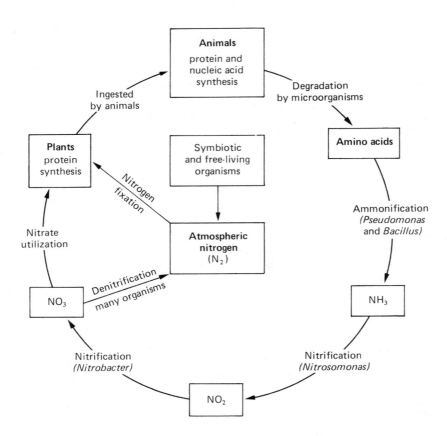

Another vital role of some soil organisms is their ability to produce metabolic end products that, in minute concentrations, are capable of inhibiting the growth of other microorganisms. These antimicrobial substances are called **antibiotics** and are used extensively in the treatment of infectious diseases (see Experiment 43).

Microbial Populations in Soil: Enumeration

PURPOSES

1. To acquaint students with the microbial soil flora.

2. To determine the number of bacteria and fungi present in a sample of soil.

PRINCIPLE

Soil contains myriads of microorganisms, including bacteria, fungi, protozoa, algae, and viruses. Despite this diversity, bacteria, including the moldlike actinomycetes, and fungi are the most prevalent, as shown at the bottom of this page.

It is essential to bear in mind the fact that the soil environment differs from one location to another and from one period of time to another. Therefore factors such as moisture, pH, temperature, gaseous oxygen content, and organic and inorganic composition of soil will be crucial in determining the specific microbial flora of a particular sample.

Just as the soil differs, microbiological methods used to analyze soil also vary. A single technique cannot be used to count all the different types of microorganisms present in a given soil sample. This is because there is no one laboratory cultivation procedure that can provide all the physical and nutritional requirements necessary for the growth of a greatly diverse microbial population. In this experiment only the relative number of bacteria, actinomycetes, and fungi are determined. The method used is the serial dilution agar plating procedure described in Experiment 13. To support the growth of these three types of microorganisms,

different media are employed: Glycerol yeast agar for the isolation of actinomycetes and Sabouraud agar for the isolation of fungi. To inhibit the growth of bacteria, both media are supplemented with 10 μg of chlortetracycline (Aureomycin) per milliliter of medium (Figure 51.1 on page 302).

MATERIALS

Soil

1-gm sample of finely pulverized, rich garden soil in a flask containing 99 ml sterile water; labeled flask 1: 1:100 dilution (10^{-2}).

Media

Per designated student group: One 75-ml flask of molten nutrient agar maintained at 45 degrees C; one 75-ml flask of molten glycerol yeast extract agar maintained at 45 degrees C, one 75-ml flask of molten Sabouraud agar maintained at 45 degrees C, and two 99-ml flasks of sterile water.

Equipment

Bunsen burner, Quebec colony counter, mechanical hand counter, sterile 1-ml pipettes, sterile Petri dishes, and wax marking pencil.

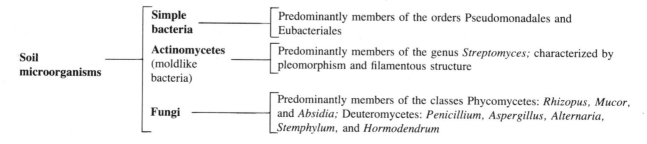

Soil microorganisms

Simple bacteria — Predominantly members of the orders Pseudomonadales and Eubacteriales

Actinomycetes (moldlike bacteria) — Predominantly members of the genus *Streptomyces;* characterized by pleomorphism and filamentous structure

Fungi — Predominantly members of the classes Phycomycetes: *Rhizopus, Mucor,* and *Absidia;* Deuteromycetes: *Penicillium, Aspergillus, Alternaria, Stemphylum,* and *Hormodendrum*

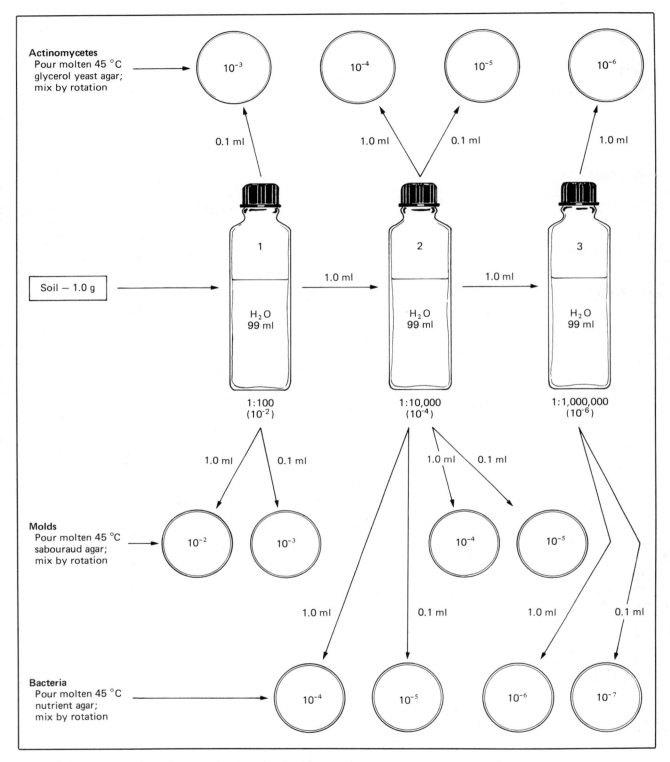

FIGURE 51.1. Procedure for enumeration of soil microorganisms

PROCEDURE

1. With a wax pencil label each set of four Petri dishes as follows:

   Nutrient agar: 10^{-4}, 10^{-5}, 10^{-6}, and 10^{-7} (to be used for enumeration of bacteria).

   Glycerol yeast extract agar: 10^{-3}, 10^{-4}, 10^{-5}, and 10^{-6} (to be used for enumeration of actinomycetes).

   Sabouraud agar: 10^{-2}, 10^{-3}, 10^{-4}, 10^{-5} (to be used for enumeration of fungi).

   Identify each plate with your initials.

2. With a wax pencil label the 99-ml sterile water flasks 2 and 3.

3. Shake the provided soil sample dilution of 1:100 (10^{-2}) vigorously approximately 30 times, with your elbow resting on the table.

4. With a sterile 1-ml pipette, transfer 1 ml of the provided soil sample dilution to flask 2 and shake vigorously as before. The final dilution is 1:10,000 (10^{-4}).

5. Using another sterile 1-ml pipette, transfer 1 ml of dilution 2 to flask 3 and shake vigrously as before. The final dilution is 1:1,000,000 (10^{-6}).

6. Using sterile 1-ml pipettes and aseptic technique, distribute the proper amount of each dilution to the appropriately labeled plates to effect the final required dilutions as follows:

   a. For molds—in plates labeled Sabouraud agar:

      1 ml of dilution 1 in plate to effect a 10^{-2} dilution.

      0.1 ml of dilution 1 in plate to effect a 10^{-3} dilution.

      1 ml of dilution 2 in plate to effect a 10^{-4} dilution.

      0.1 ml of dilution 2 in plate to effect a 10^{-5} dilution.

   b. For actinomycetes—in plates labeled glycerol yeast extract agar:

      0.1 ml of dilution 1 in plate to effect a 10^{-3} dilution.

      1 ml of dilution 2 in plate to effect a 10^{-4} dilution.

      0.1 ml of dilution 2 in plate to effect a 10^{-5} dilution.

      1 ml of dilution 3 in plate to effect a 10^{-6} dilution.

   c. For bacteria—in plates labeled nutrient agar:

      1 ml of dilution 2 in plate to effect a 10^{-4} dilution.

      0.1 ml of dilution 2 in plate to effect a 10^{-5} dilution.

      1 ml of dilution 3 in plate to effect a 10^{-6} dilution.

      0.1 ml of dilution 3 in plate to effect a 10^{-7} dilution.

7. Remove the molten nutrient agar medium from the water bath and wipe the flask dry with paper towel. Using sterile technique, pour approximately 15 ml of medium into each dilution plate for bacterial enumeration. Rotate each plate gently to ensure uniform distribution of cells in the medium.

8. Repeat step 7 using the molten Sabouraud agar for enumeration of fungi and the molten glycerol yeast extract agar for enumeration of actinomycetes.

9. Allow all the agar plate cultures to solidify.

10. Incubate the plates in an inverted position for three to seven days at 25 degrees C.

Observations and Results

1. Using a Quebec colony counter and a mechanical hand counter, observe all the colonies on each plate. Plates with more than 300 colonies cannot be counted and are designated as **too numerous to count, TNTC;** plates with fewer than 30 colonies are designated as **too few to count, TFTC.** Count only plates between 30 and 300 colonies.

2. Determine the number of organisms per milliliter of original culture by multiplying the number of colonies counted by the dilution factor. Refer to the observations and results section of Experiment 13 for examples in the calculation of cell counts.

3. Record your observation and calculated cell count per gram of sample in the chart.

Organism	*Dilution*	*Number of colonies*	*Organisms per gm of soil*
Bacteria	10^{-4} 10^{-5} 10^{-6} 10^{-7}		
Actinomycetes	10^{-3} 10^{-4} 10^{-5} 10^{-6}		
Molds	10^{-2} 10^{-3} 10^{-4} 10^{-5}		

4. Based on your results, which of the three types of soil organisms were most abundant in your sample? _____ . Least abundant? _____

REVIEW QUESTIONS

1. Would you expect to be able to duplicate your results if a soil sample were taken from the same location at a different time of the year? _____ Explain.

2. In the experiment performed, why wasn't the same medium used for enumeration of all three types of soil organisms?

3. Would you expect to be able to isolate an anaerobic organism from any of your cultures? _____ Explain.

4. Discuss the necessity of diverse microbial flora in the soil for the maintenance of life on this planet.

5. Explain why most microorganisms are present in the upper layers of the soil.

Nitrogen Cycle: Ammonification

PURPOSE

To demonstrate the liberation of ammonia from nitrogenous organic compounds.

PRINCIPLE

Ammonification, one of the essential phases of the nitrogen cycle, involves degradation of nitrogenous biopolymers and subsequent release of ammonia. This process is initiated by excretion of extracellular proteolytic enzymes that are commonly produced by such soil organisms as members of the genera *Bacillus, Clostridium,* and *Streptomyces.* These enzymes act sequentially to hydrolyze the proteins of plant and animal origins into their constituent amino acids. The amino acids are subsequently enzymatically deaminated, with the release of ammonia, as shown at the bottom of this page.

In this experiment, peptone broth, which contains an organic nitrogen substrate, is used to demonstrate the ability of some microorganisms to degrade proteins with the resultant formation of ammonia. Following incubation, the presence of ammonia, indicative of ammonification, is detectable by the yellow color when Nessler's reagent is added to samples of the test cultures. The relative amount of ammonia can be determined by differences in the degree of yellow coloration.

MATERIALS

Cultures

24-hour nutrient broth cultures of *Bacillus cereus, Pseudomonas fluorescens,* and *Proteus vulgaris.*

Soils

0.1-gm samples each of rich and poor garden soil.

Reagent

Nessler's reagent.

Media

Six 4 percent peptone broth tubes per designated student group.

Equipment

Bunsen burner, inoculating loop, spot plates, and Pasteur pipettes.

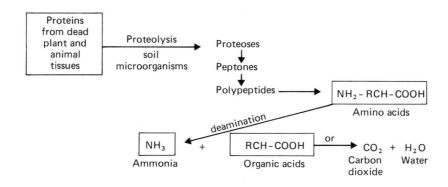

PROCEDURE

1. Label each of the peptone broth tubes with the name of the organism and soil sample to be inoculated, and identify each with your initials. The last tube of medium is labeled as a control.

2. Using sterile inoculating technique, inoculate each experimental organism and add each soil sample into its appropriately labeled tube.

3. Incubate the culture tubes for seven days at 25 degrees C.

Observations and Results EXPERIMENT **52**

1. Test each culture for the presence of ammonia on days 3, 5, and 7 following inoculation as described:

   Using a Pasteur pipette, place one drop of Nessler's reagent into six separate depressions on the spot plate. Add a loopful of each experimental culture and the control to a separate depression containing the Nessler's reagent. Mix by rotating the plate.

2. Using the chart, determine if ammonia is present and its relative amount.

Color	*Result*	
No change	0	No ammonia
Pale yellow	1+	Small amount of ammonia
Deep yellow	2+	Moderate amount of ammonia
Brown precipitate	3+	Large amount of ammonia

3. Record in the chart:

   a. The observed color of each Nessler's reagent culture mixture.

   b. The absence or presence, as 1+, 2+, or 3+, of ammonia in each test culture.

Test samples and organisms	*Day 3*		*Day 5*		*Day 7*	
	Color	*Result*	*Color*	*Result*	*Color*	*Result*
Rich garden soil						
Poor garden soil						
B. cereus						
P. fluorescens						
P. vulgaris						
Control						

4. Based on your results obtained in the soil sample cultures:

   a. Was there a difference in the amount of ammonia present in the two cultures? _____ Explain.

   b. Was the amount of ammonia present in each soil culture different from one day of observation to the next? _____
   Describe any differences observed in each of the following samples:
   Poor garden soil sample:
   Rich garden soil sample:

REVIEW QUESTIONS

1. Is ammonification restricted to solely aerobic or anaerobic organisms? _____
   Explain.

2. List sequentially the end products formed as a result of the enzymatic hydrolysis of proteins
   to amino acids.

3. Does proteolysis occur extracellularly or intracellularly? _____ Explain.

4. Explain why only some species of microorganisms are capable of ammonification.

Nitrogen Cycle: Nitrification

PURPOSE

To demonstrate the enzymatic conversion of ammonia to nitrates by soil microorganisms.

PRINCIPLE

Ammonia that is liberated during the ammonification process seldom accumulates to any appreciable extent in the soil. In an aerobic environment it undergoes bacterial oxidation to nitrates by a two-step process primarily involving the chemoautotrophic soil organisms *Nitrosomonas* and *Nitrobacter*. This process is called **nitrification** and the chemical transformations are illustrated as follows:

Step 1. **Nitrite formation**

$$NH_4^+ \ + \ O_2 \ \xrightarrow{Nitrosomonas} \ NO_2^- \ + \ H_2O$$

Ammonium   **Nitrite**
   ion   **ion**

Step 2. **Nitrate formation**

$$NO_2^- \ + \ O_2 \ \xrightarrow{Nitrobacter} \ NO_3^-$$

Nitrite   **Nitrate**
 ion   **ion**

The nitrates released in the soil are highly soluble and are assimilated by terrestrial plants and some microorganisms for the biosynthesis of cellular proteins.

In this experiment ammonium sulfate broth is used to demonstrate the ability of some microorganisms to oxidize ammonia to nitrite. The nitrite broth illustrates the further oxidation of nitrite to nitrate. Following incubation, the presence of nitrite and nitrate is determined as follows:

Determination of nitrite production in ammonium sulfate broth soil cultures:

1. Test for the presence of ammonia by use of Nessler's reagent. A yellow color indicates that am-

monia was not oxidized to nitrate. No color change indicates the absence of ammonia, and therefore nitrite should be present.

2. Test for the presence of nitrite by use of Trommsdorf's reagent and sulfuric acid. The presence of a blue-black color is indicative of the presence of nitrite; no color change is indicative of the absence of nitrite.

Determination of nitrate production in nitrite broth soil cultures:

1. Test for the presence of nitrite by use of Trommsdorf's reagent and sulfuric acid. The appearance of a blue-black color indicates that nitrites are present and that nitrates have not been formed.

2. Test for presence of nitrate by use of diphenylamine reagent and sulfuric acid if the absence of nitrites (no color change with Trommsdorf's reagent) was ascertained. The appearance of a deep blue color indicates nitrates are present.

MATERIALS

Soil

0.1-gm samples of a slightly alkaline and an acidic garden soil.

Media

Two 25-ml flasks each of ammonium sulfate broth and nitrite broth per designated student group.

Reagents

Nessler's reagent, diphenylamine reagent, Trommsdorf's reagent, concentrated sulfuric acid, and dilute sulfuric acid (one part concentrated H_2SO_4 plus three parts H_2O).

Equipment

Bunsen burner, Pasteur pipettes, spot plate, and wooden
applicator sticks.

PROCEDURE

1. Label each of the flasks of ammonium sulfate broth
   and nitrite broth with the type of soil sample to be
   inoculated and identify each with your initials.

2. Add the provided soil samples to the appropriately
   labeled flasks. Shake vigorously.

3. Incubate all of the flasks for three weeks at 25
   degrees C.

Observations and Results

Examine all flasks at seven-day intervals for nitrite and nitrate production as follows:

1. Nitrite production in ammonium sulfate soil cultures

   a. Determine the presence or absence of ammonia with Nessler's reagent. Refer to Experiment 52 for the procedure to be followed.

   b. Determine the presence or absence of nitrites by mixing three drops of Trommsdorf's reagent with one drop of dilute sulfuric acid in two separate depressions on the spot plate. With a Pasteur pipette, add one drop of each soil culture to a separate reagent mixture. Mix with a wooden applicator. Observe for a color change.

2. Nitrate production in nitrite broth soil cultures

   a. Repeat step 1b to determine the presence or absence of nitrite. Proceed to check for nitrates only if nitrites are absent, as the nitrate test will give a positive result in the presence of both nitrate and nitrite.

   b. Determine the presence or absence of nitrate by placing one drop of diphenylamine and two drops of concentrated H_2SO_4 into separate depressions on the spot plate. Add one drop of each soil culture to a separate reagent mixture. Mix and observe for color change.

3. Record in the chart the observed color change, if any, and the presence ($+$) or absence ($-$) of the compound being tested for.

	Alkaline soil sample			Acidic soil sample		
	Week			Week		
	1	2	3	1	2	3
Ammonium sulfate cultures Nessler's reaction for ammonia						
Trommsdorf's reaction for nitrite						
Nitrite cultures Trommsdorf's reaction for nitrite						
Diphenylamine reaction for nitrate						

4. Cite any observable differences in the rate at which nitrite and nitrate are formed in the two experimental soil cultures.

Do you feel that your experimental result is the same as the expected result? _____ Explain.

REVIEW QUESTIONS

1. Distinguish between ammonification and nitrification.

2. Are anaerobic organisms capable of nitrification? _____ Explain.

3. Name the major genus of organisms capable of nitrite formation: _____ . Nitrate formation: _____ .

Nitrogen Cycle: Denitrification

PURPOSE

To demonstrate the reduction of nitrates to nitrogen gas.

PRINCIPLE

Nitrates produced in the soil by microbial nitrification or deposited by the use of nitrate fertilizers, must be used rapidly or they are lost by leaching or by microbial reduction. In the latter, nitrates are reduced to nitrites, ammonia, nitrous oxide, and finally to elemental nitrogen in the form of nitrogen gas. This process is called **denitrification.**

Some of the soil organisms involved in denitrification are members of the genera *Pseudomonas, Bacillus,* and *Micrococcus*. Under anaerobic conditions, these organisms have the enzyme profile enabling them to use nitrate rather than gaseous oxygen as a final electron acceptor in their energy metabolism, thereby reducing the nitrate to gaseous nitrogen.

$$NO_3^- \longrightarrow NO_2^- \longrightarrow NO \longrightarrow N_2 \uparrow$$
Anaerobiosis

The following experiment uses a medium containing nitrate, the substrate for nitrogen gas formation, and a Durham tube for detection of the evolution of nitrogen gas. Following incubation, the evolution of gaseous nitrogen, the end product of denitrification, can be detected by the presence of an air bubble in the Durham tube. If denitrification does not occur there will not be an air bubble in the tube.

MATERIALS

Cultures

24-hour nutrient broth cultures of *Pseudomonas aeruginosa* and *Proteus vulgaris*.

Soil

0.1-gm samples of rich and poor soil.

Media

Four nitrate broth tubes containing Durham tubes per designated student group.

Equipment

Bunsen burner and inoculating loop.

PROCEDURE

1. Label each of the nitrate broth tubes with the name of the organisms and soil samples to be inoculated and identify each with your initials.

2. Using sterile inoculating technique, inoculate each test organism and soil sample into its appropriately labeled tube. **Do not shake the culture tubes during inoculation.**

3. Incubate the culture tubes for two weeks at 25 degrees C.

Observations and Results

1. At one-week intervals, examine all the test cultures for the presence or absence of an air bubble in the Durham tube. Record your observations as (+) or (−) in the chart.

2. Based on your observations, determine and record whether denitrification has taken place.

Organisms and Soils	7th day		14th day	
	Gas production	Denitrification	Gas production	Denitrification
P. aeruginosa				
P. vulgaris				
Rich soil				
Poor soil				

REVIEW QUESTIONS

1. Distinguish between nitrification and denitrification.

2. Show the chemical reactions that occur in denitrification.

3. Name two genera of soil organisms that play a role in denitrification.

Nitrogen Cycle: Nitrogen Fixation

PURPOSE

To demonstrate the fixation of atmospheric nitrogen by symbiotic and nonsymbiotic microorganisms.

PRINCIPLE

Nitrogen fixation is the phase of the nitrogen cycle during which enzymatically competent microorganisms convert atmospheric nitrogen into nitrogenous compounds, fixed nitrogen. This vital process replenishes the soil with usable nitrogen that is rapidly removed by plant use, denitrification, and leaching.

Nitrogen fixation is mediated by two microbial systems. One consists of nonsymbiotic, free-living microorganisms such as members of the genera *Azotobacter, Beijerinckia, Clostridium,* and the Cyanobacteria, which are capable of using nitrogen gas as their nitrogen source. A distinguishing characteristic of these organisms is their ability to form thick walls, thereby producing a dormant cyst that is resistant to drying and ultraviolet radiation; however, it is sensitive to heat. The second system involves symbiotic microbial forms, such as members of the genus *Rhizobium,* which, following infection, grow in a tumorlike nodule in the roots of leguminous plants. A mutually beneficial association is thus established in which the microorganisms use nutrients in the plant sap and act to fix atmospheric nitrogen to ammonia for its subsequent assimilation into plant proteins.

This procedure consists of the following two parts designed to isolate symbiotic and nonsymbiotic nitrogen fixers.

1. *Rhizobium:* A crushed leguminous root nodule is used as the source for a stained slide preparation. The *Rhizobium* organisms, called bacteroids when present in plant cells, are distinguishable by their pleomorphism; they appear in a variety of X, Y, T, V, and stellate shapes.

2. *Azotobacter:* Colonies of these species will be isolated from an alkaline soil sample on a nitrogen-free mannitol agar medium. The differentiation between the two mannitol-utilizing *Azotobacter* species is made on the basis of the following distinguishing characteristics:

Species	Production of water-insoluble pigment	Color of agar colonies	Production of water-soluble pigment	Fluorescence under ultraviolet light	Microscopic appearance
A. chroococcum	+	Brown to black with age	−	None	Gram-negative; large ovoid rods, often in pairs; cysts may be present
A. vinelandii	−	Colorless	+	Green	Gram-negative; large ovoid rods, often in pairs; cysts may be present
A. macrocytogenes*	−	Colorless	−	None	Gram-variable; large ovoid to coccoid cells, occurring singly or in pairs; no cysts

Asomonas and *Azotobacter* are closely related genera; therefore your isolate may be a member of this genus.

MATERIALS

Cultures

1-gm samples of slightly alkaline soil. Freshly picked leguminous plants (peas, clover, or alfalfa) with root nodules.

Media

One 50-ml nitrogen-free mannitol broth in screw-cap bottle and one nitrogen-free mannitol agar plate per designated student group.

Reagents

Methylene blue, crystal violet, Gram's iodine, ethyl alcohol, and safranin.

Equipment

Bunsen burner, inoculating loop, glass slides, staining tray, and ultraviolet lamp.

PROCEDURE

Azotobacter Isolation

1. Add a 1-gm soil sample to a bottle of the nitrogen-free mannitol broth that was previously labeled with your initials. Tighten the cap and shake vigorously until a uniform mixture is established.

2. Incubate the culture for four to seven days at room temperature, 25 degrees C.

3. At the end of the incubation period, examine the surface of the culture for the presence of a thin film. **Do not shake the flask or disturb the film.**

4. Using sterile inoculating technique, transfer a loopful of the surface film to a nitrogen-free mannitol agar plate labeled with your initials. Perform a four-way streak inoculation for the isolation of colonies (see Experiment 11 to review the procedure).

5. Inoculate the plate in an inverted position for four to six days at room temperature, 25 degrees C.

Rhizobium Isolation

1. Thoroughly rinse a nodule obtained from the roots of a leguminous plant.

2. Crush the nodule between two slides.

3. Spread a loopful of the material from the nodule in a thin layer on the surface of a clean slide.

4. Air dry, heat fix, and stain the smear preparation with methylene blue for one minute. Rinse the slide under running water and blot dry with bibulous paper.

Observations and Results

Azotobacter Identification

1. Select two or three isolated colonies whose appearances differ on the agar plates.

2. Observe the selected colonies for presence or absence of pigmentation.

3. Place two loopsful of each colony on a separate slide. Place each slide under ultraviolet light to determine the presence or absence of green fluorescence. **Do not look into UV source. Use of safety glasses is recommended.**

4. Prepare a Gram stain of each isolate. Examine all slides microscopically to determine the Gram reaction and the size, shape, and arrangement of the cells.

5. Based on your observations, record your results in the chart.

	Colony 1	Colony 2	Colony 3
Drawing of representative microscopic field:	◯	◯	◯
Microscopic morphology:	___ ___ ___	___ ___ ___	___ ___ ___
Color of colony on agar culture:	___	___	___
Fluorescence under ultraviolet light:	___	___	___
Identification of isolate:	___	___	___

Rhizobium Identification

1. Examine your methylene blue slide preparation under the oil-immersion objective for the presence of pleomorphic cell forms.

2. Draw a representative field of your observations.

◯ Representative field of *Rhizobium*

Microscopic cell morphology:
_____
_____
_____
_____

REVIEW QUESTIONS

1. What are the methods by which soil is depleted of useable forms of nitrogen?

2. Distinguish between symbiotic and nonsymbiotic nitrogen fixation.

3. Distinguish between denitrification and nitrogen fixation.

4. Why is the medium used for isolation of *Azotobacter* devoid of nitrogen-containing compounds?

Isolation of Antibiotic-Producing Microorganisms and Determination of Antimicrobial Spectrum of Isolates

PURPOSES

1. To isolate antibiotic-producing microorganisms.

2. To determine the spectrum of the antimicrobial activity of the isolated antibiotic.

PRINCIPLE

Soil is the major repository of microorganisms that produce **antibiotics** capable of inhibiting the growth of other microorganisms. Clinically useful antibiotics have been isolated from four groups of soil microorganisms—*Streptomyces, Bacillus, Penicillium,* and *Cephalosporium*—that represent three microbial types, namely, actinomycetes, bacteria, and molds, respectively.

Although soils from all parts of the world are continually screened in industrial laboratories for the isolation of new antibiotic-producing microorganisms, industrial microbiology is directing its energies toward chemical modification of existing antibiotic substances. This is accomplished by adding or replacing chemical side chains, reorganizing intramolecular bonding, or producing mutant microbial strains capable of excreting a more potent form of the antibiotic. The establishment of chemical congeners has been responsible for the circumventing of antibiotic resistance, minimizing adverse side effects in the host and increasing the effective spectrum of a given antibiotic.

In Part A of the experiment, students will use the **crowded-plate technique** for isolation of antibiotic-producing microorganisms from two soil samples, one of which is seeded with *Streptomyces griseus* to serve as a positive control. Figure 56.1 on page 324 illustrates the procedure to be followed. In Part B isolates exhibiting antibiotic activity will be screened against several different microorganisms to establish their effectiveness.

MATERIALS

Cultures

For Part B: 24-hour trypticase soy broth cultures of *Escherichia coli, Staphylococcus aureus, Mycobacterium smegmatis,* and *Pseudomonas aeruginosa.*

Soil Suspensions

For Part A: 1:500 dilution of soil sample suspension (0.1 gm of soil per 50 ml of tap water) to serve as an unknown; 1:500 dilution of soil sample seeded with *Streptomyces griseus* (0.1 gm of soil per 50 ml of tap water) to serve as a positive control.

Media

Per designated student group:
Part A: Six 15-ml trypticase soy agar deep tubes, and two trypticase soy agar slants.
Part B: Two trypticase soy agar slants.

Equipment

Part A: 500-ml beaker, test tubes, sterile Petri dishes, inoculating needle, hot plate, thermometer, 1-ml and 5-ml pipettes, and magnifying hand lens.
Part B: Bunsen burner and inoculating loop.

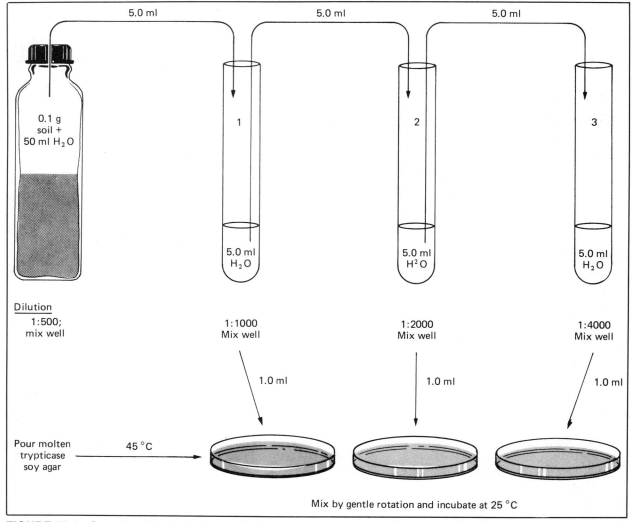

FIGURE 56.1. Crowded-plate technique for isolation of antibiotic-producing microorganisms

PART A: Isolation of Antibiotic-Producing Microorganisms

1. Label two sets of three sterile Petri dishes as to the soil sample being used and dilution (1:1000, 1:2000, and 1:4000), and identify each with your initials.

2. Place six trypticase soy agar deep tubes into a beaker of water and bring to 100 degrees C on a hot plate. Once agar is liquefied, add cool water to the water bath. Cool to 45 degrees C by checking the temperature with a thermometer.

3. Prepare a serial dilution of the unknown and positive control 1:500 soil samples as follows: (refer to Figure 56.1)

   a. Label three test tubes 1, 2, and 3. With a pipette, add 5 ml of tap water to each tube.

   b. Shake the provided 1:500 soil sample thoroughly for five minutes to effect a uniform soil-water suspension.

   c. Using a 5-ml pipette, transfer 5 ml from the 1:500 dilution to tube 1 and mix. The final dilution is 1:1000.

   d. Using another pipette, transfer 5 ml from tube 1 to tube 2 and mix. The final dilution is 1:2000.

   e. Using another pipette, transfer 5 ml from tube 2 to tube 3 and mix. The final dilution is 1:4000.

   f. Using separate 1-ml pipettes, transfer 1 ml of the 1:1000, 1:2000, and 1:4000 dilutions to their appropriately labeled Petri dishes.

   g. Pour one tube of molten trypticase soy agar, cooled to 45 degrees C, into each plate and mix by gentle rotation.

   h. Allow all plates to solidify.

4. Incubate all plates in an inverted position for two to four days at 25 degrees C.

5. After incubation, aseptically isolate one colony showing a zone of growth inhibition from each soil culture with an inoculating needle and streak onto trypticase soy agar slants labeled with the soil sample from which the isolate was obtained and identified with your initials.

6. Incubate the slants for two to four days at 25 degrees C. These will serve as stock cultures of antibiotic-producing isolates to be used in Part B.

PART B: Determination of Antimicrobial Spectrum of Isolates

1. Label the trypticase soy agar plates with the soil sample source of the isolate and identify each with your initials.

2. Using sterile technique, make a single line streak inoculation of each isolate onto the surface of an agar plate so as to divide the plate in half as shown:

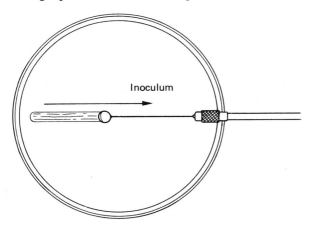

3. Incubate the plates in an inverted position for three to five days at 25 degrees C.

4. Following incubation, on the bottom of each plate draw four lines perpendicular to the growth of the antibiotic-producing isolate as shown:

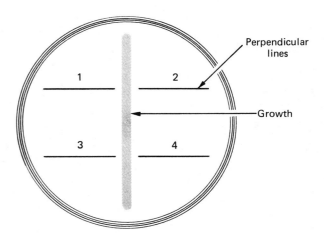

5. Aseptically make a single-line streak inoculation of each of the four test cultures following the inoculation template on each plate. Start close to, but not touching, the growth of the antibiotic-producing isolate and streak toward the edge of the plate.

6. Incubate the plates in an inverted position for 24 hours at 37 degrees C.

Observations and Results   EXPERIMENT **56**

Part A

1. Examine all crowded-plate dilutions for colonies exhibiting zones of growth inhibition. Use a hand magnifying lens if necessary.

2. Record in the chart the number of colonies showing zones of inhibition.

| | Number of colonies | | |
| | Dilutions | | |
Soil sample	1:1000	1:2000	1:4000
Unknown			
Positive control			

Part B

1. Examine all plates for inhibition of test organisms.

2. Draw a representation of your observed antibiotic activity against the test organisms.

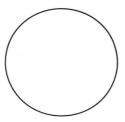

Antibiotic-producing isolate 1

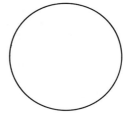

Antibiotic-producing isolate 2

3. Based on your observations, record in the chart the presence (+) or absence (−) of antibiotic activity against each of the test organisms and the spectrum of antimicrobial activity (broad or narrow).

| | Test organisms | | | | |
Soil sample	E. coli gram-negative	S. aureus gram-negative	P. aeruginosa gram-negative	M. smegmatis acid-fast	Spectrum
Unknown					
Positive control					

REVIEW QUESTIONS

1. Why is it frequently advantageous to modify antibiotics in industrial laboratories?

2. Is the ability to produce antibiotics limited only to bacterial species? _____ Explain.

3. Do you feel that a sufficient number of test organisms was used in Part B to determine fully the spectrum of activity of each isolated antibiotic? _____ Explain.

XIII

Bacterial Genetics

PURPOSE

To demonstrate the applicability of bacterial test systems in genetic-related studies. The procedures include enzyme induction, detection of potential chemical carcinogens, transfer of genetic material by means of conjugation, and isolation of streptomycin-resistant mutants.

INTRODUCTION

In recent years, bacteria have proved to be essential organisms in research into the structure and function of DNA, the universal genetic material. Their use is predicated on the following:

1. Their haploid genetic state that allows for the phenotypic, observable expression of a genetic trait in the presence of a single mutant gene.

2. Their rapid rate of growth, which permits observation of transmission of a trait through many generations.

3. The availability of large test populations that allows isolation of spontaneous mutants and their induction by chemical and physical mutagenic agents.

4. Their low cost of maintenance and propagation that make it possible to perform a large number of experimental procedures.

In the following experiments bacterial test systems are used to demonstrate enzyme induction, screening for chemical carcinogens, and the genetic phenomena of mutation and genetic transfer. The last two mechanisms introduce genetic variability, essential for evolutionary survival, in asexually reproducing bacterial populations.

Point mutations are permanent, sudden qualitative alterations in genetic material that arise as a result of the addition, deletion, or substitution of one or more bases in the region of a single gene. As a result, one or more amino acid substitutions occur during translation and a protein that may be inactive, reduced in activity, or entirely different, is synthesized. **Spontaneous mutations** are the result of the chemical and physical components in the organism's natural environment. The rate at which they occur is extremely low in all organisms. For example, in *E. coli,* the spontaneous mutation rate at a single locus (specific site of DNA) is estimated to be in the area of 1×10^{-7}, and the possibility of a mutation at any locus in the genome is approximately 1×10^{-4}. **Induced mutations** are genetic changes resulting from the organism's exposure to an artificial physical or chemical mutagen, an agent capable of inducing a mutation. The resultant mutations are of the same type that occur spontaneously, however, their rate is increased, and in some cases dramatically so.

Transfer of genetic material and its subsequent incorporation into the bacterial genome is also a source of genetic variation in some bacteria. This transfer may occur by means of the following:

1. **Conjugation,** a mating process between "sexually" differentiated bacterial strains that allows one-directional transfer of genetic material.

2. **Transduction,** a bacteriophage-mediated transfer of genetic material from one cell to another.

3. **Transformation,** a genetic alteration in a cell resulting from the introduction of free DNA from the environment across the cell membrane.

Enzyme Induction

PURPOSE

To illustrate experimentally the induction of beta-galactosidase synthesis in *Escherichia coli*.

PRINCIPLE

Despite the fact that bacteria possess a single chromosome, each cell is capable of synthesizing hundreds of different enzymes, although studies have shown that these enzymes are not present within the cells in equal concentrations. This is because some, called **constitutive enzymes,** are synthesized at a constant rate regardless of conditions in the cell's environment. Synthesis of other enzymes occurs only when necessary, and is subject to regulatory mechanisms that are dependent on the environment. One such mechanism, induction, requires the presence of a substrate, the inducer, in the environment to initiate synthesis of its specific enzyme, called an **inducible enzyme.** An extensively studied inducible enzyme in *Escherichia coli* is **beta-galactosidase,** which acts on the disaccharide lactose to yield the monosaccharides glucose and galactose. The gene for beta-galactosidase is a member of a cluster of genes, called an **operon,** and is involved with the metabolism of lactose. The member genes of the lactose operon function as a unit, all being transcribed only when the inducer, lactose, is present in the surrounding medium.

To illustrate beta-galactosidase induction, two test strains of *E. coli* will be used: A prototrophic (wild type) strain (lactose-positive) and an auxotrophic (mutant) strain (lactose-negative), which carries a mutation in the gene for beta-galactosidase as well as a mutation in the lactose operon regulatory gene. Both test strains will be grown in the following media:

1. Inorganic synthetic medium lacking an organic carbon and energy source that is required by the heterotrophic *E. coli*.

2. Inorganic synthetic medium plus glucose, which can be utilized by both strains as a carbon and energy source.

3. Inorganic synthetic medium plus lactose, which can only be utilized by the prototrophic strain.

Orthonitrophenyl-β-D-galactoside (ONPG), a colorless analog of lactose, is capable of inducing beta-galactosidase synthesis, as it is hydrolyzed to galactose and a yellow nitrophenolate ion. Following a short incubation period, growth in all the cultures will be determined by spectrophotometry. Induction of beta-galactosidase synthesis and activity will be ascertained by the appearance of a yellow color in the medium following addition of ONPG, which only occurs in the presence of the nitrophenolate ion. Absence of this macroscopically visible color change indicates absence of enzyme induction in the lactose-negative strain.

MATERIALS

Cultures

25-ml inorganic synthetic broth suspensions of 12-hour nutrient agar cultures of a lactose-positive *Escherichia coli* strain (ATCC e 23725) and a lactose-negative *Escherichia coli* strain (ATCC e 23735) adjusted to O.D. of 0.1 at 600 mμ.

Media

Per designated student group: Dropper bottles of sterile 10 percent glucose, 10 percent lactose, and water.

Reagent

Dropper-bottles of toluene and orthonitrophenyl-β-D-galactoside (ONPG).

Equipment

1-ml and 5-ml sterile pipettes, six sterile 13 × 100-mm test tubes, six sterile 25-ml Erlynmeyer flasks, Bausch and Lomb Spectronic 20 spectrophotometer, and shaking water-bath incubator.

PROCEDURE

1. Label three sterile test tubes and three sterile 25-ml Erlynmeyer flasks as ''Lac$^+$'' (lactose positive) and the name of the substrate to be added (glucose, lactose, or water). Identify each with your initials. Similarly label three sterile flasks ''Lac$^-$'' (lactose negative) for each test organism.

2. Using sterile 5-ml pipettes, aseptically transfer 5 ml of the Lac$^+$ and Lac$^-$ inorganic synthetic broth cultures to their respectively labeled test tubes.

3. Using a sterile 1-ml pipette, aseptically add 0.5 ml of the glucose and lactose solutions and 0.5 ml of sterile distilled water to the appropriately labeled tubes.

4. Determine the optical density (O.D.) of all cultures at a wavelength of 600 mμ.

5. Aseptically transfer each culture to its respectively labeled flask. (Note: if side-arm flasks are available, additions and O.D. readings may be made directly.)

6. Incubate all flasks for 2 hours in a shaking water-bath at 37 degrees C and 100 strokes per minute.

7. Following incubation, transfer all cultures back to their respectively labeled test tubes.

8. Determine the O.D. for each culture at a wavelength of 600 mμ.

9. To each culture, add 5 drops of toluene and shake vigorously (toluene ruptures the cells releasing intact enzymes).

10. To each culture, add 5 drops of ONPG solution.

11. Incubate all cultures for 40 minutes at 37 degrees C.

12. Observe all cultures for development of yellow color.

Observations and Results

1. Record the initial O.D. of each lac$^+$ and lac$^-$ *E. coli* broth culture at 600 mμ.

2. Following incubation, determine and record the O.D. of each culture. Based on your observations, indicate whether growth has occurred in each of the cultures.

3. Following the addition of ONPG, observe the cultures for the presence of a yellow coloration indicative of beta-galactosidase synthesis and activity. Record your results as to the color of the cultures and the presence ($+$) or absence ($-$) of the β-galactosidase activity.

| *Cultures* | *O.D. at 600 mμ* | | *Growth (+) or (−)* | *Color of Culture with ONPG* | *β-galactosidase (+) or (−)* |
	Prior to incubation	*Following incubation*			
Lac$^+$ *E. coli* glucose					
Lactose					
Water					
Lac$^-$ *E. coli* glucose					
Lactose					
Water					

4. Explain the absence of growth in some of the cultures.

REVIEW QUESTIONS

1. Distinguish between constitutive enzymes and inducible enzymes.

2. Explain what is meant by an operon.

3. Explain the purpose of the ONPG in the procedure.

Bacterial Conjugation

PURPOSE

To demonstrate genetic recombination in bacteria by the process of conjugation.

PRINCIPLE

Genetic variability is essential for the evolutionary success of all organisms. In diploid eukaryotes, the processes of **crossing over,** exchange of genetic material between homologous chromosomes, and **meiosis** contribute to this variability. In haploid, asexually reproducing prokaryotic organisms, genetic recombination may occur by **conjugation, transduction** and **transformation.** In this exercise only the process of conjugation is considered.

Conjugation is a mating process during which a one-directional transfer of genetic material occurs at physical contact between two "sexually" differentiated cell types. This differentiation, or existence of different mating strains in some bacteria, is determined by the presence of a **fertility factor,** or **F factor,** within the cell. Cells that lack the F factor are recipients (females) of the genetic material during conjugation and are designated as **F⁻.** Cells possessing the F factor have the ability to act as genetic donors (males) during mating. If this F factor is extrachromosomal (a **plasmid** or **episome**) the cells are designated as **F⁺**; most commonly only the F factor is transferred during conjugation. If this factor becomes incorporated into the bacterial chromosome there is a transfer of chromosomal genes, although generally not involving the entire chromosome or the F factor. The cells are designated **Hfr, high-frequency recombinants**.

In this experiment students will prepare a mixed culture representing a cross between an Hfr protrophic (wild type) strain of *E. coli* that is streptomycin-sensitive, and an F⁻ autotrophic (mutant) *E. coli* strain that requires threonine, leucine, and thiamine and is streptomycin-resistant. Following a short incubation period, isolation of only the threonine and leucine recombinants will be performed by plating the mixed culture on a minimal medium containing streptomycin and thiamine. The streptomycin is incorporated into the medium to inhibit the growth of the wild-type, streptomycin-sensitive parental Hfr cells. The thiamine is required as an essential growth factor for the thiamine⁻ recombinant cells. Because of its distant location on the chromosome, this marker will not be transferred during the short mating period. A genetic map showing the site of origin of transfer and the locations of the relevant markers is given in Figure 58.1.

MATERIALS

Cultures

12-hour nutrient broth cultures of F⁻ *Escherichia coli* strain thr⁻, leu⁻, thi⁻, and Str-r (ATCC e 23724); and Hfr *Escherichia coli* strain Str-s (ATCC e 23740).

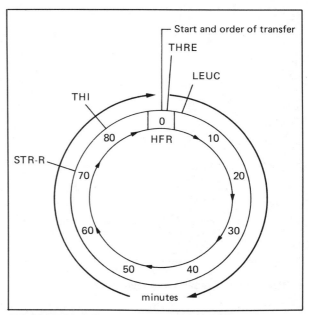

FIGURE 58.1. Genetic map of *Escherichia coli*

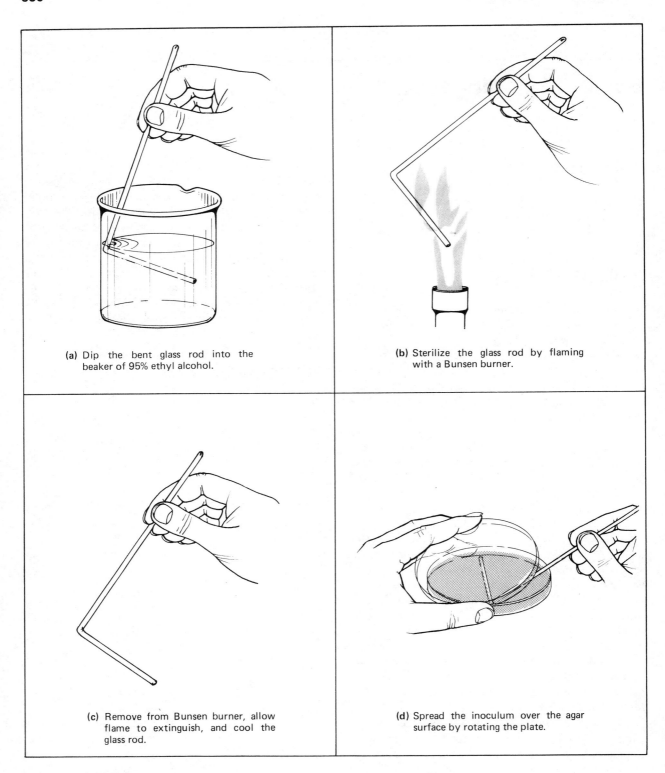

(a) Dip the bent glass rod into the beaker of 95% ethyl alcohol.

(b) Sterilize the glass rod by flaming with a Bunsen burner.

(c) Remove from Bunsen burner, allow flame to extinguish, and cool the glass rod.

(d) Spread the inoculum over the agar surface by rotating the plate.

FIGURE 58.2. Spread-plate technique

Media

Three plates of minimal medium plus streptomycin and thiamine per designated student group.

Equipment

Bunsen burner, beaker with 70 percent ethyl alcohol, bent glass rod, 1-ml sterile pipettes, sterile 13 × 100-mm test tube, and wax pencil.

PROCEDURE

1. With separate sterile 1-ml pipettes, aseptically transfer 1 ml of the F⁻ *E. coli* culture and 0.3 ml of the Hfr *E. coli* culture into the sterile 13 × 100 mm test tube.

2. Mix by gently rotating the culture between the palms of the hands.

3. Incubate the culture for 30 minutes at 37 degrees C.

4. Prepare control plates of the parental Hfr and F⁻ *E. coli* strains on two appropriately labeled minimal plus streptomycin and thiamine agar plates using the spread-plate technique as illustrated in Figure 58.2.

   a. Aseptically add 0.1 ml of each *E. coli* strain to its appropriately labeled agar plate.

   b. With an alcohol-dipped and flamed glass rod, spread the inoculum over the entire surface of the agar plate.

5. Following incubation, **vigorously** agitate the mixed culture to terminate the genetic transfer.

6. Aseptically add 0.1 ml of the mixed culture to an appropriately labeled minimal plus streptomycin and thiamine plate and spread the inoculum over the entire surface with a sterile glass rod.

7. Incubate all plates in an inverted position for 48 hours at 37 degrees C.

Observations and Results

1. Observe all plates for the presence (+) or absence (−) of colonies. Record your results in the chart.

	Hfr E. coli plate	*F⁻ E. coli plate*	*Mixed culture plate*
Growth (+) or (−)			

2. Do you expect any growth to be present on the two parental *E. coli* minimal agar plates? _____ Explain.

3. Did genetic recombination occur? _____ Explain how your observations support your answer.

REVIEW QUESTIONS

1. Explain how genetic variations may be introduced in eukaryotic and prokaryotic cells.

2. Explain the significance of the F factor.

3. Distinguish between F^+ and Hfr bacterial strains.

4. Explain the importance of the streptomycin marker in the parental *E. coli* strains.

Isolation of Streptomycin-Resistant Mutant

PURPOSE

To isolate a streptomycin-resistant mutant in a proto-trophic bacterial population by means of the gradient-plate technique.

PRINCIPLE

Mutation, a change in the base sequence of a single gene, although infrequent, is one of the sources of genetic variability in cells. In some instances, these changes enable the cell to survive in an otherwise deleterious environment. An example of such a genetic adaptation is the development of **antibiotic resistance** in a small population of microorganisms prior to the advent and large-scale use of these agents. More recently, this microbial characteristic of antibiotic resistance has become of major clinical importance, as the number of drug-resistant microbial strains continues to increase due to their extensive use and frequent misuse, which over the years have selected for the drug-resistant strains by their microbicidal effects on the sensitive cell forms. These agents select for the resistant mutant and do not act as inducers of the mutation.

In a drug-resistant organism, the mutated gene enables the cell to circumvent the antimicrobial effect of the drug by a variety of mechanisms, including:

1. An enzymatic alteration in the chemical structure of the antibiotic as in penicillin resistance.

2. A change in the selective permeability of the cell membrane as in streptomycin resistance.

3. A decrease in the sensitivity of their enzymes to inhibiting mechanisms as in the resistance to streptomycin, which interferes with the translation process at the ribosomes.

4. An overproduction of a natural substrate (metabolite) to compete effectively with the drug (anti-metabolite) as in the resistance to sulfonamides, which produce their antimicrobial effect by competitive inhibition.

The following procedure is designed to allow students to isolate a streptomycin-resistant mutant from a prototrophic (wild type, streptomycin-sensitive) *Escherichia coli* culture by means of the **gradient-plate technique.** This requires preparation of a double-layered agar plate as illustrated in Figure 59.1. The lower, slanted agar medium layer lacks streptomycin. When poured over the lower slanted layer, the molten agar medium containing the antibiotic will produce a streptomycin concentration gradient in the surface layer. Following a spread-plate inoculation of the prototrophic test culture and incubation, the appearance of colonies in a region of high streptomycin concentration is indicative of streptomycin-resistant mutants.

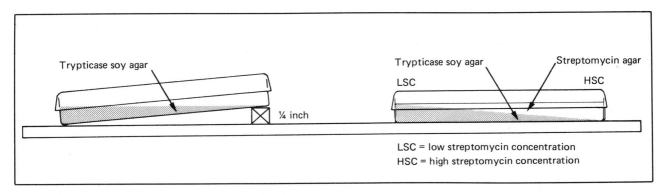

FIGURE 59.1. Preparation of a streptomycin gradient plate

MATERIALS

Cultures

24-hour nutrient broth cultures of *Escherichia coli*.

Media

Two 10-ml trypticase soy agar deep tubes per designated student group.

Reagent

Stock streptomycin solution (10 mg per 100 ml of water)

Equipment

Sterile Petri dish, sterile 1-ml pipettes, bent glass rod, beaker with 70 percent ethanol, and water bath.

PROCEDURE

1. In a hot water bath, melt two trypticase soy agar tubes. Cool and maintain at 45 degrees C.

2. Place a pencil under one end of a sterile Petri dish, pour a sufficient amount of the molten agar medium to cover the entire bottom surface and allow to solidify in the slanted position.

3. Using a sterile 1-ml pipette, add 0.1 ml of the streptomycin solution to a second tube of molten trypticase soy agar. Mix by rotating the tube between the palms of your hands.

4. Place the dish in a horizontal position, pour a sufficient amount of the molten agar medium containing streptomycin to cover the gradient agar layer and allow to solidify.

5. With a sterile 1-ml pipette add 0.2 ml of the *E. coli* test culture. With an alcohol-dipped and flamed bent glass rod, spread the culture over the entire agar surface as illustrated in Figure 58.2.

6. Label the plate with your initials and incubate it in an inverted position for 48 hours at 37 degrees C.

7. Following incubation select one or two isolated colonies present in the middle of the streptomycin concentration gradient. With a sterile inoculating loop streak the selected colonies toward the high-concentration end of the plate.

8. Incubate the plate in an inverted position for 48 hours at 37 degrees C.

Observations and Results

1. Following the initial incubation, observe the plate for the appearance of discrete colonies and indicate their position in the diagram.

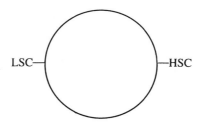

2. Following the second incubation period, observe for a line of growth from the streaked colonies into the area of high streptomycin concentration. Growth in this area is indicative of streptomycin-resistant mutants.

3. Indicate the observed line(s) of growth in the diagram.

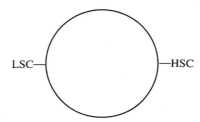

REVIEW QUESTIONS

1. What is the significance of streptomycin resistance today?

2. List the mechanisms responsible for antibiotic resistance.

3. Why is it necessary to use an antibiotic gradient-plate preparation for isolation of mutants?

4. Why has there been an increase in drug-resistant bacterial strains in recent years?

5. Does the streptomycin in the medium cause the mutations? _____ Explain.

The Ames Test: A Bacterial Test System for Chemical Carcinogenicity

PURPOSE

To perform a screening procedure for the detection of potential chemical carcinogens using a bacterial test system.

PRINCIPLE

Our exposure to a wide variety of chemical compounds has been markedly increasing over the past decades. Oncological epidemiologists strongly suspect that the intrusion of these chemicals in the form of industrial pollutants, pesticides, food additives, hair dyes, and cigarette smoke, among others, may play a significant role in the induction of malignant transformations in humans. From a genetic aspect there is strong evidence linking **carcinogenicity** to **mutagenicity.** Research indicates that approximately 90 percent of the chemicals proved to be carcinogens are mutagens; they cause cancer by inducing mutations in somatic cells. These mutations are most frequently a result of base substitutions, the substitution of one base for another in the DNA molecule, and frame-shift mutations, a shift in the reading frame of a gene due to the addition or deletion of a base.

In view of the rapid advent of new products and new industrial processes with their resultant pollutants, it is essential to determine their potential genetic hazards. Despite the fact that mammalian cell structure and human enzymatic pathways differ from those in bacteria, the chemical nature of DNA is common to all organisms; this permits the use of bacterial test systems for the rapid detection of possible mutagens and therefore possible carcinogens.

The **Ames test** is a simple and inexpensive procedure that uses a bacterial test organism to screen for mutagens. The test organism is a histidine-negative (his$^-$) auxotrophic strain of *Salmonella typhimurium* that will not grow on a medium deficient in histidine unless a back mutation to his$^+$ (histidine-positive) has occurred. It is recognized that the mutagenic effect of a chemical is frequently influenced by the enzymatic

pathways of an organism, whereby nonmutagens are transformed into mutagens and vice versa when introduced into human systems. In mammals, this toxification or detoxification frequently occurs in the liver. The Ames test generally requires the addition of a liver homogenate S-9, which serves as a source of activating enzymes, to make this bacterial system more comparable to a mammalian test system.

To perform the Ames test by the spot method (see Figure 60.1 on page 346), molten agar containing the test organism, S-9 mix, and a trace of histidine to allow the bacteria to undergo several cell divisions necessary for mutation to occur, is poured on a minimal agar plate. A disc impregnated with the test chemical is then placed in the center of the test plate. Following diffusion of the test compound from the disc, a decreasing concentration of the chemical is established. Following incubation, a qualitative indication of the mutagenicity of the test chemical can be determined by noting the number of colonies present on the plate. Each colony represents a his$^-$ $\rightarrow$ his$^+$ revertant. A positive result, indicating mutagenicity, is obtained when an obvious increase in the number of colonies is evident when compared to the number of spontaneous revertants on the negative control plate.

In the following procedure students will perform a modified Ames test, not using the S-9 mix to test for the mutagenicity of nitro compounds, which, as in humans, are activated by the bacterial nitro reductases. Four minimal agar plates are inoculated with the *S. typhimurium* test organism. One plate, the negative control, is not exposed to a test chemical. Any colonies developing on this plate are representative of spontaneous his$^-$ $\rightarrow$ his$^+$ mutations. The second plate, the positive control is exposed to a known nitrocarcinogen, 2-nitrofluorene. The remaining two plates are used to determine the mutagenicity of two commercial hair dyes.

MATERIALS

Cultures

24-hour nutrient broth cultures of *Salmonella typhimurium,* strain TA 1538 (ATCC e 29631)

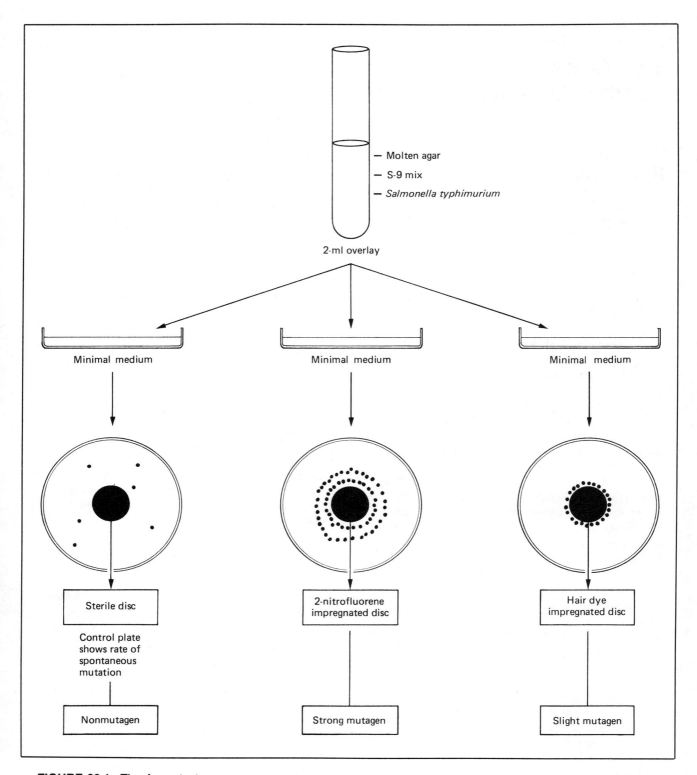

FIGURE 60.1. The Ames test

Media

Four minimal agar plates and four 2-ml top agar tubes per designated student group.

Reagents

Sterile biotin-histidine solution, 2-nitrofluorene dissolved in alcohol, two mixtures of commercial hair dye and hydrogen peroxide.

Equipment

1-ml sterile pipettes, sterile discs, forceps, water bath, and Bunsen burner.

PROCEDURE

1. Label three minimal agar plates with the name of the test chemical to be used and identify each with your initials. The fourth plate is labeled as a negative control.

2. Melt four tubes of top agar in a hot-water bath and maintain the molten agar at 45 degrees C.

3. To each molten top agar tube, aseptically add 0.2 ml of the sterile biotin-histidine solution and 0.1 ml of the *S. typhimurium* test culture. Mix by rotating the test tube between the palms of your hands.

4. Aseptically pour the top agar cultures onto the minimal agar plates and allow to solidify.

5. Using a sterile forceps, dip each disc into its respective test chemical solution and drain by touching the side of the container.

6. Place the chemical-impregnated discs in the center of the respectively labeled minimal agar plates. Place a sterile disc on the plate labeled negative control. With the sterile forceps, **gently** press down on the discs so that they adhere to the surface of the agar.

7. Incubate all plates in an inverted position for 24 hours at 37 degrees C.

Observations and Results EXPERIMENT 60

1. Count the number of large colonies present on each plate and record on the chart.

2. Determine and record the number of chemically induced mutations by subtracting the number of colonies on the negative control plate, representative of spontaneous mutations, from the number of colonies on each test plate.

3. Determine and record the relative mutagenicity of the test compounds on the basis of the number of induced mutations: If below 10 ($-$); if greater than 10 ($1+$); if greater than 100 ($2+$); and if greater than 500 ($3+$).

Test chemical	Number of colonies	Number of induced mutations	Degree of mutagenicity ($-$), ($1+$), ($2+$), or ($3+$)
Negative control			
2-nitrofluorene			
Hair dye-H_2O_2 mixture 1			
Hair dye-H_2O_2 mixture 2			

REVIEW QUESTIONS

1. What is the purpose of the S-9 in the Ames test?

2. What is the purpose of the biotin-histidine solution in the Ames test?

3. Explain the relationship between chemical carcinogenicity and mutagenicity.

4. What are the advantages of using bacterial systems instead of mammalian systems to test for chemical carcinogenicity?

5. What are the disadvantages of using bacterial systems instead of mammalian systems to test for chemical carcinogenicity?

Medical Microbiology

PURPOSES

To familiarize students with:

1. The characteristics and methodology for isolating and identifying selected pathogenic microorganisms.

2. The indigenous microbial flora of selected human anatomical sites.

INTRODUCTION

Although it is true that microorganisms are ubiquitous and their benefit to humans has been recorded, a diminutive population of organisms remains an enigma: They are the pathogens, whose existence makes medical or clinical microbiology an especially important science.

That living agents are capable of inducing infections (contagium viva) was first put forward by the monk Fracastoro in Verona about 500 years ago. In 1659 Kircher reported the presence of minute motile organisms in the blood of plague victims. Two hundred years following Fracastoro's initial concept, the germ theory of disease was formulated by Plenciz based on Leeuwenhoek's revolutionary microscopic observation of microorganisms. Perhaps the most important contributions to microbiology were made by Pasteur, Koch, and Lister during the Golden Era of microbiology from 1870 to 1920. During this time these investigators and their students recorded the observations and discoveries that cemented the cornerstone to medical microbiology. The body of knowledge that has accrued since these early years has been instrumental in making clinical microbiology a major component of laboratory or diagnostic medicine. The major responsibility of this science is isolating and identifying infectious pathogens to enable physicians to treat patients with infectious diseases prudently, intelligently, and rapidly.

Many of the experiments so far described have application in the field of clinical microbiology. Among these are isolation and identification of unknown cultures, the use of selective and differential media, and biochemical tests used to separate and identify various microorganisms. Although studying all of the bacterial pathogens responsible for human illness is not possible here, routine experiments for isolating and identifying some of the most frequently encountered infectious organisms and microorganisms which constitute the indigenous flora of the human body, are included. The pathogens chosen are pyogenic cocci, members of the genera *Staphylococcus* and *Streptococcus*, the Enterobacteriaceae, and the organisms suspected in formation of dental caries. Experimental procedures designed for the detection and presumptive identification of microorganisms in blood and urine, which are normally sterile body fluids, have also been incorporated into this section. Organisms that naturally reside in or on body surfaces and constitute its **normal flora** are also examined.

The need for the expeditious detection and identification of pathogens has led to the development of rapid testing methods. These are microbiologically and immunologically based and can be performed quickly and without the need for sophisticated and expensive equipment. Some prototypic experiments using these rapid methods are included along with the traditional procedures.

Many of the organisms that are used, although attenuated by having been subcultured on artificial media for many generations, must be viewed as potential pathogens and therefore handled with respect. All necessary precautions must be taken to prevent infection of oneself or others.

Microbial Flora of the Mouth: Determination of Susceptibility to Dental Caries

PURPOSES

1. To acquaint students with the organisms responsible for dental caries.

2. To perform experiments that demonstrate a host's susceptibility to caries formation.

PRINCIPLE

A variety of microorganisms is known to be involved in the formation of dental caries, including *Lactobacillus acidophilus, Streptococcus mutans,* and *Actinomyces odontolyticus.* These organisms in the oral flora produce organic acids, particularly lactic, derived from fermentable carbohydrates that adhere to the surface of the teeth. In the continued presence of lactic acid, dental enamel undergoes decalcification and softening, which result in the formation of tiny perforations called dental caries.

The actual mechanism of action of these organisms is still unclear. It has been noted that *Streptococcus mutans* excretes an enzyme called **dextransucrase** (glycosyl transferase), which is capable of polymerizing sucroses into a large polymer, glucan, plus the monosaccharide fructose. This polysaccharide clings tenaciously to the teeth and forms dental plaque, in which streptococci reside and ferment fructose with the formation of lactic acid.

Similarly, *Lactobacillus acidophilus* produces lactic acid as an end product of carbohydrate fermentation. Some members of this genus are capable of producing acid environments in the range of a pH of 2 to 3. Oral lactobacilli are capable of metabolizing glucose found in the mouth, producing organic acids that reduce the oral acid concentration to a pH of less than

5. At this pH, decalcification occurs and dental decay begins.

One of the best microbiological methods for determining susceptibility to dental caries is the **Snyder test,** which measures the amount of acid produced by the action of the lactobacilli on glucose. The test employs a differential medium, Snyder agar (pH 4.7), which contains glucose and the pH indicator brom cresol green that gives the medium a green color.

Following incubation, Snyder agar cultures containing lactobacilli in the saliva will show glucose fermentation with the production of acid, which tends to lower the pH to 4.4, the level of acidity at which dental caries form. At this pH the green medium turns yellow. A culture demonstrating a yellow color within 24 to 48 hours is suggestive of the host's susceptibility to the formation of dental caries. A culture that does not change color is indicative of lower susceptibility.

MATERIALS

Cultures

Organisms of the normal oral flora present in saliva.

Media

Two Snyder test agar deep tubes per designated student group.

Equipment

Bunsen burner, 1-inch square blocks of paraffin, sterile 1-ml pipettes, and sterile test tubes.

PROCEDURE

1. Label the two tubes of Snyder agar with your initials.

2. Melt the two Snyder agar deep tubes and cool to 45 degrees C.

3. Chew one square of paraffin for a period of three minutes **without swallowing the saliva.** As saliva develops, collect it in a sterile test tube.

4. Vigorously shake the collected saliva sample and transfer 0.2 ml of saliva with a sterile pipette into one of the Snyder test medium tubes that has been cooled to 45 degrees C. **Caution: pipette should not touch sides of tubes or agar.**

5. Mix the contents of the tube thoroughly by rolling the tube between the palms of the hands or by tapping it with your finger.

6. Rapidly cool the inoculated tube of Snyder agar in an ice-water bath.

7. Repeat steps 4 through 6 to inoculate the second tube.

8. Incubate both tubes for 72 hours at 37 degrees C.

Observations and Results

1. Examine the Snyder test cultures daily during the 72-hour incubation period for a change in the color of the culture medium. Use an uninoculated tube of the medium as a control. Record the color of the cultures in the chart.

2. Using Table 61.1 to interpret your observations, record your results as to caries susceptibility in the chart.

TABLE 61.1. Assessment of Susceptibility to Caries

	Hours of incubation		
Caries activity	*24*	*48*	*72*
Marked	Positive	. . .	. . .
Moderate	Negative	Positive	. . .
Slight	Negative	Negative	Positive
Negative	Negative	Negative	Negative

Source: Courtesy of Difco Laboratories, Inc., Detroit, Michigan.
Positive: Complete color change, green is no longer dominant.
Negative: No color change or a slight color change; medium retains green color throughout.

Tube number	*Color of Synder test cultures*			*Caries susceptibility (Yes) or (No)*
	24 hrs.	*48 hrs.*	*72 hrs.*	

REVIEW QUESTIONS

1. Explain the differential nature of the Snyder agar medium as used for the detection of dental caries.

2. Explain the mechanism responsible for the formation of dental caries by resident microorganisms.

3. What is the function of the paraffin in this procedure?

4. Based on your results, what is your tendency to form dental caries?

Is this result consistent with your past dental history?

Normal Microbial Flora of the Throat and Skin

PURPOSE

To identify microorganisms that normally reside in the throat and on the skin.

PRINCIPLE

The so-called normal flora are regularly found in specific areas of the body. This specificity is far from arbitrary and depends on environmental factors such as pH, oxygen concentration, amount of moisture present, and types of secretions associated with each anatomical site. Native microbial flora are broadly located as follows:

1. Skin: Staphylococci (predominantly *S. epidermidis*), streptococci (α-hemolytic, nonhemolytic, and enterococci), diphtheroid bacilli, yeasts, and fungi.

2. Eye conjunctiva: Staphylococci, streptococci, diphtheroids, and neisseriae.

3. Upper respiratory tract: Staphylococci, streptococci (α-hemolytic, nonhemolytic, enterococci, and *S. pneumoniae*), diphtheroids, spirochetes, members of the genera *Branhamella, Neisseria,* and *Hemophilus.*

4. Mouth and teeth: Anaerobic spirochetes and vibrios, fusiform bacteria, staphylococci, anaerobic levan-producing and dextran-producing streptococci responsible for dental caries.

5. Intestinal tract: In the upper intestine, predominantly lactobacilli and enterococci. In the lower intestine and colon 96 percent to 99 percent anaerobes such as members of the genera *Bacteroides, Lactobacillus, Clostridium,* and *Streptococcus;* 1 percent to 4 percent aerobes including coliforms, enterococci, and a small number of *Proteus, Pseudomonas,* and *Candida* species.

6. Genitourinary tract: Staphylococci, streptococci, lactobacilli, gram-negative enteric bacilli, clostridia, spirochetes, yeasts, and protozoa such as *Trichomonas* species.

In this exercise the resident flora of the throat and skin will be studied. As these sites represent sources of mixed microbial populations, streak-plate inoculations, as outlined in Experiment 11, will be performed to effect their separations. The discrete colonies, thus formed, can then be studied morphologically, biochemically, and microscopically to identify the individual genera of these mixed flora.

The procedure used to identify the native flora of the throat involves the following steps:

1. A **blood agar plate** inoculated to demonstrate the alpha-hemolytic and beta-hemolytic reactions of some **streptococci** and **staphylococci:** A distinction between these two genera can be made based on their colonial and microscopic appearance. The streptococci typically form pinpoint colonies on blood agar, whereas the staphylococci form larger, pinhead colonies that might show a golden coloration. When viewed under a microscope, the streptococcal cells form chains of varying lengths, whereas the staphylococci arrange themselves in clusters.

2. A **chocolate agar plate** inoculated to detect **Neisseria** sp. by means of the oxidase test: Members of this genus are recognized when the colonies develop coloration that is pink to dark purple upon addition of tetramethyl-p-phenylenediamine dihydrochloride following incubation.

3. A **Mueller-Hinton tellurite plate** inoculated to demonstrate the presence of **diptheroids,** which appear as black, pinpoint colonies on this medium: This coloration is due to the diffusion of the tellurite ions into the bacterial cells and their subsequent reduction to tellurium metal, which precipitates inside the cells.

The procedure used to identify the native flora of the skin involves the following steps:

1. A **blood agar plate** inoculated to determine the presence of hemolytic microorganisms, specifically the **staphylocci** and **streptococci:** Differentiation between these two genera may be made as previously described.

2. A **mannitol salt agar plate** inoculated for the isolation of the **staphylocci:** The generally avirulent staphylococcal species can be differentiated from the pathogenic *S. aureus* because the latter is able to ferment mannitol, causing yellow coloration of this medium surrounding the growth.

3. A **Sabouraud agar plate** inoculated to detect **yeasts** and **molds:** Yeast cells will present pigmented or nonpigmented colonies that are elevated, moist, and glistening. Mold colonies will appear as fuzzy, powdery growths arising from a mycelial mat in the agar medium.

MATERIALS

Media

Per designated student group: two blood agar plates, one mannitol salt agar plate, one chocolate agar plate, one Mueller-Hinton tellurite plate, one Sabouraud agar plate, and two 5-ml sterile saline tubes.

Reagents

Crystal violet, Gram's iodine, 95 percent ethanol, safranin, tetramethyl-p-phenylenediamine dihydrochloride, and lactophenol cotton blue.

Equipment

Sterile cotton swabs, desiccator jar with candle, microscope, glass slides, Bunsen burner, and wax pencil.

PROCEDURE

1. Using a sterile cotton swab, obtain a specimen from the pharyngeal tonsil by rotating the swab vigorously over its surface without touching the tongue.

2. Inoculate a tube of sterile saline with the swab and mix to effect a uniform suspension.

3. Using a sterile inoculating loop, inoculate one plate each of blood agar, chocolate agar, and Mueller-Hinton tellurite, all previously labeled with the source of the specimen and identified with your initials, by means of a four-way streak inoculation as described in Experiment 11.

4. Using a sterile cotton swab moistened in sterile saline, obtain a specimen from the skin by rubbing the swab vigorously against the palm of the hand.

5. Inoculate a tube of sterile saline with the swab and mix.

6. Inoculate one plate each of blood agar, mannitol salt agar, and Sabouraud agar, as described in step 3.

7. Incubate the inverted chocolate agar plate in a CO_2 incubator, in a CO_2 incubation bag, or a candle jar. If the candle jar is used, place a lighted candle into a desiccator jar and cover the jar tightly to effect a 5 percent to 10 percent CO_2 environment required for the growth of the *Neisseria*. Incubate the jar for 48 hours at 37 degrees C.

8. Incubate the inverted Sabouraud agar plate for 48 hours at 25 degrees C and the remaining plates at 37 degrees C.

Observations and Results   EXPERIMENT **62**

1. Examine the blood agar plate cultures for zones of hemolysis (refer to Experiment 15 for an explanation of the hemolytic reactions).

2. Add the tetramethyl-p-phenylenediamine dihydrochloride to the surface of the growth on the chocolate agar plate. Observe for the appearance of a pink-to-purple color on the surface of any of the colonies.

3. Examine the Mueller-Hinton tellurite plate as to the presence of black colonies.

4. Examine the Sabouraud agar plate for the appearance of moldlike growth.

5. Examine the mannitol salt agar plate for the presence of growth that is indicative of staphylococci. Then examine the color of the medium surrounding the growth. A yellow color is indicative of *S. aureus* (refer to Experiment 15 for the differentiation of staphylococcal species on this medium).

6. Record your observations in the chart and indicate the types of organisms that may be present in each specimen.

Culture	*Throat specimen*	*Skin specimen*
Blood agar (+) or (−) hemolysis		
Chocolate agar (+) or (−) pink-to-purple colonies		
Mueller-Hinton tellurite (+) or (−) black colonies		
Sabouraud agar (+) or (−) fungal colonies		
Mannitol salt agar (+) or (−) growth		— Growth
Color of medium		red
Types of organisms present		

7. Prepare two Gram-stained smears from each of the blood agar cultures, choosing well-isolated colonies that differ in their cultural appearance and demonstrate hemolytic activity. Observe microscopically for the Gram reaction and the size, shape, and arrangement of the cells. In the chart, draw a representative field, describe the morphology of the cells, and attempt to identify each isolate.

8. Prepare two lactophenol cotton blue-stained smears of organisms obtained from discrete colonies that differ in appearance on the Sabouraud agar culture (refer to Experiment 34). Observe microscopically, draw a representative field, and attempt to identify the fungi by referring to Experiment 36.

Skin specimen

Isolate 1

Isolate 2

Gram reaction: _____ _____
Morphology: _____ _____
_____ _____
Organism: _____ _____

Throat specimen

Isolate 1

Isolate 2

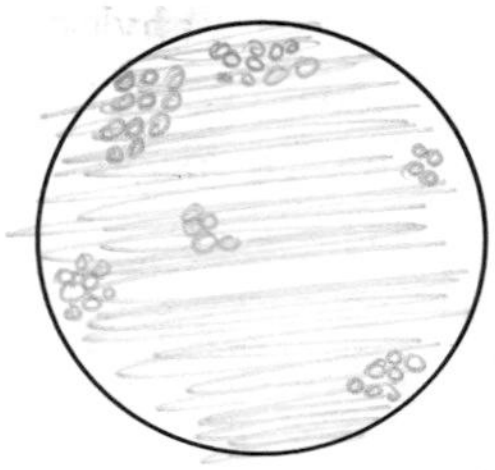

Gram reaction: _____ _____
Morphology: _____ _____
_____ _____
Organism: _____ _____

Sabouraud agar culture specimen

Isolate 1

Isolate 2

Morphology: _____ _____
Organism: _____ _____

REVIEW QUESTIONS

1. Account for the differences in microbial flora found in different parts of the body.

Identification of Human Staphylococcal Pathogens

DNA is positive
mannitol well turn yellow

PURPOSE

To familiarize students with:

1. The medical significance of the staphylococci.

2. Selected laboratory procedures designed to differentiate among the major staphylococcal species.

PRINCIPLE

The genus *Staphylococcus* is comprised of both pathogenic and nonpathogenic organisms. They are gram-positive cocci and occur most commonly as irregular clusters of spherical cells. They are mesophilic non-spore-formers; however, they are generally highly resistant to drying, especially when sequestered in organic matter such as blood, pus, and tissue fluids. They are capable of surviving outside of the body for extended periods of time, even up to several months. Many staphylococci are indigenous to skin surfaces and mucous membranes of the upper respiratory tract. Breaks in the skin and mucous linings may serve as portals of entry into underlying tissues, with the possibility of infection by virulent strains.

The three major species include *S. aureus, S. saprophyticus,* and *S. epidermidis.* Strains of the last two species are generally avirulent; however, under special circumstances where a suitable portal of entry is pro-vided, *S. epidermidis* may be the etiological agent for **skin lesions** and **endocarditis,** and *S. saprophyticus* has been implicated in some **urinary tract infections.**

Infections are primarily associated with *S. aureus,* pathogenic strains that are often responsible for the formation of **abscesses,** localized pus-producing lesions. These lesions most commonly occur in the skin and its associated structures, resulting in **boils, carbuncles, acne,** and **impetigo.** Infection of internal organs and tissues is not uncommon, however, and include **pneumonia, osteomyelitis** (abscesses in bone and bone marrow), **endocarditis** (inflammation of the endocardium), **cystitis** (inflammation of the urinary bladder), **pyelonephritis** (inflammation of the kidneys), **staphylococcal enteritis** due to enterotoxin contamination of foods, and on occasion, **septicemia.**

Strains of *S. aureus* produce a variety of metabolic end products, some of which may play a role in the organisms' pathogenicity. Included among these are **coagulase,** which causes clot formation; **leukocidin,** which lyses white blood cells; **hemolysins,** which are active against red blood cells; and **enterotoxin,** which is responsible for a type of gastroenteritis. Additional metabolites of a nontoxic nature are **DNase, lipase, gelatinase,** and the fibrinolysin **staphylokinase.**

When there is a possibility of staphylococcal infection isolation of *S. aureus* is of clinical importance. These virulent strains can be differentiated from other staphylococci and identified by a variety of laboratory tests, some of which are illustrated below.

Test	S. aureus	S. epidermidis	S. saprophyticus
Mannitol salt agar			
Growth	+	+	+
Fermentation	+	−	−
Colonial pigmentation	Generally golden yellow	White	White
Coagulase	+	−	−
DNase	+	−	−
Hemolysis	Generally beta	−	−
Novobiocin resistance	Resistant	Resistant	Sensitive

In this exercise students will distinguish among the staphylococcal species by performing a computer-assisted multitest procedure, traditional test procedures, or both. The traditional and computer-based procedures are described below.

The traditional procedures involve the following steps:

1. **Mannitol salt agar:** This medium is selective for salt-tolerant organisms such as staphylococci. Differentiation among the staphylococci is predicated on their ability to ferment mannitol. Following incubation, mannitol-fermenting organisms, typically *S. aureus* strains, exhibit a yellow halo surrounding their growth, and nonfermenting strains do not. (Refer to Experiment 15 for a more detailed discussion on the use of this medium.)

2. **Coagulase test:** Production of coagulase is indicative of a *S. aureus* strain. The enzyme acts within host tissues to convert fibrinogen to fibrin. It is theorized that the fibrin meshwork that is formed surrounds the bacterial cells or infected tissues, protecting the organisms from nonspecific host resistance mechanisms such as phagocytosis and the antistaphylococcal activity of normal serum. In the coagulase tube test for bound and free coagulase, a suspension of the test organism in citrated plasma is prepared and the inoculated plasma is then periodically examined for fibrin formation, or coagulation. Clot formation within four hours is interpreted as a positive result and indicative of a virulent *S. aureus* strain. The absence of coagulation after 24 hours of incubation is a negative result, indicative of an avirulent strain.

3. **Deoxyribonuclease (DNase) test:** Generally, coagulase-positive staphylococci also produce the hydrolytic enzyme DNase; thus this test can be used to reconfirm the identification of *S. aureus.* The test organism is grown on an agar medium containing DNA. Following incubation, DNase activity is determined by the addition of 0.1 percent toluidine blue to the surface of the agar. DNase-positive cultures capable of DNA hydrolysis will show a rose pink halo around the area of growth. The absence of this halo is indicative of a negative result and the inability of the organism to produce DNase.

4. **Novobiocin resistance:** This test is used to distinguish *S. epidermidis* from *S. saprophyticus.* It requires inoculation of a Mueller-Hinton agar plate with the test organism and application of a 30-μg novobiocin antibiotic disc to the surface of the agar.

Following incubation, the sensitivity of an organism to the antibiotic is determined by the Kirby-Bauer method as described in Experiment 43.

Computer-assisted procedures involve the STAPH-IDENT system (developed by Analytab Products, Division of Sherwood Medical, Plainview, New York). STAPH-IDENT is a rapid, computer-based micromethod for the separation and identification of the newly proposed thirteen species of staphylococci. The system consists of ten microcupules containing dehydrated substrates for the performance of conventional and chromagenic tests. The addition of a suspension of the test organism serves to hydrate the media and to initiate the biochemical reactions. The identification of the staphylococcal species is made with the aid of the differential chart and the API STAPH-IDENT Profile Register that is part of the system (Table 63.1), or both.

MATERIALS

Cultures

24-hour trypticase soy agar slant cultures of *Staphylococcus epidermidis*, *Staphylococcus saprophyticus* (ATCC 15305), and *Staphylococcus aureus* (ATCC e 27660). Number-coded, 24-hour blood agar cultures of the above organisms for the STAPH-IDENT system.

Media

Per designated student group: one mannitol salt agar plate, one deoxyribonucleic acid agar plate, three Mueller-Hinton agar plates, and the STAPH-IDENT system.

Reagents

Citrated human or rabbit plasma and 0.1 percent toluidine blue and McFarland barium sulfate standards.

Equipment

Bunsen burner, inoculating loop, 13 × 100-mm test tubes, sterile Pasteur pipettes, 1-ml sterile pipettes, sterile cotton swabs, 30-μg novobiocin antibiotic discs, wax marking pencil, forceps, and beaker with 95 percent alcohol.

Table 63.1 API STAPH-IDENT Profile Register

Profile	Identification			Profile	Identification		
0 040	STAPH CAPITIS			4 700	STAPH AUREUS	COAG +	
0 060	STAPH HAEMOLYTICUS				STAPH SCIURI	COAG −	
0 100	STAPH CAPITIS			4 710	STAPH SCIURI		
0 140	STAPH CAPITIS						
0 200	STAPH COHNII			5 040	STAPH EPIDERMIDIS		
0 240	STAPH CAPITIS			5 200	STAPH SCIURI		
0 300	STAPH CAPITIS			5 210	STAPH SCIURI	COAG +	
0 340	STAPH CAPITIS			5 300	STAPH AUREUS	COAG −	
0 440	STAPH HAEMOLYTICUS				STAPH SCIURI		
0 460	STAPH HAEMOLYTICUS			5 310	STAPH SCIURI		
0 600	STAPH COHNII			5 600	STAPH SCIURI		
0 620	STAPH HAEMOLYTICUS			5 610	STAPH SCIURI	COAG +	
0 640	STAPH HAEMOLYTICUS			5 700	STAPH AUREUS	COAG −	
0 660	STAPH HAEMOLYTICUS				STAPH SCIURI		
				5 710	STAPH SCIURI		
1 000	STAPH EPIDERMIDIS			5 740	STAPH AUREUS		
1 040	STAPH EPIDERMIDIS						
1 300	STAPH AUREUS			6 001	STAPH XYLOSUS	XYL +	ARA +
1 540	STAPH HYICUS (An)				STAPH SAPROPHYTICUS	XYL −	ARA −
1 560	STAPH HYICUS (An)			6 011	STAPH XYLOSUS		
				6 021	STAPH XYLOSUS		
2 000	STAPH SAPROPHYTICUS	NOVO	R	6 101	STAPH XYLOSUS		
	STAPH HOMINIS	NOVO	S	6 121	STAPH XYLOSUS		
2 001	STAPH SAPROPHYTICUS			6 221	STAPH XYLOSUS		
2 040	STAPH SAPROPHYTICUS	NOVO	R	6 300	STAPH AUREUS		
	STAPH HOMINIS	NOVO	S	6 301	STAPH XYLOSUS		
2 041	STAPH SIMULANS			6 311	STAPH XYLOSUS		
2 061	STAPH SIMULANS			6 321	STAPH XYLOSUS		
2 141	STAPH SIMULANS			6 340	STAPH AUREUS	COAG +	
2 161	STAPH SIMULANS				STAPH WARNERI	COAG −	
2 201	STAPH SAPROPHYTICUS			6 400	STAPH WARNERI		
2 241	STAPH SIMULANS			6 401	STAPH XYLOSUS	XYL +	ARA +
2 261	STAPH SIMULANS				STAPH SAPROPHYTICUS	XYL −	ARA −
2 341	STAPH SIMULANS			6 421	STAPH XYLOSUS		
2 361	STAPH SIMULANS			6 460	STAPH WARNERI		
2 400	STAPH HOMINIS	NOVO	S	6 501	STAPH XYLOSUS		
	STAPH SAPROPHYTICUS	NOVO	R	6 521	STAPH XYLOSUS		
2 401	STAPH SAPROPHYTICUS			6 600	STAPH WARNERI		
2 421	STAPH SIMULANS			6 601	STAPH SAPROPHYTICUS	XYL −	ARA −
2 441	STAPH SIMULANS				STAPH XYLOSUS	XYL +	ARA +
2 461	STAPH SIMULANS			6 611	STAPH XYLOSUS		
2 541	STAPH SIMULANS			6 621	STAPH XYLOSUS		
2 561	STAPH SIMULANS			6 700	STAPH AUREUS		
2 601	STAPH SAPROPHYTICUS			6 701	STAPH XYLOSUS		
2 611	STAPH SAPROPHYTICUS			6 721	STAPH XYLOSUS		
2 661	STAPH SIMULANS			6 731	STAPH XYLOSUS		
2 721	STAPH COHNII (SSP1)						
2 741	STAPH SIMULANS			7 000	STAPH EPIDERMIDIS		
2 761	STAPH SIMULANS			7 021	STAPH XYLOSUS		
				7 040	STAPH EPIDERMIDIS		
3 000	STAPH EPIDERMIDIS			7 141	STAPH INTERMEDIUS (An)		
3 040	STAPH EPIDERMIDIS			7 300	STAPH AUREUS		
3 140	STAPH EPIDERMIDIS			7 321	STAPH XYLOSUS		
3 540	STAPH HYICUS (An)			7 340	STAPH AUREUS		
3 541	STAPH INTERMEDIUS (An)			7 401	STAPH XYLOSUS		

Profile	Identification			Profile	Identification	
3 560	STAPH HYICUS (An)			7 421	STAPH XYLOSUS	
3 601	STAPH SIMULANS	NOVO	S	7 501	STAPH INTERMEDIUS (An)	COAG+
	STAPH SAPROPHYTICUS	NOVO	R		STAPH XYLOSUS	COAG−
				7 521	STAPH XYLOSUS	
4 060	STAPH HAEMOLYTICUS			7 541	STAPH INTERMEDIUS (An)	
4 210	STAPH SCIURI			7 560	STAPH HYICUS (An)	
4 310	STAPH SCIURI			7 601	STAPH XYLOSUS	
4 420	STAPH HAEMOLYTICUS			7 621	STAPH XYLOSUS	
4 440	STAPH HAEMOLYTICUS			7 631	STAPH XYLOSUS	
4 460	STAPH HAEMOLYTICUS			7 700	STAPH AUREUS	
4 610	STAPH SCIURI			7 701	STAPH XYLOSUS	
4 620	STAPH HAEMOLYTICUS			7 721	STAPH XYLOSUS	
4 660	STAPH HAEMOLYTICUS			7 740	STAPH AUREUS	

Source: Analytab Products, Plainview, New York.

PROCEDURE

1. Preparation of mannitol salt agar and DNA agar plate cultures:

   a. With a wax marking pencil, divide the bottoms of the two plates into three sections. Label each section with the name of the organism to be inoculated and identify the covers with your initials.

   b. Aseptically make a single line of inoculation of each test organism in its respective sector on both agar plates.

2. Preparation of agar plate cultures for novobiocin-resistance test:

   a. Label the three Mueller-Hinton agar plates with the name of the test organism to be inoculated and identify each with your initials. Inoculate each plate with its respective organism according to the Kirby-Bauer procedure as outlined in Experiment 43.

   b. Using an alcohol-dipped and flamed forceps, aseptically apply a novobiocin antibiotic disc to the surface of each inoculated plate. **Gently** press the discs down with a sterile forceps to ensure that they adhere to the agar surface.

3. Incubate all plate cultures in an inverted position for 24 to 48 hours at 37 degrees C.

4. Coagulase test procedure:

   a. Label three 13 × 100-mm test tubes with the name of the organism to be inoculated and identify each with your initials.

   b. Aseptically add 0.5 ml of a 1:4 dilution of citrated rabbit or human plasma and 0.1 ml of each test culture to its respectively labeled test tube.

   c. At the end of the laboratory session incubate all tubes that are coagulase-negative for 20 hours at 37 degrees C.

5. STAPH-IDENT system procedure:

   a. Preparation of strip:

      i. Dispense 5 ml of tap water into incubation tray.

      ii. Place API strip into incubation tray.

   b. Preparation of inoculum:

      i. Add 2 ml of 0.85 percent saline (pH 5.5–7.0) into a sterile 15 × 150-mm test tube.

      ii. Using a sterile swab, pick up a sufficient amount of inoculum to prepare a saline suspension with a final turbidity that is equivalent to a No. 3 McFarland ($BaSO_4$) turbidity standard. **Caution: suspension must be used within fifteen minutes of preparation.**

   c. With a sterile Pasteur pipette, add 2 to 3 drops of the inoculum to each microcupule.

   d. Place plastic lid on tray and incubate for 5 hours at 37 degrees C.

Observations and Results EXPERIMENT **63**

1. Examine the bacterial plasma suspensions for clot formation 5 minutes, 20 minutes, 1 hour, and 4 hours after inoculation by holding the test tubes in a slanted position. Place all coagulase-negative suspensions in a 37 degrees C incubator for observation 24 hours after inoculation. Record your observations and results in the chart.

Bacterial species	Appearance of plasma clotted (+) or unclotted (−)					Coagulase (+) or (−)
	5 min.	20 min.	1 hr.	4 hrs.	24 hrs.	
S. aureus						
S. epidermidis						
S. saprophyticus						

2. Examine the mannitol salt agar plate. Note and record the following in the chart:

   a. Presence (+) or absence (−) of growth of each test organism.

   b. Color of the medium surrounding the growth of each test organism.

   c. Whether each test organism is a mannitol fermenter (+) or nonfermenter (−).

3. Flood the DNA agar plate with 0.1 percent toluidine blue. Observe for the delayed development of a rose pink coloration surrounding the growth of each test organism. Record your color observation and indicate the presence (+) or absence (−) of DNase activity in the chart.

4. With a metric ruler, measure the size of the zone of inhibition, if present, surrounding each of the novobiocin discs on the agar plates. A zone of inhibition of 17 mm or less is indicative of novobiocin resistance, whereas a zone greater than 17 mm indicates that the organism is sensitive to this antibiotic. Record in the chart the susceptibility of each test organism to novobiocin as sensitive (S), or resistant (R).

Procedure	S. aureus	S. epidermidis	S. saprophyticus
Mannitol salt agar Growth Color of medium Fermentation	*fermentega yellow*	*Pink*	*pink*
DNA agar Color of medium DNase activity	*Pink* *rose + Pink*	*blue*	*blue*
Novobiocin resistance Growth inhibition in mm Susceptibility (R) or (S)	*S*	*S*	*R*

5. Interpret your STAPH-IDENT system reactions on the basis of the observed color changes in each of the microcupules described in the chart below. Record your color observation and result as (+) or (−) for each test in this chart.

Microcupule			Interpretation of Reactions		Reaction Results	
No.		Substrate	Positive	Negative	Color	(+) or (−)
1	PHS	p-nitrophenyl-phosphate, disodium salt	Yellow	Clear or straw-colored		
2	URE	Urea	Purple to red-orange	Yellow or yellow-orange		
3	GLS	p-nitrophenyl-β-D-glucopyranoside	Yellow	Clear or straw-colored		
4	MNE	Mannose	Yellow or yellow-orange	Red or orange		
5	MAN	Mannitol				
6	TRE	Trehalose				
7	SAL	Salicin				
8	GLC	p-nitrophenyl-β-D-glucuronide	Yellow	Clear or straw-colored		
9	ARG	Arginine	Purple to red-orange	Yellow or yellow-orange		
10	NGP	2-naphthyl-β-D-galactopyranoside	Add 1-2 drops of STAPH-IDENT reagent			
			Plum-purple (mauve)	Yellow or colorless		

6. Construct a four-digit profile for your unknown organism as follows: A four-digit profile is derived from the results obtained with STAPH-IDENT. The 10 biochemical tests are divided into four groups, as follows:

PHS	MNE	SAL	NGP
URE	MAN	GLC	
GLS	TRE	ARG	

Only positive reactions are assigned a numerical value. The value depends on the location within the group.

A value of one for the first biochemical in each group (i.e., PHS, MNE . . .)
A value of two for the second biochemical in each group (i.e., URE, MAN . . .)
A value of four for the third biochemical in each group (i.e., GLS, TRE . . .)
A value of zero for all negative reactions.
A four-digit number is obtained by totaling the values of each of the groups.

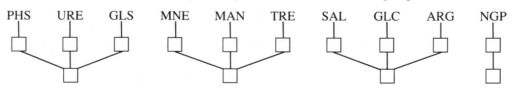

Using Table 63.1 and your four-digit profile number, identify your organism.

Unknown Organism: _____

Staphylococcal Phage Typing

PURPOSE

To identify pathogenic *Staphylococcus aureus* strains by phage typing.

PRINCIPLE

Phage typing is a useful procedure that allows differentiation of pathogenic strains within one species that cannot be separated on the basis of biochemical and morphological differences. This microbiological tool has demonstrated significant merit in epidemiology when it is necessary to determine the causative agent of a disease and its source or carriers. The basis for phage typing is the fact that following continuous reproduction in a particular bacterial host strain, the phages become modified and therefore adapted specifically for that singular strain.

Many pathogenic strains of *S. aureus* can now be identified by this method. Specific staphylophages are known to interact with approximately 70 percent of the coagulase-positive staphylococci; no interaction is noted with coagulase-negative members of this species. At the present time, more than 20 *S. aureus* strains have been classified into five lytic groups, four assigned and one unassigned, on the basis of staphylophage susceptibility. The staphylophages have been coded by number, and the different strains are identified by the number of the phage responsible for its lysis. For instance,

hospital-acquired staphylococcal infections have been related to the staphylophage types numbered 52, 52A, and 80, all of which belong to lytic group I; the organism responsible for skin lesions interacts with phage type 71, found in lytic group II. Table 64.1 shows the known members of each phage group and the lytic group to which they belong.

Phage typing first requires the seeding of an isolated pathogen on an appropriate agar plate medium. This is followed by applying specific phage cultures to the agar surface at designated positions. Following incubation, **plaque formation** (clear zones) will develop only in the areas where phage infection of host cells has occurred; this identifies the phage type capable of lysing the susceptible bacterial host.

MATERIALS

Cultures

24-hour nutrient broth cultures of coagulase-positive *Staphylococcus aureus* strains (ATCC e 27691, e 27693, and e 27697) and staphylophage culture types 47, 52A, 71, and 81.

Media

Three tryptone agar plates per designated student group.

Equipment

Bunsen burner, inoculating loop, sterile cotton swabs, and wax pencil.

PROCEDURE

1. With a wax pencil, rule the bottom of each of the three tryptone agar plates into nine equal squares.

TABLE 64.1. Lytic Groups and Phages of *Staphylococcus aureus*

Lytic Group	Phages
I	29, 52, 52A, 79, 80
II	3A, 3B, 3C, 65, 71
III	6, 7, 42E, 47, 53, 54, 75, 77, 83
IV	42D
Unassigned	81, 187

Label four separated squares with the phage type numbers, 47, 52A, 71, and 81 as illustrated.

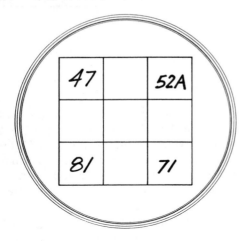

2. Label the cover of each agar plate with the *S. aureus* strain to be inoculated and identify each with your initials.

3. Using sterile cotton swabs, inoculate the entire surface of each agar plate with its respective test organism by streaking the plate first in a horizontal direction and then vertically.

4. Aseptically place a loopful of each phage culture in the center of its appropriately labeled square on each of the plates.

5. Allow agar surfaces to dry.

6. Incubate all plate cultures in an inverted position for 24 hours at 37 degrees C.

Observations and Results EXPERIMENT **64**

1. Observe all plates for zones of lysis in the staphylococcal lawn.

2. Indicate in the chart the positions of the phage types used and the presence (+) or absence (−) of lysis.

3. Based on your observations, indicate for each *S. aureus* strain its phage type susceptibility and lytic group classification.

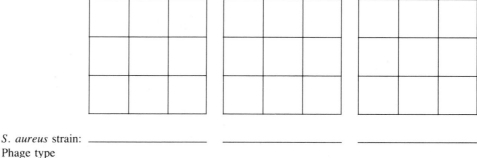

S. aureus strain: _____ _____ _____
Phage type
susceptibility: _____ _____ _____
Lytic group
classification: _____ _____ _____

REVIEW QUESTIONS

1. Describe the relationship of phage typing to the field of epidemiology.

2. How do bacteriophages become specific for host cell strains?

3. Describe the phage typing procedure for bacterial strain classification.

Identification of Human Streptococcal Pathogens

PURPOSE

To familiarize students with:

1. The medical significance of streptococci.

2. Selected laboratory procedures designed to differentiate streptococci on the basis of their hemolytic activity and biochemical patterns associated with the Lancefield group classifications.

PRINCIPLE

Members of the genus *Streptococcus* are perhaps responsible for a greater number of infectious diseases than any other group of microorganisms. Morphologically, they are cocci that divide in a single plane forming chains. They form circular, translucent to opaque, pinpoint colonies on solid media. All members of this group are gram-positive and many are nutritionally fastidious, requiring enriched media such as blood for growth.

The streptococci are classified by two major methods, (1) their **hemolytic activity** and (2) the **serologic classification of Lancefield.** The observed hemolytic reactions on blood agar are of the following three types:

1. (**α**) **Alpha-hemolysis,** an incomplete form of hemolysis produces a green zone around the colony. Alpha-hemolytic streptococci, the *Streptococcus viridans* species, are usually nonpathogenic opportunists. In some instances, however, they are capable of inducing human infections such as **subacute endocarditis,** which may precipitate valvular damage and heart failure if untreated. *Streptococcus pneumoniae,* the causative agent of **lobar pneumonia,** will be studied in a separate experiment.

2. (**β**) **Beta-hemolysis,** a complete destruction of red blood cells, exhibits a clear zone of approximately two to four times the diameter of the colony. The streptococci capable of producing beta-hemolysins are most frequently associated with pathogenicity.

3. (**γ**) **Gamma-hemolysis** is indicative of the absence of any hemolysis around the colony. Most commonly, gamma-hemolytic streptococci are avirulent.

Lancefield classified the streptococci into **groups,** designated **A** through **0,** based on the presence of an antigenic group-specific hapten called the **C-substance.** This method of classification generally implicates the members of groups A, B, C, and D in human infectious processes.

Beta-hemolytic streptococci belonging to **group A,** and collectively referred to as *Streptococcus pyogenes,* are the human pathogens of prime importance. Members of this group are the main etiological agents of human respiratory infections such as **tonsillitis, bronchopneumonia,** and **scarlet fever** as well as skin disorders such as **erysipelas** and **cellulitis.** In addition, these organisms are responsible for the development of complicating infections, namely **glomerulonephritis** and **rheumatic fever,** which may surface when primary streptococcal infections either go untreated or are not completely eradicated by antibiotics. The beta-hemolytic streptococci found in **group B** are indigenous to the vaginal mucosa and have been shown to be responsible for **puerperal fever** (childbirth fever), a sometimes-fatal **neonatal meningitis,** and **endocarditis.** Members of **group C** are also beta-hemolytic and have been implicated in **erysipelas, puerperal fever,** and **throat infections.** *Group D* streptococci generally exhibit alpha or gamma hemolysis on blood agar plates. This group includes the **enterococci** such as *S. faecalis,* which may be the etiological agent of **urinary tract infections,** and the **nonenterococci** such as *S. bovis,* which is of lesser medical significance in humans.

The virulence of the streptococci is associated with their ability to produce a wide variety of extracellular metabolites. Included among these are the **hemolysins** (alpha and beta), **leucocidins** that destroy phagocytes, and the **erythrogenic toxin** responsible for the rash of scarlet fever. Also of medical significance are the metabolites **hyaluronidase** (the spreading factor), which hydrolyzes the tissue cement hyaluronic acid; **strep-**

TABLE 65.1.   Laboratory differentiation of streptococci

Gamma   *for non group D*

Group	A	B	C	D		K, H, N
Organisms	S. pyogenes	S. agalactiae	S. equi	S. faecalis enterococci	S. bovis nonentero-cocci	S. salivarius S. sanguis S. mitis
Hemolysis	β	β	β	α → γ	α → γ	α
Bacitracin test	+	−	−	−	−	−
~~Camp test~~	−	+	−	−	−	−
Bile esculin hydrolysis	−	−	−	+	(+)	(−)
6.5% NaCl medium	NG	NG	NG	G	NG	NG
Growth at 10° C	NG	NG	NG	G	NG	NG
Growth at 45° C	NG	NG	NG	G	NG or G	NG

NG = no growth.
G = growth.

U TI

tokinase, a fibrinolysin; and the **nucleases,** ribonuclease and deoxyribonuclease, which destroy viscous tissue debris. The last three metabolic end products facilitate the spread of the organisms, thereby initiating secondary sites of streptococcal infection.

Although the different groups of streptococci have similar colonial morphology and microscopic appearance, they can be separated and identified by the performance of a variety of laboratory tests. Toward this end, students will perform laboratory procedures to differentiate among the medically significant streptococci on the basis of their Lancefield group classification and their hemolytic patterns. Table 65.1 will aid in this separation.

Identification of group A streptococci involves the following procedures:

1. **Bacitracin test:** A filter-paper disc impregnated with 0.04 units of bacitracin is applied to the surface of a blood agar plate previously streaked with the organism to be identified. Following incubation, the appearance of a zone of growth inhibition surrounding the disc is indicative of group A streptococci. Absence of this zone suggests a non-group A organism.

2. **Directigen test:** A rapid, non-growth-dependent immunological procedure for the detection of the group A antigen developed by Becton Dickinson and Company, Cockeysville, Maryland. In this test, a clinical specimen is subjected to reagents de-

signed to extract the group A antigen, which is then mixed with a reactive and a negative control latex. Agglutination with the reactive latex is indicative of group A streptococci.

Identification of group B streptococci is performed with the **CAMP test** (named for Christie, Atkins, and Munch-Peterson). Group B streptococci produce a peptide, the CAMP substance, that acts in concert with the beta-hemolysins produced by some strains of *Staphylococcus aureus,* causing an increased hemolytic effect. Following inoculation and incubation, the resultant effect appears as an arrow-shaped zone of hemolysis adjacent to the central streak of *S. aureus* growth. The non-group B streptococci do not produce this reaction. Figure 65.1 illustrates the CAMP reactions.

Identification of group D streptococci involves the following procedures:

1. **Bile esculin test:** In the presence of bile, group D streptococci hydrolyze the glycoside esculin to 6,7-dehydroxy-coumarin that reacts with the iron salts in the medium to produce a brown-to-black coloration of the medium following incubation. Lack of this dark coloration is indicative of a non-group D organism.

2. **6.5 percent sodium chloride broth:** The group D enterococci can be separated from the nonenterococci by the ability of the former to grow in this medium.

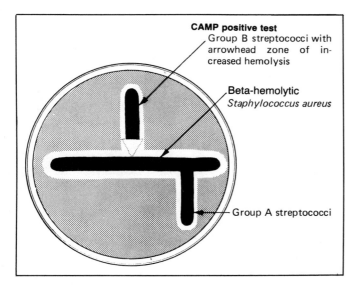

CAMP positive test
Group B streptococci with arrowhead zone of increased hemolysis

Beta-hemolytic
Staphylococcus aureus

Group A streptococci

FIGURE 65.1. CAMP reactions

Hemolytic activity is identified with a blood agar medium. The pathogenic streptococci, primarily the beta-hemolytic, can be separated from the generally avirulent alpha and gamma-hemolytic streptococci by the type of hemolysis produced on blood agar as previously described.

MATERIALS

Cultures

24-hour blood agar slant cultures of *Streptococcus pyogenes*, *Streptococcus faecalis*, *Streptococcus bovis*, *Streptococcus agalactiae*, *Streptococcus mitis*, and *Staphylococcus aureus* (ATCC 25923).

Media

Per designated student group: five blood agar plates, three bile esculin agar plates, and three 6.5 percent sodium chloride broth.

Reagents

Directigen Rapid Group A Strep Test (Becton Dickinson and Company, Cockeysville, Maryland). Crystal violet, Gram's iodine, ethyl alcohol, safranin, and Taxo A discs (0.04 unit of bacitracin).

Equipment

Bunsen burner, inoculating loop, staining tray, lens paper, bibulous paper, microscope, sterile cotton swabs, wax pencil, sterile 12 × 75-mm test tubes, sterile Pasteur pipettes, sterile applicators, and mechanical rotator.

PROCEDURE

1. Prepare a Gram stain preparation of each streptococcal culture and observe under oil immersion. Record your observations in the chart provided as to cell morphology and Gram reaction.

2. Prepare the blood agar plate cultures to identify the type of hemolysis as follows:

   a. With a wax marking pencil divide the bottoms of two blood agar plates to accommodate the five test organisms. Label each section with the name of the culture to be inoculated and identify each plate with your initials.

   b. Using sterile inoculating technique, make a single line streak of inoculation of each organism in its respective sector on the blood plates.

3. Prepare the blood agar plate cultures for the bacitracin test as follows:

   a. With a wax marking pencil label the covers of two blood agar plates with your initials and the organisms to be inoculated, *S. pyogenes* and *S. agalactiae*.

   b. Using a sterile cotton swab inoculate the agar surface of each plate with its respective test organism by streaking first in a horizontal direction, then vertically to ensure a heavy growth over the entire surface.

   c. Using an alcohol-dipped and flamed forceps, apply a single 0.04-unit bacitracin disc to the surface of each plate. Gently, touch each disc to ensure its adherence to the agar surface.

4. Prepare a blood agar plate culture for the CAMP test as follows:

   a. Using a sterile inoculating loop, make a single line of inoculation along the center of the plate using the *S. aureus* culture.

   b. With a sterile loop, inoculate *S. pyogenes* on one side and perpendicular to the central *S. aureus* streak, starting about 5 mm from the central streak and extending toward the periphery of the agar plate.

c. On the opposite side of the central streak, but not directly opposite the *S. pyogenes* line of inoculation, repeat step 4b using *S. agalactiae*.

5. Prepare the bile esculin agar plate cultures as follows:

   a. Label the three bile esculin plates with your initials and *S. bovis*, *S. mitis*, and *S. faecalis*, respectively.

   b. Aseptically inoculate each plate with its test organism by making several lines of inoculation on the agar surface.

6. Prepare 6.5 percent sodium chloride broth cultures as follows:

   a. Label three tubes of 6.5 percent sodium chloride broth with your initials and *S. bovis*, *S. faecalis*, and *S. mitis*, respectively.

   b. With a sterile loop, inoculate each tube with its organism.

7. Directigen test procedure:

   a. Label two sterile 12 × 75-mm test tubes as *S. pyogenes* and *S. agalactiae*.

   b. Add 0.3 ml of Reagent 1 to both test tubes.

   c. Using a sterile cotton swab, transfer the test organisms into their respectively labeled test tubes. Note: these samples will emulate the throat swabs obtained in a clinical situation.

   d. Add one drop of Reagent 2 to each test tube. Mix by rotating the swab against the side of

the tube. Allow the swabs to remain in the test tubes for three minutes.

   e. Add one drop of Reagent 3 to both tubes and mix.

   f. Remove swabs, extracting as much liquid as possible by rolling swab against the side of the tube.

   g. Place one drop of negative antigen control onto both circles in column A of test slide.

   h. Place one drop of positive antigen control onto both circles in column B of test slide.

   i. Dispense one drop of each streptococcal sample onto both circles in columns C and D respectively.

   j. Using a new, sterile applicator, spread each specimen within the confines of both circles in columns A, B, C, and D.

   k. Add one drop of reactive latex to the top row of circles.

   l. Add one drop of control latex to the bottom row of circles.

   m. Place the slide on a mechanical rotator for four minutes under a moistened humidifying cover.

   n. Compare the agglutination seen in the upper "reactive latex" circles with the consistency of the latex in the bottom "control latex" circle. Any agglutination in the top circles distinct from any background granules seen in the bottom circles is indicative of group A streptococci.

8. Incubate all tubes and plates in an inverted position for 24 hours at 37 degrees C.

Observations and Results EXPERIMENT **65**

1. Prepare a Gram-stained smear of the test organisms. Microscopically identify cellular morphology and Gram reaction and record your observations.

2. Examine the two blood agar plates for bacitracin activity. Record your observations as to the presence (+) or absence (−) of a zone of inhibition of any size surrounding the discs.

3. Examine the blood agar plate for the CAMP reaction. Record your observations as to the presence (+) or absence (−) of increased arrow-shaped hemolysis.

4. Examine the bile esculin plates for the presence (+) or absence (−) of a brown-black coloration in the medium and record your observations.

5. Observe the 6.5 percent sodium chloride broth cultures for the presence (+) or absence (−) of growth and record your observations.

6. Examine the two blood agar plates for the presence and type of hemolysis produced by each of the test organisms. Record your observations as to the appearance of the medium surrounding the growth and the type of hemolytic reaction that has occurred—alpha, beta, or gamma.

7. Observe the Directigen test slide for the presence (+) or absence (−) of agglutination in the reactive and control latex circles. Based on your observations, indicate the Lancefield group classification of the test organisms. Record your results.

8. Based on your observations, classify each test organism according to its Lancefield group.

Procedure	S. pyogenes	S. agalactiae	S. bovis	S. faecalis	S. mitis

Gram stain
   Morphology
   Reaction
	_____
	_____
		_____
		_____
			_____
			_____
				_____
				_____
					_____
					_____

Bacitracin test Zone of inhibition	_____	_____	_____	_____	_____
CAMP test Increased hemolysis	_____	_____	_____	_____	_____

Bile esculin test
   Color of medium
   Result (+) or (−)
	_____
	_____
		_____
		_____
			_____
			_____
				_____
				_____
					_____
					_____

6.5 percent NaCl broth Growth	_____	_____	_____	_____	_____

Hemolytic activity
   Appearance of
   medium
   Type of hemolysis
	_____
	_____
		_____
		_____
			_____
			_____
				_____
				_____
					_____
					_____

Directigen test:
   agglutination (+)
   or (−)
   Reactive circle
   Control circle
   Lancefield group
	_____
	_____
	_____
		_____
		_____
		_____
			_____
			_____
			_____
				_____
				_____
				_____
					_____
					_____
					_____

Group classification	_____	_____	_____	_____	_____

REVIEW QUESTIONS

1. Distinguish between the purposes of the bacitracin and CAMP tests.

2. What is the mechanism of the bile esculin test?

3. Why is it important medically to distinguish between the enterococci and the nonenterococci?

4. Account for the ability of some streptococci to produce secondary sites of infection.

5. What is the clinical justification for the use of a rapid test procedure such as the Directigen test, for the identification of Group A streptococci?

Identification of *Streptococcus pneumoniae*

PURPOSE

To familiarize students with laboratory procedures to differentiate between *Streptococcus pneumoniae* and other alpha-hemolytic streptococci.

PRINCIPLE

The pneumococcus *Streptococcus pneumoniae* is the major etiological agent of **lobar pneumonia,** an infection characterized by acute inflammation of the bronchial and alveolar membranes. They are gram-positive cocci, tapered or lancet-shaped at their edges, and occur in pairs or as short, tight chains. The large, thick **capsules** formed *in vivo* are responsible for antiphagocytic activity, which is believed to enhance the organisms' virulence. In addition, the pneumococci produce **alpha-hemolysis** on blood agar plates. Because of these properties (short chain formation, alpha-hemolysis, and failure of the capsule to stain on Gram staining), the organisms closely resemble *Streptococcus viridans* species. The *S. pneumoniae* can be differentiated from other alpha-hemolytic streptococci on the basis of the following laboratory tests:

Test	S. pneumoniae	S. mitis
Hemolysis	α	α
Bile solubility	+	−
Optochin	+	−
Inulin fermentation	+	−
Quellung reaction	+	−
Mouse virulence	+	−

The following is a brief description of the tests and their mechanisms:

1. **Bile solubility test:** In the presence of surface-active agents such as **bile** and **bile salts** (sodium desoxycholate or sodium dodecyl sulfate) the cell wall of the pneumococcus undergoes lysis. Other members of the alpha-hemolytic streptococci will not be lysed by these agents and are bile-insoluble. Following incubation, bile-soluble cultures will appear clear; bile-insoluble cultures will be turbid.

2. **Optochin test:** This is a growth inhibition test in which 6-mm filter-paper discs impregnated with 5 mg **ethylhydrocupreine hydrochloride** (optochin) and called P-discs, are applied to the surface of a blood agar plate streaked with the test organism. The *S. pneumoniae,* being sensitive to this surface-active agent, are lysed with the resultant formation of a zone of inhibition greater than 15 mm surrounding the P-disc. Nonpneumococcal alpha-hemolytic streptococci are resistant to optochin and fail to show a zone of inhibition or produce a zone less than 15 mm.

3. **Inulin fermentation:** The pneumococci are capable of fermenting inulin, while most other alpha-hemolytic streptococci are inulin-nonfermenters. Following incubation, the **acid** resulting from inulin fermentation will change the color of the culture from red to yellow. Cultures that are not capable of fermenting inulin will not exhibit a color change, which is a negative test result.

4. **Quellung** (Neufield) **reaction:** This **capsular swelling** reaction is a sensitive and accurate method of determining the presence of *S. pneumoniae* in sputum. The reaction of the pneumococcal capsular polysaccharide, a hapten antigen, with an omnivalent capsular antiserum produces a microscopically visible swollen capsule surrounding the *S. pneumoniae* organisms.

5. **Mouse virulence test:** Laboratory white mice are highly susceptible to infection by *S. pneumoniae* and resistant to other streptococcal infections. Intraperitoneal injection of 0.1 ml of pneumococcus-infected sputum will kill the mouse. Examination of the peritoneal fluid by Gram stain and culture will reveal the presence of *S. pneumoniae*.

In the following experiment, students will use hemolytic patterns, bile solubility, the Quellung reaction, the optochin test, and inulin fermentation for laboratory differentiation of *S. pneumoniae* from other alpha-hemolytic streptococci.

MATERIALS

Cultures

24-hour blood agar slant cultures of *Streptococcus pneumoniae* and *Streptococcus mitis*.

Media

Per designated student group: one blood agar plate, two phenol red inulin broth tubes, and four 13 × 75-mm tubes containing 1 ml of nutrient broth.

Reagents

Crystal violet, Gram's iodine, ethyl alcohol, safranin, methylene blue, 10 percent sodium desoxycholate, commercially available Taxo P-discs (5 mg optochin), and omnivalent pneumococcal antiserum.

Equipment

Bunsen burner, inoculating loop, cotton swabs, and sterile 1-ml serological pipettes.

PROCEDURE

1. **Bile solubility test**

   a. Label two nutrient broth tubes *S. pneumoniae* and two remaining tubes as *S. mitis*, respectively, and mark each with your initials.

   b. Aseptically add two loopsful of the test organisms to the appropriately labeled sterile test tubes to effect a heavy suspension.

   c. Aseptically add 0.5 ml of sodium desoxycholate to one tube of each test culture. The remaining two cultures will serve as controls.

   d. Incubate the tubes in a water bath at 37 degrees C for one hour.

2. **Optochin test**

   a. With a wax pencil divide the bottom of a blood agar plate into two equal sections and label one section *S. pneumoniae* and the other *S. mitis*, and mark with your initials.

   b. Using a sterile cotton swab, heavily inoculate the surface of each section with its respective test organism in a horizontal and then vertical direction, being careful to stay within the limits of each section.

   c. Using an alcohol-dipped and flamed forceps, apply a single Taxo P-disc (optochin) to the surface of the agar in each section of the inoculated plate. Touch each disc slightly to ensure its adherence to the agar surface.

   d. Incubate the plate in an inverted position for 24 to 48 hours at 37 degrees C.

3. **Inulin fermentation test**

   a. Label two phenol red inulin broth tubes with the name of each test organism to be inoculated and identify both with your initials.

   b. Using sterile technique, inoculate each experimental organism into its appropriately labeled tube of medium by loop inoculation.

   c. Incubate the tube cultures for 24 to 48 hours at 37 degrees C.

4. **Quellung reaction**

   a. Spread a loopful of each test culture on labeled clean glass slides and allow the slides to air dry.

   b. Place a loopful of the omnivalent capsular antiserum and a loopful of methylene blue to each of two coverslips.

   c. Place the coverslips over the dried bacterial smears.

5. Prepare a Gram stain preparation of each test organism and observe under oil immersion. Record your observations in the chart as to cell morphology and Gram reaction.

Observations and Results EXPERIMENT **66**

1. Examine blood agar plates for the presence of hemolysis and optochin activity by measuring the zone of inhibition, if present, surrounding the disc. Record your results in the chart as to the presence (+) or absence (−) of the 15-mm or greater zone of inhibition surrounding the optochin discs and indicate if each organism is optochin-sensitive or optochin-resistant.

2. For the bile solubility test, examine the tubes after one hour of incubation for the presence or absence of turbidity in each culture. Record your observations as to appearance (clear or turbid) and bile solubility of each test organism.

3. Observe the phenol red inulin broth cultures and record the color of each culture and whether it is indicative of a positive (+) or negative (−) result.

4. Examine slides of the Quellung reaction under oil immersion and indicate in the chart the presence (+) or absence (−) of capsular swelling surrounding the blue-stained cells.

Procedure	*S. pneumoniae*	*S. mitis*
Gram stain Morphology Reaction		
Optochin test Zone of inhibition in mm Resistant or sensitive		
Bile solubility test Appearance of culture Bile solubility		
Inulin fermentation Color of medium Fermentation		
Quellung reaction Capsular swelling		

REVIEW QUESTIONS

1. Why is it clinically important to distinguish *S. pneumoniae* from other alpha-hemolytic streptococci?

2. What is the function of a bacterial capsule, and what role does it play in infection?

3. Describe how you would separate *S. pneumoniae* from other alpha-hemolytic streptococci.

Identification of Enteric Microorganisms Using Computer-Assisted Multitest Microsystems

PURPOSES

1. To acquaint students with the members of the family Enterobacteriaceae.

2. To familiarize students with laboratory procedures designed to identify enteric pathogens using commercial multitest microsystems.

INTRODUCTION

The Enterobacteriaceae are a significant group of bacteria that are endogenous to the intestinal tract or may gain access to this site with ingestion of contaminated food and water. The family consists of a number of genera whose members vary in their capacity to produce disease. The *Salmonella* and *Shigella* are considered to be pathogenic, and members of other genera, particularly *Escherichia* and *Enterobacter* and to a lesser extent *Klebsiella* and *Proteus,* constitute the natural flora of the intestines and are generally considered to be avirulent. It must be remembered, however, that all can produce disease under appropriate conditions.

The Enterobacteriaceae are gram-negative, short rods. They are mesophilic, nonfastidious organisms that multiply in many foods and water sources. They are all nonspore-formers and susceptible to destruction by common physical and chemical agents. They are resistant to destruction by low temperatures and can therefore frequently survive in soil, sewage, water, and many foods for extended periods of time.

From a medical point of view, the pathogenic Enterobacteriaceae are salmonellae and shigellae. Salmonellae are responsible for enteric fevers, **typhoid** and milder **paratyphoid,** and **gastroenteritis.** In typhoid the *Salmonella typhi* penetrate the intestinal mucosa and enter the blood stream, thus infecting organs such as the gall bladder, intestines, liver, kidney, spleen, and heart. Ulceration of the intestinal wall, a long febrile period due to the release of the lipopolysaccharide endotoxin into the blood, and enteric symptoms are common. **Gastroenteritis** is caused by a number of *Salmonella* species. Symptoms associated with this type of food poisoning include abdominal pain, nausea, vomiting, and diarrhea, which develop within 24 hours of ingestion of contaminated food and last for several days.

Several shigellae are responsible for **shigellosis,** a bacillary dysentery that varies in severity. In the more severe cases there is ulceration of the large intestine, explosive diarrhea, fever, and dehydration.

Isolation and identification of enteric bacteria from feces, urine, blood, and fecally contaminated materials is of major importance in diagnosis of enteric infections. Although the Enterobacteriaceae are morphologically alike and in many ways metabolically similar, laboratory procedures for the identification of these bacteria are based on differences in biochemical activities (Figure 67.1 on page 384).

In the past decade several **multitest systems** have been developed for differentiation and identification of members of the Enterobacteriaceae. They use microtechniques that incorporate a number of media in a single unit. At least six multitest systems are commercially available. The obvious advantages of these units are the need for minimal storage space, the use of fewer media, the rapidity with which results may be obtained, and the applicability of the results to a computerized system for identification of organisms. There are also certain disadvantages with these systems, including difficulty in obtaining the proper inoculum size since some media require heavy inoculation while others need to be lightly inoculated, the possibility of media carryover from one compartment to another, and the possibility of using inoculum of improper age. Despite these difficulties, when properly correlated with other properties such as Gram stain and colonial morphology on specialized solid media, these systems are acceptable for the identification of Enterobacteriaceae. The most frequently used systems are discussed.

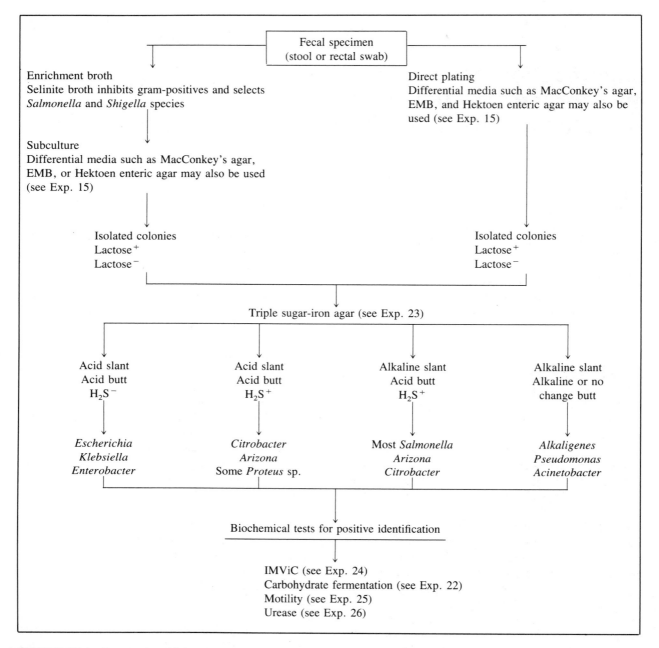

FIGURE 67.1. Conventional laboratory procedures for isolation and identification of enteric microorganisms

ENTEROTUBE MULTITEST SYSTEM AND ENCISE II

The Enterotube Multitest System (Roche Diagnostics, Hoffmann-La Roche, Inc., Nutley, New Jersey) consists of a single tube containing 12 compartments (Figure 67.2) and a self-enclosed inoculating needle. This needle can touch a single isolated colony and then in one operation be drawn through all 12 compartments, thereby inoculating all of the test media. In this manner, 15 standard biochemical tests can be performed in one inoculating procedure. Following incubation, the color changes that occur in each of the compartments are interpreted according to the manufacturer's instructions to identify the organisms. This method has been further refined to permit identification of the enteric bacteria by means of a computer-assisted system called **ENCISE** (Enterobacteriaceae numerical coding and identification system for enterotube).

API (ANALYTICAL PROFILE INDEX) SYSTEM

The API-20E (Analytab Products Inc., Plainview, New York) employs a plastic strip composed of 20 individual microtubes, each containing a dehydrated medium in the bottom and an upper cupule as shown in Figure 67.3. The media become hydrated during inoculation of a suspension of the test organism, and the strip is then incubated in a plastic-covered tray to prevent evaporation. In this manner 22 biochemical tests are performed. Following incubation, identification of the organism is made by using differential charts supplied by the manufacturer or by means of a computer-assisted system called **PRS** (profile recognition system). This includes an API coder, profile register, and selector.

In the following experiment students will inoculate an Enterotube and an API strip with an unknown en-

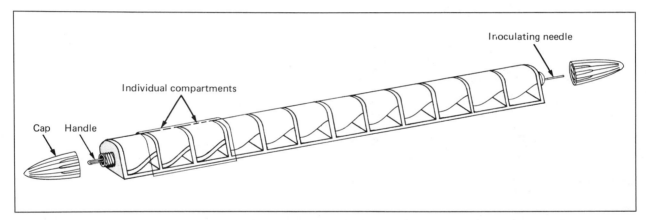

FIGURE 67.2. Enterotube multitest system (courtesy of Roche Diagnostics, Division of Hoffmann-La Roche, Inc.)

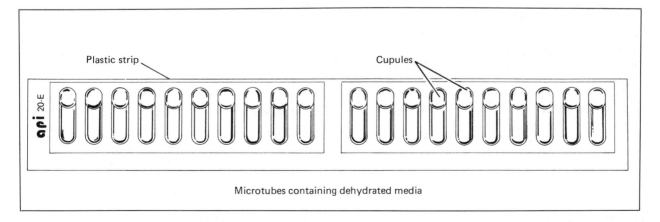

FIGURE 67.3. The API-20E system (Courtesy of Analytab Products Division of Sherwood Medical, Plainview, N.Y.)

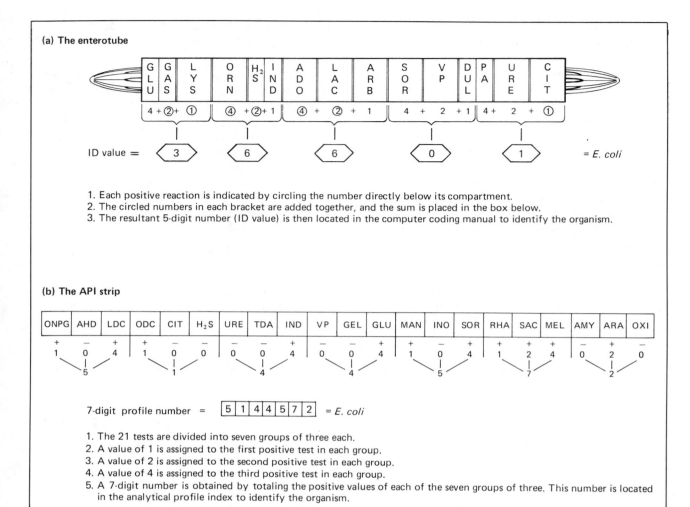

FIGURE 67.4. Computer-assisted techniques for the identification of Enterobacteriaceae

teric organism. Following incubation, identification will be made by two methods: (1) the traditional method of noting the characteristic color changes and interpreting them according to manufacturers' instructions and (2) the computer-assisted method illustrated in Figure 67.4.

MATERIALS

Cultures

Number-coded, 24-hour trypticase soy agar streak plates of *Escherichia coli*, *Salmonella typhimurium*, *Klebsiella pneumoniae*, *Enterobacter aerogenes*, *Shigella dysenteriae*, and *Proteus vulgaris*.

Media

Per designated student group: one Enterotube II, one API-20E strip, and one 5-ml tube of 0.85 percent sterile saline.

Reagents

Sterile mineral oil, 10 percent ferric chloride, Kovac's reagent, V-P reagent for API system, nitrate reduction reagents, Barritt's reagent (V-P test reagent for Enterotube system), 1.5 percent hydrogen peroxide, and 1 percent tetramethyl-p-phenylenediamine dihydrochloride (oxidase reagent).

Equipment

Bunsen burner, inoculating loop, 5-ml pipette, sterile Pasteur pipettes, wax pencil, API profile recognition system and differential identification charts, and Enterotube ENCISE pads and color reaction charts.

PROCEDURE

API-20E System

1. Familiarize yourself with the components of the system: Incubation tray, lid, and the strip with its 20 microtubes.

2. Label the elongated flap on the incubation tray with your name and the number of the unknown culture supplied by the instructor.

3. With a pipette, add approximately 5 ml of tap water to the incubation tray.

4. Using a sterilized loop, touch an isolated colony on the provided streak-plate culture, transfer the inoculum to a 5-ml tube of sterile saline, and mix well to effect a uniform suspension.

5. Remove the API strip from its sterile envelope and place it in the incubation tray.

6. Tilt the incubation tray. Using a sterile Pasteur pipette containing the bacterial saline suspension, fill the tube section of each compartment by placing the tip of the pipette against the side of the cupule. Fill the cupules in the CIT, VP, and GEL microtubes with the bacterial suspension.

7. Using a sterile Pasteur pipette, completely fill the cupules of the ADH, LDC, ODC, and URE microtubes with sterile mineral oil to provide an anaerobic environment.

8. Cover the inoculated strip with the tray lid and incubate for 18 to 24 hours at 37 degrees C.

Enterotube System

1. Familiarize yourself with the components of the system: Screw caps at both ends, medium-containing compartments, self-enclosed inoculating needle, plastic slide bar, and blue-taped section.

2. Label the Enterotube with your name and the number of the unknown culture supplied by the instructor.

3. Remove the screw caps from both ends of the Enterotube. Using the inoculating needle contained in the Enterotube, aseptically pick some inoculum from an isolated colony on the provided streak-plate culture.

4. Inoculate the Enterotube as follows:

    a. Twist the needle in a rotary motion and withdraw it slowly through all 12 compartments.

    b. Replace the needle in the tube and with a rotary motion push the needle into the first three compartments (GLU, LYS, and ORN). The point of the needle should be visible in the H_2S/IND compartment.

    c. Break the needle at the exposed notch by bending, discard the needle remnant, and replace the caps at both ends. The presence of the needle in the three compartments maintains anaerobiosis, which is necessary for dextrose fermentation, CO_2 production, and the decarboxylation of lysine and ornithine.

5. Remove the blue tape covering the ADO, LAC, ARB, SOR, VP, DUL/PA, URE, and CIT compartments. Beneath this tape are tiny air vents that provide aerobic conditions in these compartments.

6. Place the clear plastic slide band over the GLU compartment to contain the wax, which may be spilled by the excessive gas production of some organisms.

7. Incubate the tube on a flat surface for 24 hours at 37 degrees C.

Observations and Results

API-20E System

1. Observe all reactions in the API strip that do not require addition of a test reagent, and interpret your observations using the manufacturer's instructions. Record your observations and results in the chart.

2. Add the required test reagents in the following order: Kovac's reagent to IND, VP reagent to VP (read the result after 15 minutes), ferric chloride to TDA, nitrate reagents to GLU, and oxidase reagent to OXI. Note color changes and interpret your observations according to the manufacturer's instructions. Record your observations and results in the chart.

3. Based on your results, identify your unknown organism using the differential identification chart.

4. Determine and record in the chart the seven-digit profile number as described in Figure 67.4. Identify your unknown organism by referring to the profile recognition system.

Enterotube II System

1. Observe all reactions in the Enterotube except IND and VP, and interpret your observations using the manufacturer's instructions. Record your observations and results in the chart.

2. Perform the IND and VP tests as follows:

   a. Place the Enterotube in a rack with the GLU and VP compartments facing downward.

   b. With a needle and a syringe gently pierce the plastic film of the H_2S/IND compartment and add two to three drops of Kovac's reagent. Read the results after one minute.

   c. As in step 2b, add two drops of Barritt's reagent to the VP compartment and read the result after 20 minutes.

   d. Record your IND and VP observations and results.

3. Based on your results, identify your unknown organism using the manufacturer's color identification charts.

4. In the chart, determine and record the five-digit ID value as described in Figure 67.4. Identify your unknown organism by referring to the computer-coding manual.

		API-20E		Enterotube	
Code	**Name**	Appearance (color)	Result (+) or (−)	Appearance (color)	Result (+) or (−)
OPNG	Beta-galactosidase				
ADH	Arginine dihydrolase				
LDC	Lysine decarboxylase				
ODC	Ornithine decarboxylase				
CIT	Citrate				
H₂S	Hydrogen sulfide				
URE	Urease				
TDA	Tryptophan deaminase				
IND	Indole				
VP	Acetonin				
GEL	Gelatin				
GLU	Glucose				
MAN	Mannitol				
INO	Inositol				
SOR	Sorbitol				
RHA	Rhamnose				
SAC	Sucrose				
MEL	Melibiose				
AMY	Amygdalin				
ARA	Arabinose				
OXI	Oxidase				
NO₃	Nitrate reduction				
GAS	Gas production				
PHE	Phenylalanine				
LAC	Lactose				
DUL	Dulcitol				
Organism					

Determination of API-20E seven-digit profile number

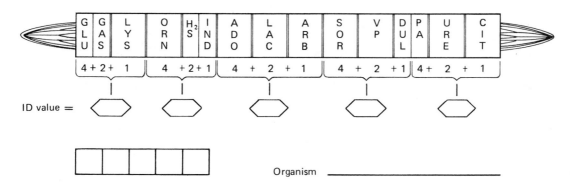

ONPG	AHD	LDC	ODC	CIT	H₂S	URE	TDA	IND	VP	GEL	GLU	MAN	INO	SOR	RHA	SAC	MEL	AMY	ARA	OXI

Organism _____

Determination of Enterotube five-digit identification number

GLU	GAS	LYS	ORN	H₂S	IND	ADO	LAC	ARB	SOR	VP	DUL	PA	URE	CIT

4 + 2 + 1   4 + 2 + 1   4 + 2 + 1   4 + 2 + 1   4 + 2 + 1

ID value =   ⬡   ⬡   ⬡   ⬡   ⬡

Organism _____

REVIEW QUESTIONS

1. Cite the advantages of multitest systems:

Disadvantages:

2. What Enterobacteriaceae are of medical significance?

   List and describe the infections caused by these organisms.

3. Were similar results obtained by use of the computer-assisted method and the traditional color change method?

Isolation and Presumptive Identification of *Campylobacter*

PURPOSE

To demonstrate the laboratory procedures required for isolation, cultivation, and presumptive identification of members of the genus *Campylobacter*.

PRINCIPLE

Clinicians are aware of the medical significance of *Campylobacter* strains as the etiological agents of enteric infections. The incidence of enteritis caused by *Campylobacter fetus* subspecies *jejuni* has been found to equal or exceed that of salmonellosis or shigellosis. The clinical syndrome, although varying in severity, is generally characterized by acute gastroenteritis accompanied by the rapid onset of fever, headache, muscular pain, malaise, nausea, and vomiting. Twenty-four hours following this acute phase, diarrhea develops that may be mucoid, bloody, bile-stained, and watery. The precise epidemiology of the infection is not clear; however, contact with animals, water-borne organisms, and oral-fecal transmission remain suspect.

The organisms (*campylo,* curved; *bacter,* rod) were formerly called vibrios because of their curved and spiral morphology. In the early 1970s they were reclassified in the genus *Campylobacter.* They are gram-negative, curved or spiral, and have a single flagellum located at one or both poles of the cell. In pure culture, two types of colonies have been recognized and designated as types I and II. The more commonly observed type I colonies are large, flat, and spreading with uneven margins. They are nonhemolytic, watery, and grayish. Type II colonies are also nonhemolytic but they are smaller (one to two mm), with unbroken edges. They are convex and glistening.

Initially, the isolation of *Campylobacter* organisms from fecal specimens was difficult because of their microaerophilic nature and their 42-degree C optimal growth temperature. Furthermore, in the absence of selective media, their growth was masked by the overgrowth of other enteric organisms, and they were often overlooked on primary isolation. This situation has been rectified with the development of selective media that are designed specifically for isolating *Campylobacter* species and that inhibit the growth of other enteric organisms. These media are nutritionally enriched and supplemented with 5 percent to 10 percent sheep or horse blood. In addition, they contain three to five antimicrobial agents depending on the medium. For example, cephalosporins, one of the antimicrobial agents present in the Campy BAP medium (BBL, Cockeysville, Md.), is selective for *C. fetus* subspecies *jejuni,* and inhibits the subspecies *intestinalis,* which is rarely responsible for enteric infections.

The most essential requirement for cultivating campylobacteria is a microaerophilic incubation atmosphere. High concentrations of oxygen are toxic to these organisms and an atmosphere of 3 percent to 10 percent carbon dioxide and 5 percent to 10 percent oxygen is optimal for their growth. The incubation temperature for *C. fetus* subsp. *jejuni* has been found to be 42 degrees C. At this temperature the organism grows optimally, while growth of *C. fetus* subsp. *intestinalis* is inhibited.

In the experiment to follow, a simulated fecal specimen, a culture containing *C. fetus* subsp. *jejuni,* and other enteric organisms is used. Students will attempt to isolate the *Campylobacter* organisms by using the following two procedures:

1. A conventional method uses MacConkey's agar directly, circumventing enrichment procedures, as a mixed simulated fecal population is the test culture.

2. A special method employs one of the preceding media and cultural conditions and uses the Campy Pak and GasPak jar (BBL), which are illustrated in Figure 68.1 on page 394.

Presumptive identification is made on the basis of colonial morphology and the microscopic appearance of the organisms obtained from a typical isolated colony. Students may perform the catalase and oxidase tests as described in Experiments 29 and 30 for further presumptive identification. In the case of *C. fetus* subsp. *jejuni,* both tests should be positive.

FIGURE 68.1. Campy Pak and GasPak Jar. Courtesy of BBL Microbiology Systems, Division of Becton Dickinson and Company. (GasPak and Campy Pak are trademarks of Becton Dickinson and Company.)

MATERIALS

Cultures

Mixed saline suspensions of *Campylobacter fetus* subspecies *jejuni* cultured on a sheep blood-enriched medium, *Salmonella typhimurium,* and *Escherichia coli.*

Media

Per designated student group: one Campy BAP agar plate and one MacConkey's agar plate.

Reagents

Crystal violet, Gram's iodine, 95 percent ethyl alcohol, and 0.8 percent carbol fuschin.

Equipment

Bunsen burner, inoculating loop, wax pencil, Campy Pak and GasPak jars, and 10-ml pipettes.

PROCEDURE

1. Label the covers of the Campy BAP agar and MacConkey's agar plates with your initials.

2. Aseptically perform a four-way streak inoculation as described in Experiment 11 for the isolation of discrete colonies on both agar plates.

3. Place the inoculated Campy BAP agar plate into the GasPak jar in an inverted position. Following the manufacturer's instructions, open the Campy Pak envelope and place it into the jar. With a pipette, add 10 ml of water to the envelope and immediately seal the jar to establish a microaerophilic environment.

4. Incubate the jar for 48 hours at 42 degrees C.

5. Incubate the MacConkey's agar plate culture in an inverted position for 48 hours at 37 degrees C.

Observations and Results

1. Observe both plate cultures for the presence of discrete colonies. In the chart, diagram the appearance of representative colonies on both plates and describe their colonial characteristics. Also, note and record the color of the medium surrounding the representative colonies on the MacConkey's plate. (Refer to Experiment 15 for an explanation of the selective and differential nature of MacConkey's agar.)

Plate culture	Diagram of colonies	Colonial characteristics	Color of medium
Campy BAP agar			
MacConkey's agar			

2. Prepare a Gram stain preparation using 0.8 percent carbol fuschin as the counterstain of each representative colony on both agar plate cultures. Observe microscopically, note, and record the microscopic morphology and Gram reaction of each preparation.

	Campy BAP plate isolate	MacConkey's agar plate	
		Isolate 1	Isolate 2
Drawing of representative field			
Microscopic morphology			
Gram reaction			

3. Based on your observations, identify your isolates:

Campy BAP agar culture isolate: _____

MacConkey's agar culture isolate 1: _____

MacConkey's agar culture isolate 2: _____

4. Optional: Perform the catalase and oxidase tests on the representative isolates.

REVIEW QUESTIONS

1. Explain why members of *Campylobacter* may not be isolated from a stool specimen in a diagnostic laboratory.

2. Describe the clinical syndrome induced by *C. fetus* subsp. *jejuni*.

3. Explain the purposes of the antimicrobial agents present in the selective media used for the isolation of *Campylobacter*.

4. Describe how *C. fetus* subsp. *jejuni* may be separated from *C. fetus* subsp. *intestinalis*.

Microbiological Analysis of Urine Specimens

PURPOSES

To acquaint students with:

1. The organisms commonly responsible for infections of the genitourinary tract.

2. Laboratory methods for detection of bacteriuria and identification of microorganisms associated with the urinary tract.

PRINCIPLE

The anatomical structure of the mammalian urinary system is such that the external genitalia and the lower aspects of the urethra are normally contaminated with a diverse population of microorganisms. The tissues and organs that comprise the remainder of the urinary system, the bladder, ureters, and kidneys, are sterile, and therefore urine that passes through these structures is also sterile. When pathogens gain access to this system, they can establish infection. Some etiological agents of urinary-tract diseases are illustrated in the right-hand column.

Urinary-tract infections may be limited to a single tissue or organ or they may spread upward and involve the entire system. Infections such as **cystitis** initially involve the bladder and spread through the ureters to the kidneys. Infections limited to the ureters and kidneys are called **pyelitis. Glomerulonephritis** is an inflammation that results in the destruction of renal corpuscles; **pyelonephritis** results in the destruction of renal tubules. Organisms other than bacteria may also act as etiological agents of urogenital infections. *Trichomonas vaginalis,* a pathogenic flagellated protozoan, is commonly found in the vagina, and under appropriate conditions is responsible for a severe inflammatory **vaginitis.** *Candida albicans,* a pathogenic yeast, is normally found in low numbers in the intestines. Under suitable conditions, it can enter the urogenital system where it gives rise to vaginal infections. *Schistosoma haematobium* is a pathogenic fluke, a helminth, responsible for severe bladder infections.

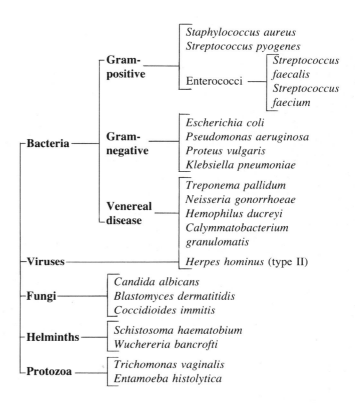

The initial step in diagnosis of a possible urinary-tract infection is laboratory examination of a urine specimen. The sample must be collected midstream in a sterile container following adequate cleansing of the external genitalia. It is imperative to culture the freshly voided, unrefrigerated urine sample immediately to avoid growth of normal indigenous organisms, which may overtake the growth of the more slowly growing pathogens. In this event the infectious organism might be overlooked, resulting in an erroneous diagnosis.

Clinical evaluation of the specimen requires a quantitative determination of the microorganisms per ml of urine. Urine in which the bacterial count per ml exceeds 100,000 (10^5) denotes significant bacteriuria and is indicative of a urinary-tract infection. Urine in which counts range from 0 to 1000 per ml are normal.

In the conventional method, a urine sample is streaked over the surface of an agar medium with a special loop calibrated to deliver a known volume. Following incubation, the number of isolated colonies present on the plate is determined and multiplied by a factor that converts the volume of urine to 1 ml. The final calculation is then equal to the number of organisms per ml of sample.

Example: 25 colonies were present on a plate inoculated with a loop calibrated to deliver 0.01 ml of a urine specimen.

$$\begin{array}{l} \text{Number of} \\ \text{colonies} \end{array} \times \begin{array}{l} \text{Factor that} \\ \text{converts 0.01 ml} \\ \text{to 1 ml} \end{array} = \begin{array}{l} \text{organisms} \\ \text{per ml} \end{array}$$

$$25 \times 100 \qquad = 2500 \text{ organisms per ml}$$

If the specimen is turbid, dilution is necessary prior to culturing. In this case, conventional tenfold dilutions are prepared in physiological saline to effect a final dilution of 1:1000 (see Experiment 13). Each of the dilutions—10^{-1}, 10^{-2}, and 10^{-3}—is then streaked on the surface of a suitable agar plate medium for isolation of colonies. Following incubation, the number of microorganisms per ml of sample is determined by the following formula:

$$\begin{array}{l} \text{Organisms per ml} = \\ \begin{array}{l} \text{Number of} \\ \text{colonies} \end{array} \times \begin{array}{l} \text{Factor that converts the} \\ \text{volume of urine to 1 ml} \end{array} \times \begin{array}{l} \text{Dilution} \\ \text{factor} \end{array} \end{array}$$

Example: 25 colonies were counted on a 10^{-2} dilution plate inoculated with a loop calibrated to deliver 0.01 ml of urine. Therefore $25 \times 100 \times 100 = 250,000$ organisms per ml.

On determination of bacteriuria, identification of the infectious organism can be accomplished by the laboratory procedures outlined in Figure 69.1.

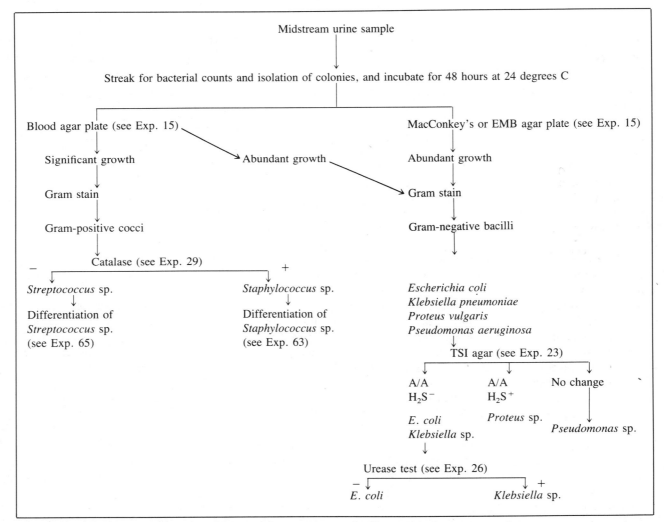

FIGURE 69.1. Laboratory procedures for the isolation and identification of urinary tract pathogens

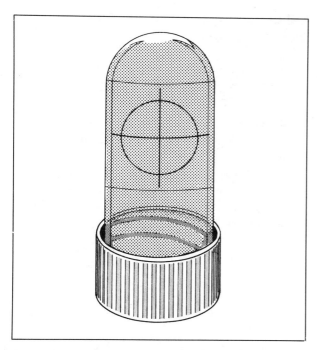

FIGURE 69.2. Bacturcult culture tube (Reproduced with permission from Wampole Laboratories, a Division of Carter-Wallace, Inc.)

A newer, less conventional, and less time-consuming method uses a diagnostic urine-culture tube, Bacturcult, devised by Wampole Laboratories, Dist. (Figure 69.2). Bacturcult is a sterile, disposable plastic tube coated on the interior with a special medium that allows detection of the bacteriuria and a presumptive class identification of urinary bacteria.

Following incubation of the Bacturcult urine culture, bacteriuria can be detected with a bacterial count. This is performed by placing the counting strip around the Bacturcult tube over an area of even colony distribution, and counting the number of colonies within the circle. The average number of colonies counted is interpreted in the chart shown below.

For the presumptive identification of bacteria, the medium contains the substrates lactose and urea, and the pH indicator phenol red. Depending on the organism's enzymatic action on these substrates, differentiation of urinary bacteria into three groups following incubation is possible based on observable color changes that occur in the culture:

Group I: *E. coli* and *Enterococcus*—yellow.

Group II: *Klebsiella, Staphylococcus,* and *Streptococcus*—rose to orange.

Group III: *Proteus* and *Pseudomonas*—purplish-red.

It must be realized that mixed cultures do not always produce clear-cut color changes. Therefore if additional testing is required, the discrete colonies that develop on the medium can be used as the source for subculturing into other media.

In this experiment, the conventional procedure using the calibrated loop establishes the number of cells in seeded urine specimens. The Bacturcult tube will be used for enumeration and presumptive group identification. Should the instructor desire to emulate more closely a clinical evaluation of urine, a mixed seeded culture must be used. Representative colonies isolated from the blood agar streak-plate culture for detection of bacteriuria can be then identified following the schema in Figure 69.1.

MATERIALS

Cultures

Six urine samples, each seeded with one of the following 24-hour cultures: *Streptococcus faecalis, Staphylococcus aureus, Proteus vulgaris, Escherichia coli, Pseudomonas aeruginosa,* and *Klebsiella pneumoniae.* Optional: Urine sample seeded with a gram-positive and a gram-negative organism.

Average number of colonies within circle	*Approximate number of bacteria per ml*	*Diagnostic significance*
Fewer than 25	Fewer than 25,000	Negative bacteriuria
25 to 50	25,000 to 100,000	Suspicious*
More than 50	Greater than 100,000	Positive bacteriuria

Source: Courtesy of Wampole Laboratories.
*Additional testing recommended.

Media

Per designated student group: six blood agar plates, three sterile 9-ml tubes of saline, and six Bacturcult culture tubes.

Equipment

Bunsen burner, calibrated 0.01-ml platinum loop, wax marking pencil, and sterile 1-ml pipettes.

PROCEDURE

Bacturcult Procedure

1. Label the Bacturcult tubes with the name of the bacterial organism present in the urine sample and identify each with your initials.

2. Fill each tube almost to the top with urine.

3. Immediately pour the urine out of each tube, allowing all of the fluid to drain for several seconds. Replace the screw cap securely.

4. Immediately prior to incubation, loosen the cap on each tube by turning the screw cap counterclockwise for one-half turn.

5. Incubate the tubes with the caps down for 24 hours at 37 degrees C.

Calibrated Loop Procedure for Bacterial Counts

1. Label the three 9-ml sterile saline tubes and the three blood agar plates 10^{-1}, 10^{-2}, and 10^{-3}, respectively.

2. Using the three 9-ml saline blanks, aseptically prepare a tenfold dilution of the urine sample to effect 10^{-1}, 10^{-2}, and 10^{-3} dilutions.

3. With a calibrated loop, aseptically add 0.01 ml of the 10^{-1} urine dilution to the appropriately labeled blood agar plate and streak for isolation of colonies as illustrated.

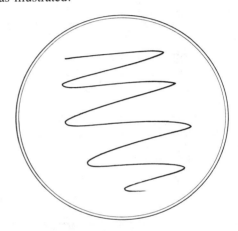

4. Repeat step 3 to inoculate the remaining urine sample dilutions.

5. Incubate all plates in an inverted position for 24 hours at 37 degrees C.

Observations and Results

1. Determine the number of colonies in each of the Bacturcult urine cultures as follows:

   a. Place the counting strip around the tube over an area of even colony distribution and count the number of colonies within the circle.

   b. Repeat the count in another area of the tube.

   c. Average the two counts.

   d. Record in the chart the average number of colonies counted within the circle.

2. Based on your colony count, determine and record in the chart the approximate number of bacteria per ml of each sample and its diagnostic significance as negative bacteriuria, suspicious, or positive bacteriuria.

3. Observe and record in the chart the color of the medium in each of the urine cultures and the presumptive bacterial group.

Urine culture	Number of colonies	Number of bacteria per ml	Diagnostic significance	Color of medium	Presumptive group
S. faecalis					
S. aureus	< 25	Fewer than 25000	− bacteriuria	red	
K. pneumoniae					
P. vulgaris	7 50	7 100 000	+ bacteriuria	purple	
P. aeruginosa	7 50	7 100 000	+ bacteriuria	orange	
E. coli	< 25	fewer than 25000	− bacteria	yellow	

4. Determine the number of colonies on each blood agar culture plate and calculate the number of organisms per ml of the urine. Record your results in the chart.

Urine sample dilution	Number of colonies	Organisms per ml of sample	Bacteriuria (+) or (−)
10^{-1}			
10^{-2}			
10^{-3}			

REVIEW QUESTIONS

1. Discuss the types of urinary infections that may be caused by different microorganisms.

2. How accurate is a laboratory analysis of a 24-hour, unrefrigerated, non-midstream urine sample? _____ Explain.

3. Describe the Bacturcult tube and indicate the basis on which presumptive identification of microorganisms is made.

4. Explain how a clinical diagnosis of a bacteriuria is established.

5. If five colonies were counted on a 10^{-3} dilution plate streaked with 0.01 ml of urine, what is the number of organisms per ml of the original specimen, and is this count indicative of bacteriuria? _____ Explain.

Microbiological Analysis of Blood Specimens

PURPOSES

To acquaint students with:

1. The microorganisms most frequently associated with septicemia.

2. Laboratory methods for the isolation and presumptive identification of the etiological agents of septicemia.

PRINCIPLE

Blood is normally a sterile body fluid. This sterility may be breached, however, when microorganisms gain access into the bloodstream during the course of an infectious process. The transient occurrence of bacteria in the blood is designated as **bacteremia** and implies the presence of nonmultiplying organisms in this body fluid.

Bacteremias may be encountered in the course of some bacterial infections such as pneumonia, meningitis, typhoid fever, and urinary tract infections. A bacteremia of this nature does not present a life-threatening situation, as the bacteria are present in low numbers and the activity of the host's nonspecific immune system is generally capable of preventing further systemic invasion of tissues. A more dangerous and clinically alarming syndrome is **septicemia,** a condition characterized by the rapid multiplication of microorganisms with the possible elaboration of their toxins into the bloodstream. The clinical picture frequently present in septicemia is that of septic shock, which is recognized by a severe febrile episode with chills, prostration, and a drop in blood pressure.

A large and diverse microbial population has been implicated in septicemia. The major offenders include:

1. Gram-negative bacteria, because of their endotoxic properties are the most frequently encountered etiological agents that present the more serious complications of septicemia. Included among these agents are *Hemophilus influenzae, Neisseria meningiditis, Serratia marcescens, Escherichia coli, Pseudomonas aeruginosa* and *Salmonella* sp. Less frequently implicated are *Francisella tularensis* and members of the genera *Campylobacter* and *Brucella.*

2. Gram-positive bacteria which generally do not produce the presenting signs of septic shock. They primarily include members of the genera *Streptococcus* and *Staphylococcus.*

3. *Candida albicans* is the major fungal invader of the bloodstream.

In the clinical setting, to facilitate the rapid initiation of effective chemotherapy, a culture of the suspect blood sample is required for the isolation and identification of the offending organisms. A blood sample is drawn and cultured in an appropriate medium under both aerobic and anaerobic conditions. Over a period of three to seven days, the cultures are observed for turbidity and Gram-stained smears are prepared to ascertain the presence of microorganisms in the blood. Upon detection of microbial growth in the cultures, transfers onto a variety of specialized agar media are made for the identification of the infectious agent. The schema for this protocol is shown in Figure 70.1 on page 404.

In this exercise two methods are outlined. Either method or both methods may be used for the isolation and presumptive identification of the microorganisms in the experimental culture. Both procedures use a simulated blood specimen, a prepared culture containing blood previously seeded with selected microorganisms. The traditional method is a modification of the schema shown in Figure 70.1. This procedure requires the preparation of Gram-stained smears for the morphological study of the organisms and the inoculation of selected agar media for their isolation and preliminary identification. The alternative method uses the commercially available **Septi-Chek System** (Roche Diagnostic Systems, Division of Hoffmann-La Roche Inc., Nutley, New Jersey), a single unit composed of the Septi-Chek culture bottle and the Septi-Chek slide as

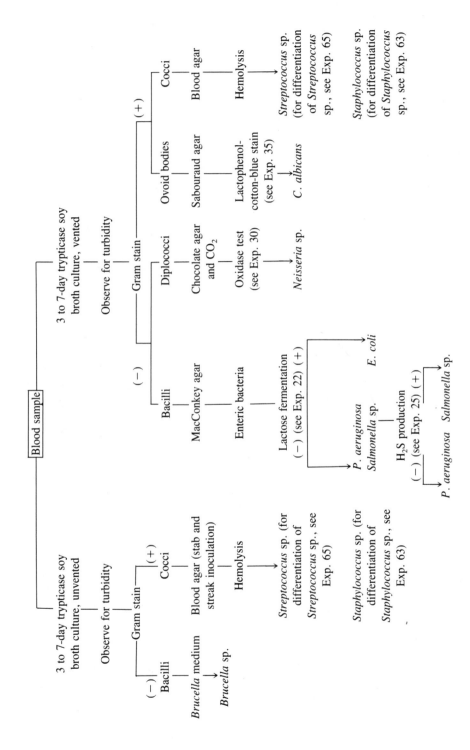

FIGURE 70.1. Schema for the Isolation and Identification of the Etiological Agents of Septicemia

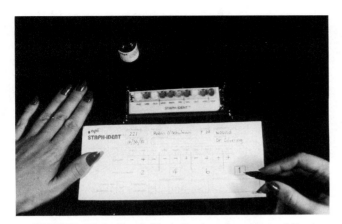

FIGURE 70.2. Septi-Chek System (courtesy of Roche Diagnostic Systems, Nutley, N.J.)

illustrated in Figure 70.2. The culture bottle component allows for the qualitative determination of the presence of microorganisms in the blood sample, and the slide component is designed for the simultaneous subculturing of the organisms onto a plastic slide containing three differential media (chocolate, MacConkey, and malt agar). Differential growth on these media provides preliminary information as to the nature of the infectious agent and isolated colonies for further study.

MATERIALS

Culture

48-hour to 72-hour blood culture prepared as follows: 10 ml of citrated blood seeded with two drops of *Escherichia coli, Neisseria perflava,* and *Saccharomyces cerivisiae,* each adjusted to an O.D. of 0.1 mμ, in 90 ml of trypticase soy broth containing 0.05 percent of sodium polyanetholesulfonate (PSP) used to prevent clotting of the blood sample.

Media

Per designated student group: one each of blood agar plate, MacConkey agar plate, chocolate agar plate, Sabouraud agar plate, and Septi-Chek system.

Reagents

Crystal violet, Gram's iodine, 95 percent ethyl alcohol, safranin, lactophenol-cotton-blue stain, 70 percent iso-propyl alcohol, and tetramethyl-para-phenylenediamine dihydrochloride.

Equipment

Sterile 20-gauge, 1½-inch needles; sterile 1-ml and 10-ml syringes, Bunsen burner; staining tray; inoculating loop; lens paper; bibulous paper; microscope; and wax pencil.

PROCEDURE

1. Using a wax pencil, label all agar plates with your initials.

2. Swab the rubber stopper of the blood culture bottle with 70 percent isopropyl alcohol and allow to air dry.

3. Using a sterile needle and 1-ml syringe, aseptically remove 0.5 ml of the blood culture by penetrating the rubber stopper.

4. To prepare a smear, place a small drop of the culture on a clean glass slide and spread evenly with an inoculating loop.

5. Place one drop of culture in one corner of the blood agar plate and prepare a four-way streak inoculation as described in Experiment 11.

6. Repeat step 5 to inoculate the MacConkey, chocolate, and Sabouraud agar plates.

7. Incubate the agar plate cultures in an inverted position for 24 to 48 hours as follows: Sabouraud agar culture at 25 degrees C, chocolate agar culture in a 10 percent CO_2 atmosphere at 37 degrees C, and the remaining cultures at 37 degrees C.

8. Septi-Chek system procedure:

   a. Label the blood culture bottle with your initials.

   b. Remove the protective top of the screw cap of the culture bottle, disinfect the rubber stopper with 70 percent isopropyl alcohol, and allow to air dry.

   c. Using the 10-ml syringe, aseptically transfer 10 ml of the experimental culture to the Septi-Chek culture bottle.

d. Aseptically vent the bottle for aerobic incubation.

e. Replace the protective top of the screw cap on the bottle.

f. Gently invert the bottle two or three times to disperse the blood evenly throughout the medium.

g. Incubate the culture for 4 to 6 hours at 37 degrees C.

h. Attach the Septi-Chek slide according to the manufacturer's instructions.

i. Tilt the combined system to a horizontal position and hold until the liquid medium enters the slide chamber and floods the agar surfaces. While maintaining this position, rotate the entire system one complete turn to ensure that all agar surfaces have come in contact with the liquid medium. Return the system to an upright position.

j. Incubate the system in an upright position at 37 degrees C.

k. Check the culture bottle daily for turbidity and the slide for visible colony formation.

Observations and Results   EXPERIMENT **70**

1. Examine the blood agar plate culture for the presence (+) or absence (−) of hemolytic activity. If hemolysis is present, determine the type observed. Record your observations in the chart.

2. For the performance of the oxidase test, add tetramethyl-p-phenylenediamine dihydrochloride to the surface of the growth on the chocolate agar plate. The presence of pink-to-purple colonies is indicative of *Neisseria* sp. Record your observations and oxidase test result in the chart.

3. Examine the MacConkey agar plate culture for determination of lactose fermentation. Lactose fermenters exhibit a pink-to-red halo in the medium, a red coloration on the surface of their growth, or both a halo and red coloration. Record your observations and results concerning the presence of lactose fermenters in the chart.

4. Examine the Sabouraud agar plate culture for the presence of growth. Prepare a lactophenol-cotton-blue-stained smear from an isolated colony (see Experiment 35). Examine the smear microscopically for the presence of large ovoid bodies indicative of yeast cells. Record your morphological observations in the chart.

5. Observe the Septi-Chek slide system for the presence of growth on the three agar surfaces. If growth is present on:

   a. Medium 1 (MacConkey agar), examine for fermentative patterns as described in step 3 and record your observations in the chart.

   b. Medium 2 (chocolate agar), perform the oxidase test as described in step 2 and record your observations in the chart.

   c. Medium 3 (Malt agar), prepare and examine microscopically a lactophenol-cotton-blue-stained smear. Record your observations in the chart.

Culture	*Traditional procedure*	*Septi-Chek system*
Blood agar		
Hemolysis: (+) or (−)	_____	_____
Type of hemolysis:	_____	_____
Chocolate agar		
Color of colonies:	_____	_____
Oxidase test: (+) or (−)	_____	_____
MacConkey agar		
Color of colonies:	_____	_____
Color of medium:	_____	_____
Lactose fermentation: (+) or (−)	_____	_____
Sabouraud or malt agar		
Cell morphology:	_____	_____
Presumptive identification of organisms present:		

REVIEW QUESTIONS

1. What are the advantages of the Septi-Chek system over the traditional procedures for the isolation and identification of microorganisms in the blood?

2. Differentiate between septicemia and bacteremia and the medical significance of each.

3. Discuss why blood samples are cultured in both vented and unvented systems.

XV

Immunology

PURPOSE

To acquaint students with:

1. The basic principles of nonspecific (innate) and specific (acquired) immunity.

2. Serological procedures that demonstrate immunological reactions of agglutination, precipitin formation, and complement fixation.

3. Rapid diagnostic screening procedures for syphilis and infectious mononucleosis.

4. The fluorescent antibody technique for identification of antigens and/or antibodies.

INTRODUCTION

Immunity or **resistance** is a state in which a person, either naturally or by some acquired mechanism, is protected from contracting certain diseases or infections. The ability to resist disease may be hereditary (nonspecific) or it may be acquired (specific) when the disease state is emulated in the host.

Nonspecific immunity is native or natural. It is hereditary, and provides the basic mechanisms that defend the host against intrusion of foreign substances or agents of disease. This defense is not restricted to a single or specific foreign agent, but provides the body with the ability to resist many pathological conditions. The mechanisms responsible for this native immunity include the **mechanical barriers,** such as the skin and mucous membranes; **biochemical factors,** such as antimicrobial substances present in the body fluids; and the more sophisticated process of **phagocytosis** and action of the **reticuloendothelial system.**

Specific immunity, **cell-mediated and humoral,** is acquired by the host in response to the presence of a single or particular foreign substance, usually protein, called **antigen.** In the latter, antigens have the capacity to stimulate within the host the formation of homologous substances known as **antibodies.** Their function is to bind to the specific antigens responsible for their production and thereby inactivate or destroy them. Once formed, these antibodies "take up residence" in the host's serum, and as serum antibodies (proteins) are referred to as **immunoglobulins.**

Sophistication in chemistry, especially immunochemistry, has enabled us to study the interaction of antigens and immunoglobulins outside the body, in a laboratory setting. These advances have provided an immunological discipline known as **serology,** which studies these *in vitro* reactions that have diagnostic, therapeutic, and epidemiological implications.

409

In the procedures that follow, students will study the serological techniques that use the principles of:

1. **Agglutination:** This type of reaction uses specific antibodies, **agglutinins,** that are formed in response to the introduction of particulate antigens into host tissues. When these particulate antigens combine with a homologous antiserum, a three-dimensional mosaic complex occurs. This is called an agglutination reaction and can be visualized microscopically and in some cases macroscopically.

2. **Precipitin formation:** This reaction requires specific antibodies, **precipitins,** that are formed in response to the introduction of soluble, nonparticulate antigens into host tissues. These antibodies, when present in serum, form a complex with the specific homologous nonparticulate antigen and result in a visible precipitate.

3. **Complement fixation:** The binding of **complement** (a complex of serum proteins) occurs in some antigen-antibody reactions. This reaction is not visible and requires the use of an indicator, which consists of sensitized sheep red blood cells, to obviate this reaction *in vitro* by the occurrence of hemolysis.

Agglutination Reaction:
The Febrile Antibody Test

PURPOSE

To demonstrate the agglutination reaction by means of the febrile antibody slide test and an antibody titer determination.

PRINCIPLE

The **febrile antibody test** is used in the diagnosis of diseases that produce febrile (fever) symptoms. Some of the microorganisms responsible for febrile conditions are salmonellae, brucellae, and rickettsiae. **Febrile antigens,** such as endotoxins, enzymes, and other toxic end products elaborated by these organisms are used specifically to detect or exclude the homologous antibodies that develop in response to these antigens during the infection.

In this procedure, the antigen is mixed on a slide with the serum being observed. Cellular clumping is indicative of the presence of homologous antibodies in the serum; the absence of homologous antibodies is indicated when there is no visible clumping. Only the febrile antigens and antibodies of *Salmonella* sp. will be used.

The second part of this experiment is designed to illustrate that agglutination reactions such as the febrile antibody test can be used to identify an unknown microorganism through serotyping. A specific antiserum prepared in a susceptible, immunologically competent laboratory animal is mixed with a variety of unknown bacterial antigen preparations on slides. The bacterial antigen that is agglutinated by the antiserum is identified and confirmed to be the agent of infection.

These tests are strictly qualitative. A quantitative result can be obtained by performing the **antibody titer,** which measures the concentration of an antibody in the serum and thus allows the physician to follow the course of an infection. The patient's serum is titrated (diluted) and the decreasing concentrations of the antiserum are mixed with a constant concentration of homologous antigen. The end point of the test will occur in the test tube containing the serum having the highest dilution showing agglutination.

MATERIALS

Cultures

Number-coded, washed saline suspensions of *Escherichia coli*, *Proteus vulgaris*, *Salmonella typhimurium*, and *Shigella dysenteriae*.

Reagents

Physiological saline (0.85 percent NaCl), commercial preparations of *Salmonella typhimurium* H antigen and *Salmonella typhimurium* H antiserum.

Equipment

Bunsen burner, inoculating loop, glass microscope slides, 13×100-mm test tubes, sterile 1-ml pipettes, applicator sticks, wax pencil, and microscope.

PROCEDURE

Febrile Antibody Slide Test

1. With a wax pencil, make two circular areas about one-half inch in diameter on microscopic slide. Label the circles A and B.

2. To area A, add one drop of *S. typhimurium* H antigen and one drop of 0.85 percent saline. Mix the two with an applicator stick.

3. To area B, add one drop *S. typhimurium* H antigen and one drop of *S. typhimurium* H antiserum. Mix the two with a clean applicator stick.

4. Pick up the slide, and with two fingers of one hand rock the slide back and forth.

5. Observe the slide both macroscopically and microscopically, under low power, for cellular clumping (agglutination).

Serological Identification of an Unknown Organism

1. Prepare two microscope slides as in the previous procedure. Label the four areas on the slides with the numbers of your four unknown cultures.

2. Into each area on both slides place one drop of *S. typhimurium* H antiserum.

3. With a sterile inoculating loop, suspend a loopful of each number-coded unknown culture in the drop of antiserum in its appropriately labeled area on the slides.

4. Pick up the slides and slowly rock them back and forth.

5. Observe both slides macroscopically and microscopically, under low power, for agglutination.

Determination of Antibody Titer

1. Place a row of 10 test tubes (13 × 100-mm) in a rack and number the tubes 1 through 10.

2. Pipette 1.8 ml of 0.85 percent saline into the first tube and 1 ml into each of the remaining nine tubes.

3. Into tube 1, pipette 0.2 ml of *Salmonella typhimurium* H antiserum. Mix thoroughly by pulling the fluid up and down in the pipette. **Avoid vigorous washing.** The antiserum has now been diluted 10 times (1:10).

4. Using a clean pipette, transfer 1 ml from tube 1 to tube 2 and mix thoroughly as described. Using the same pipette, transfer 1 ml from tube 2 to tube 3. Continue this procedure through tube 9.

5. Discard 1 ml from tube 9. Tube 10 will serve as the antigen control and therefore will not contain any antiserum.

6. The antiserum has been diluted during this twofold dilution to give final dilutions of 1:10, 1:20, 1:40, 1:80, 1:160, 1:320, 1:640, 1:1280, and 1:2560.

7. Add 1 ml of the *Salmonella typhimurium* H antigen suspension adjusted to an optical density of 0.5 at 600 mμ to all tubes.

8. Mix the contents of the test tubes by shaking the rack vigorously.

9. Incubate the test tubes in 55 degrees C water bath for two to three hours.

NAME_____

Observations and Results   EXPERIMENT **71**

Modified Febrile Agglutination Test

In the chart:

1. Indicate the presence or absence of macroscopic and microscopic agglutination.

2. Draw a representative field of areas A and B following microscopic observation for agglutination.

	Area A	Area B
Appearance of antibody-antigen mixture	 Saline *S. typhimurium* H antigen	 *S. typhimurium* H antiserum *S. typhimurium* H antigen
Macroscopic agglutination (+) or (−)	_____	_____
Microscopic agglutination (+) or (−)	_____	_____

Serological Identification of an Unknown Organism

In the chart:

1. Indicate the presence or absence of macroscopic and microscopic agglutination in each of the suspensions.

2. Indicate the suspension that is indicative of a homologous antigen-antibody reaction.

		Agglutination		
Cell antigen	*Antiserum*	*Macroscopic* *(+) or (−)*	*Microscopic* *(+) or (−)*	*Homologous antigen-antibody reaction*
Unknown #: ___	*S. typhimurium* H			
Unknown #: ___	*S. typhimurium* H			
Unknown #: ___	*S. typhimurium* H			
Unknown #: ___	*S. typhimurium* H			

Copyright © 1987 by Benjamin/Cummings Publishing Co., Inc. All rights reserved.

413

Determination of Antibody Titer

In the chart:

1. Indicate the presence or absence of agglutination in each of the antiserum dilutions.

2. Indicate the end point of the reaction.

Tube	Dilution	Agglutination	Titer
1	1:10		
2	1:20		
3	1:40		
4	1:80		
5	1:160		
6	1:320		
7	1:640		
8	1:1280		
9	1:2560		
10	Antigen control		

REVIEW QUESTIONS

1. What are febrile antibodies?

   Discuss their clinical significance.

2. Discuss the purpose of performing an antibody titer determination.

3. Explain why the antibody titer determination uses twofold dilutions of the antiserum rather than tenfold dilutions.

Agglutination Reaction: Mono-Test for Infectious Mononucleosis

PURPOSE

To detect infectious mononucleosis heterophile antibodies.

PRINCIPLE

Infectious mononucleosis (IM) is an acute, self-limiting infectious disease characterized by mild fever, sore throat, and significant enlargement of lymph nodes and spleen. Hematologically, there is marked leukocytosis in which the number of monocytes is elevated with a concomitant decrease in the number of neutrophils. Clinically, IM may mimic other diseases such as Hodgkin's disease, hepatitis, diphtheria, and lymphatic leukemia. The etiological agent is believed to be the Epstein-Barr virus (EBV).

Diagnosis is made on the basis of the clinical symptoms and the **heterophile antibody test.** Heterophile (heterogenetic) antigens are genetically unrelated, but are extremely similar in their chemical composition. Although not identical *per se,* antibodies prepared against one antigen will cross-react with others. The most prominent heterophile antigen is the one described by Forssman, who determined that when horse or guinea pig tissues were inoculated into rabbits, the resultant antibodies in the rabbit serum agglutinated sheep red blood cells. It was evident from this result that horse, cat, and guinea pig antigens were all chemically similar to those of sheep. Such antigens have been called **Forssman antigens** and the resultant antibodies are **Forssman heterophile antibodies.** Although all heterophile antibodies react with sheep red blood cells, the IM antibodies are discernible from other heterophile antibodies because of their inability to be removed from a serum sample by absorption with guinea pig antigen. Other heterophile antibodies are easily removed by this method. On this basis, the IM antibodies are **non-Forssman heterophile antibodies.**

Originally, complex presumptive and differential absorption tests were required to detect IM antibodies. The former used a titration procedure to detect anti-sheep red blood cell antibodies. A titer greater than 1:56 was indicative of a presumptive diagnosis; a titer of 1:224 with clinical symptoms was considered to be a confirmed diagnosis. A titer higher than 1:224 without clinical symptoms was suggestive of serum sickness resulting from the use of horse antisera. Positive identification of IM antibodies was determined by the **differential absorption test,** devised by Davidsohn, in which boiled guinea pig kidney was used to absorb all horse and Forssman antibodies, with the exception of the IM antibodies.

In recent years, methods have been devised that do not require this laborious test. In this experiment a two-minute rapid screening system, called **Mono-Test** (Wampole Laboratories, Cranbury, New Jersey), will be used to detect IM antibodies. In this one-step procedure, stabilized horse red blood cells will react with IM antibodies, if they are present in the patient's serum, to give a positive agglutination reaction as shown in Figure 72.1.

MATERIALS

Reagents

Human serum sample (inactivated at 56 degrees C).

Equipment

Mono-Test kit and wooden applicator sticks.

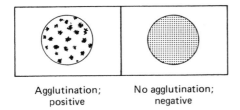

Agglutination;   No agglutination;
positive   negative

FIGURE 72.1. Mono-Test reactions

PROCEDURE

1. Using the calibrated capillary tube, draw the human serum sample up to the mark, being careful not to draw the serum into the rubber bulb, and expel gently into the center circle on the slide.

2. Place one drop of negative control serum into the circle on the right and one drop of positive control serum into the circle on the left.

3. Add one drop of Mono-Test reagent to each of the three circles.

4. Using separate wooden applicator sticks, mix and spread each mixture evenly over the entire surface of each circle.

5. Gently rock the card slide in a back and forth motion for one minute and let stand for another minute.

6. Observe immediately for agglutination.

Observations and Results   EXPERIMENT **72**

1. Observe each reaction for the presence or absence of agglutination in each circle. Diagram each reaction and record in the chart your result as positive or negative.

Positive control serum	Test serum	Negative control serum
◯	◯	◯

Agglutination: _____   _____   _____

REVIEW QUESTIONS

1. Discuss how IM antibodies can be distinguished from other heterophile antibodies.

2. Why are IM antibodies called non-Forssman antibodies?

3. What is the major advantage of the Mono-Test over the Davidsohn test?

4. Discuss the clinical syndrome seen in patients with infectious mononucleosis.

Precipitin Reaction: The Ring Test

PURPOSE

To demonstrate a precipitin reaction by means of the ring test.

PRINCIPLE

The ring or **interfacial test** is a simple serological technique that illustrates the precipitin reaction in solution. This antibody-antigen reaction can be demonstrated by the formation of a visible precipitate, a flocculent or granular turbidity, in the test fluid. Antiserum is introduced into a small-diameter test tube, and the antigen is then carefully added to form a distinct upper layer. Following a period of incubation of up to four hours, a ring of precipitate forms at the point of contact (interface) in the presence of the antigen-antibody reaction. The rate at which the visible ring forms depends on the concentration of antibodies in the serum and the concentration of the antigen.

To detect the precipitin reaction, a series of dilutions of the antigen is used, as both insufficient and excessive amounts of antigen will prevent the formation of a visible precipitate. In addition, students will be able to determine the optimal antibody:antigen ratio by the presence of a pronounced layer of granulation at the interface of the antiserum and antigen solution. This immunological reaction is illustrated in Figure 73.1.

MATERIALS

Reagents

Physiological saline (0.85 percent NaCl), and commercially available bovine globulin antiserum and normal bovine serum diluted to 1:25, 1:50, and 1:75 with physiological saline.

Equipment

Serological test tubes (8 × 75 mm), 0.5-ml pipettes, serological test tube rack, and 37 degree C incubator.

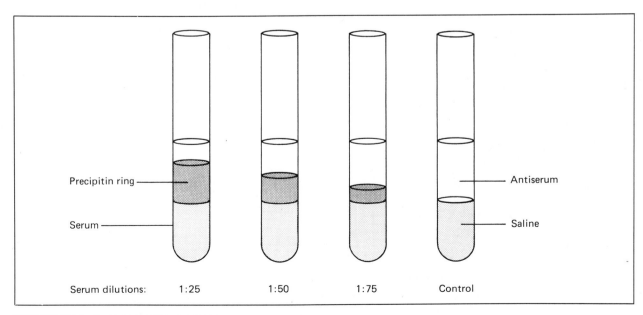

FIGURE 73.1. Test precipitin reactions

PROCEDURE

1. Label three serological test tubes according to the antigen dilution to be used (1:25, 1:50, and 1:75) and the fourth test tube as a saline control.

2. Using a different 0.5-ml pipette, transfer 0.3 ml of each of the normal bovine serum dilutions into its respectively labeled test tube.

3. Using a clean 0.5-ml pipette, transfer 0.3 ml of saline into the test tube labeled as control.

4. Carefully overlay all four test tubes with 0.3 ml of bovine globulin antiserum. To prevent mixing of the sera, tilt the test tube and allow the antiserum to run down the side of the test tube.

5. Incubate all test tubes for 30 minutes at 37 degrees C.

Observations and Results

1. Examine all test tubes for the development of a ring of precipitation at the interface. Indicate the presence or absence of a ring in the chart.

2. Determine and indicate the antigen dilution that produced the greatest degree of precipitation that is indicative of the optimal antibody:antigen ratio.

	Antigen dilutions			
	1:25	*1:50*	*1:75*	*Saline control*
Presence of interfacial ring (+) or (−)				

Dilution showing optimal antibody:antigen ratio: _____

REVIEW QUESTIONS

1. Distinguish between precipitin and agglutination reactions.

2. Explain how you would determine the optimal antigen:antibody ratio by means of the ring test.

3. Why is it essential to use a series of antigen dilutions in this procedure?

4. How would you explain the absence of visible precipitate?

Precipitin Reaction: Immunodiffusion

PURPOSE

To perform a gel diffusion experiment to:

1. Demonstrate the characteristic band precipitation patterns on an Ouchterlony plate.

2. Determine the optimal concentration (equivalence) of antibody and antigen necessary for gel diffusion.

PRINCIPLE

Immunodiffusion is an immunological procedure in which optimum concentrations of antigen and antibody join to produce visible bands of precipitation following their diffusion through a non-nutritional, clarified gel medium. The principal applications of this technique include:

1. Identification of the individual antigen and antibody components in the antigen-antibody reaction.

2. Identification of serologically related antigens.

3. Diagnosis of infectious diseases such as hepatitis, some types of typhus, and histoplasmosis.

Gel diffusion analyses may be carried out by two separate methods. The **Oudin tube method** is a one-dimensional system that allows diffusion of the antigen when it is layered onto the surface of an antibody-containing agar column in a small-bore precipitin tube. The **Ouchterlony plate method** is a two-dimensional system in which both components, antigen and antibody, are able to diffuse through the gel medium in a Petri dish. In the latter procedure, which is used in this exercise, a layer of clarified agar, ¼-inch thick, is poured into a Petri dish and allowed to solidify. When hardened, wells are cut into the agar and filled with the proper dilutions of antibody and antigen. During incubation, these components diffuse toward each other

until they reach optimal concentrations, forming precipitin bands in characteristically identifiable patterns as illustrated in Figure 74.1 on page 424.

In this experiment, students will prepare two Ouchterlony plates. One will be used to determine the optimum antigen and antibody concentrations (equivalence point). The second will be used to demonstrate the precipitin reaction patterns.

MATERIALS

Antigens

Monkey, rabbit, horse, and human sera.

Antiserum

Antihuman serum.

Media

Per designated student group: two diffusion gel plates, four 9-ml 0.85 percent saline blanks, and one 25-ml tube of 0.85 percent saline.

Equipment

No. 2 cork borer, suction apparatus, 1-ml and 5-ml pipettes, Pasteur pipettes with rubber bulbs, wax pencil, and 13 × 100-mm test tubes.

PROCEDURE

1. Using the template, prepare two Ouchterlony plates by boring five wells 7 to 10 mm apart with a No. 2 cork borer. With a suction apparatus, remove the

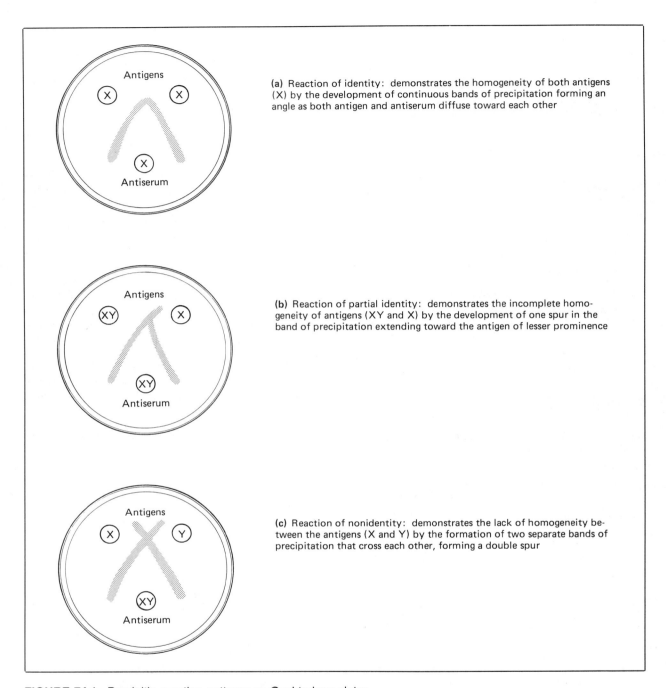

(a) Reaction of identity: demonstrates the homogeneity of both antigens (X) by the development of continuous bands of precipitation forming an angle as both antigen and antiserum diffuse toward each other

(b) Reaction of partial identity: demonstrates the incomplete homogeneity of antigens (XY and X) by the development of one spur in the band of precipitation extending toward the antigen of lesser prominence

(c) Reaction of nonidentity: demonstrates the lack of homogeneity between the antigens (X and Y) by the formation of two separate bands of precipitation that cross each other, forming a double spur

FIGURE 74.1. Precipitin reaction patterns on Ouchterlony plates

agar plugs. Label the wells on the bottom of the plate as indicated on the template.

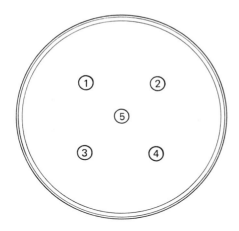

2. Prepare the plate for equivalency determination as follows:

   a. Using separate 1-ml pipettes and the four 9-ml saline blanks, perform a tenfold serial dilution of the human serum antigen to effect 10^{-1}, 10^{-2}, 10^{-3}, and 10^{-4} dilutions.

   b. Label the top of the plate "equivalency" and identify with your initials. On the bottom of the plate, label wells 1 through 4 with the antigen dilutions.

   c. Using separate Pasteur pipettes, add the antigen dilutions to their respective wells and the human antiserum to well 5. **Caution: Do not overfill the wells.**

3. Prepare the plate for demonstration of precipitin reaction patterns as follows:

   a. Prepare 1:50 0.85 percent saline dilutions of each of the four antigens using 0.1 ml of serum and 4.9 ml of 0.85 percent saline.

   b. Label the top of the plate "precipitin reactions" and identify with your initials. On the bottom of the plate, label wells 1 through 4 with the serum antigen to be used.

   c. Using separate Pasteur pipettes, add the diluted serum antigens to their respectively labeled wells and the human antiserum to well 5, being careful not to over-fill the wells.

4. Place the prepared plates, top side up, in a large beaker containing water-saturated filter paper. Cover the top of the beaker loosely with foil to retain moisture.

5. Place the beaker in a refrigerator for seven days.

Observations and Results

1. Observe the plate for equivalence determination for development of precipitin bands. Draw the observed bands in the plate diagram and indicate the antigen dilution(s)-producing zone(s) of equivalence.

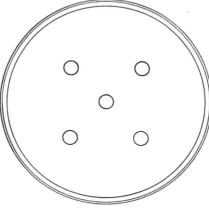

Equivalence determination
ouchterlony plate

Antigen dilution(s)-producing equivalence zone(s): _____
_____
_____

2. Observe the plate for precipitin band patterns. Draw and identify the observed precipitin bands.

Monkey serum: _____

Rabbit serum: _____

Horse serum: _____

Human serum: _____

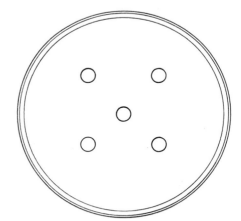

REVIEW QUESTIONS

1. What is the equivalence zone?

2. Why does the immunodiffusion medium resist contamination with bacteria?

3. What is the basic difference between gel diffusion occurring in Oudin tubes and on Ouch-terlony plates?

4. Why do you think it is necessary to use clarified gels to demonstrate immunodiffusion?

Complement Fixation Test

PURPOSE

To acquaint students with a serological test in which a serum component, called complement, plus a red blood cell indicator is used to detect an antigen-antibody reaction.

PRINCIPLE

Complement is a complex of nine major proteins found in the serum of humans and other mammals. These proteins participate in various immunological reactions. In serological reactions, in which immunoglobulins such as IgM and IgG play a role, complement becomes bound to the antigen-antibody complex. Upon binding, cytotoxic effects result on the antigens such as bacteria, causing bacteriolysis, or on red blood cells, causing hemolysis.

The **complement fixation test** is based on the measurement of the disappearance of complement during antigen and antibody reactions. It is essential to remember that in some antigen-antibody reactions, complement, if present, will be incorporated (fixed) during the reaction. The reaction, however, is not visible. Sheep red blood cells (antigen) are used as an indicator system together with antisheep red cell antibody (hemolysin). In the presence of complement, the sheep erythrocytes and hemolysin form a complex that results in lysis of blood cells. In the absence of complement, the sensitized red blood cells do not undergo lysis.

The complement fixation test will be performed in two phases—the actual test phase and the indicator phase. The test portion requires patient serum, antigen, and complement. The patient's serum is first heated to 56 degrees C for one hour to inactivate the complement normally present. The second phase, the indicator system, requires the use of sheep red blood cells and the antisheep hemolysin. Controls are included to rule out interference with reagents.

A twofold serial dilution of the patient's serum is prepared, and each of the other components is added to these dilutions in a constant and fixed concentration. The test tubes are then incubated at 37 degrees C and observed at 5-minute to 10-minute intervals for the presence or absence of hemolysis in the indicator system.

The test is positive when an antigen and antibody combine and fix complement, thereby inactivating it. On addition of the indicator system, no lysis occurs because all of the complement has been fixed. A negative test is demonstrated when the antigen-antibody reaction does not fix complement, but the free complement is fixed by the indicator system with the resultant hemolysis of the sheep red blood cells. Figure 75.1 on page 431 illustrates the complement fixation test and its interpretation.

MATERIALS

Reagents

Per designated student group: bovine serum albumin (BSA) 1:50,000 stock solution; 7-ml antibovine serum albumin antiserum (rabbit) inactivated at 56 degrees C for one hour; 15 ml guinea pig complement (2 units per ml); 10 ml hemolysin (antisheep RBC antiserum) 2 units per ml, 10 ml sheep red blood cells, 2 percent; and 10 ml 0.85 percent physiological saline.

Equipment

1-ml and 5-ml pipettes, 13 × 100-mm test tubes, test tube rack, and 37 degrees C water bath.

PROCEDURE

1. Set up a test tube rack with eleven 13 × 100-mm test tubes. Label the tubes 1 through 11.

2. Prepare a twofold serial dilution (1:2, 1:4, 1:8, 1:16, 1:32, 1:64) of the inactivated anti-BSA antiserum.

3. Using the protocol that follows, make the reagent additions.

Reagents	Test tubes										
	1	*2*	*3*	*4*	*5*	*6*	*7*	*8*	*9*	*10*	*11*
Anti-BSA antiserum (ml) undiluted	0.5	—	—	—	—	—	—	0.5	—	—	—
1:2	—	0.5	—	—	—	—	—	—	—	—	—
1:4	—	—	0.5	—	—	—	—	—	—	—	—
1:8	—	—	—	0.5	—	—	—	—	—	—	—
1:16	—	—	—	—	0.5	—	—	—	—	—	—
1:32	—	—	—	—	—	0.5	—	—	—	—	—
1:64	—	—	—	—	—	—	0.5	—	—	—	—
BSA antigen (ml)	0.5	0.5	0.5	0.5	0.5	0.5	0.5	—	0.5	—	—
Saline (ml)	—	—	—	—	—	—	—	0.5	0.5	1.0	2.5

4. Shake contents of all tubes by shaking the rack. Incubate for 10 to 30 minutes at room temperature.

Complement (ml)	*1.0*	*1.0*	*1.0*	*1.0*	*1.0*	*1.0*	*1.0*	*1.0*	*1.0*	*1.0*	—

5. Incubate all tubes at 6 to 10 degrees C for 18 hours.

Indicator system

Hemolysin (ml)	0.5	0.5	0.5	0.5	0.5	0.5	0.5	0.5	0.5	0.5	—
Sheep RBC (ml)	0.5	0.5	0.5	0.5	0.5	0.5	0.5	0.5	0.5	0.5	0.5

6. Shake the rack well and incubate at 37 degrees C. Observe every five minutes until tubes 9 and 10 become clear. Tube 9 is the antigen control and tube 10 is the RBC control.

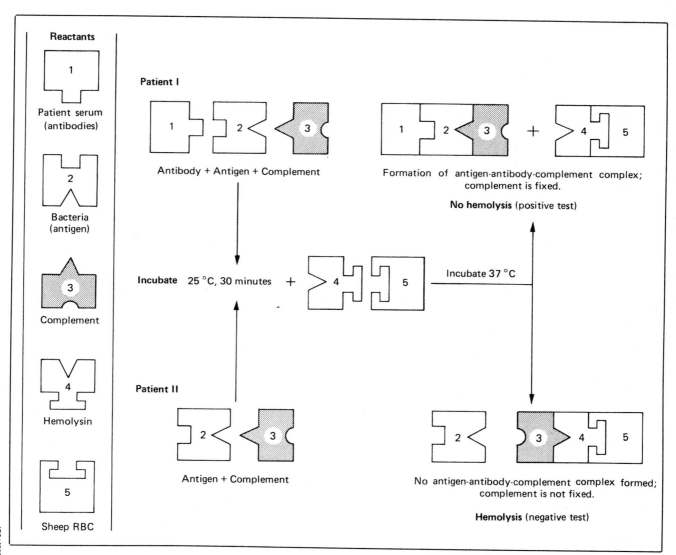

FIGURE 75.1. Complement fixation test

Observations and Results

1. Observe all tubes for the presence or absence of RBC lysis.

2. Record your results in the chart as follows:
   Complete RBC lysis = negative.
   Absence of RBC lysis = positive.

3. Indicate activity, positive or negative, in control tubes 8, 9, 10, and 11.

Tube	Antibody dilution	Lysis (+) or (−)	Result
1			
2			
3			
4			
5			
6			
7			
8			
9			
10			

REVIEW QUESTIONS

1. Explain the purpose of sheep red blood cells and antiserum in the procedure.

2. Explain the effect on the test results of using an unheated sample of the patient's serum for the performance of the complement fixation test.

3. Discuss the role of complement in serological reactions.

Immunofluorescence

PURPOSE

To demonstrate an antigen-antibody reaction.

PRINCIPLE

The **fluorescent antibody technique,** introduced by Coons in the mid-1950s, is a rapid and reliable procedure to demonstrate agglutination reactions. It has significant merit for use in the identification of microorganisms or their resultant antibodies. For example, the **fluorescent-treponemal–antibody-absorption (FTA-ABS) test** is used to diagnose syphilis using *Treponemia pallidum* as the antigen for detection of syphilitic antibodies in the patient's serum.

This technique requires the use of a specific antibody that has been tagged with a fluorescent dye such as fluorescein isocyanate or isothiocyanate. When the tagged antibody is applied to the antigen preparation, as in the direct method, or to an antigen-antibody complex, as in the indirect method, a microprecipitate forms at the site of the antigen and exhibits a yellow-green fluorescence when viewed under a fluorescence microscope.

In this experiment students will use the direct immunofluorescence method (Figure 76.1) to demonstrate the antigen-antibody reaction and to identify an unknown antigen. The fluorescein-tagged antibody is added to a heat-fixed bacterial smear. A physiologically buffered saline wash removes any uncombined fluorescent antibody. The resultant agglutination reaction, if present, will be demonstrated by a green fluorescence when viewed under a fluorescence microscope.

MATERIALS

Cultures

24-hour brain-heart infusion broth cultures of group A *Streptococcus pyogenes* and group D *Streptococcus faecalis;* numbered, unknown mixed broth cultures of group A *S. pyogenes/Escherichia coli* and group D *S. faecalis/Escherichia coli*.

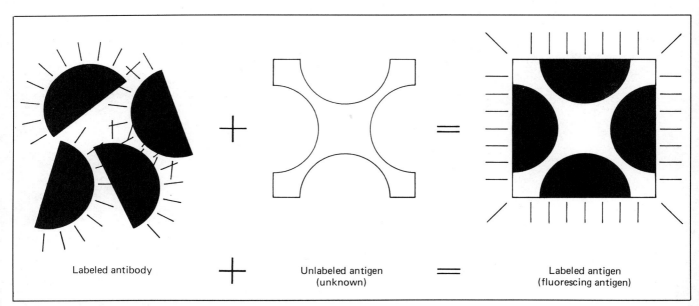

Labeled antibody + Unlabeled antigen (unknown) = Labeled antigen (fluorescing antigen)

FIGURE 76.1. Fluorescent antibody technique—direct method

Reagents

Fluorescent antibody *Streptococcus* group A (Difco #2318-56-6), and fluorescent antibody *Streptococcus* group D (Difco #2319-56-5); phosphate-buffered saline, and buffered glycerol.

Equipment

Microscope slides, Petri dishes, U-shaped glass rods to fit into Petri dishes, filter paper, Coplin jar, wax pencil, and fluorescence microscope.

PROCEDURE

1. With a wax pencil, label two slides *S. pyogenes* and *S. faecalis,* respectively. Divide the third slide in half and label as mixed unknown.

2. Prepare a heat-fixed smear of each known test organism on its respectively labeled slide. On the slide labeled mixed unknown, make a smear on each half of the slide using the unknown mixed culture.

3. On slides labeled *S. pyogenes* and *S. faecalis,* add one drop of each respective fluorescent antibody and spread gently over the surface of the smear. On the slide of the mixed unknown smears, label one side FA-A and the other side FA-D, add one drop of each fluorescent antibody to its respectively labeled smear, and allow to spread evenly over the smears.

4. Place moistened filter paper in the Petri dishes, insert the U-shaped glass rod (slide support), and place the prepared slides on the slide supports. Cover the Petri dishes and incubate for 35 minutes at 25 degrees C.

5. Remove the slides from the Petri dishes and wash away excess antibody with 1 percent phosphate-buffered saline.

6. Immerse the slides into a Coplin jar containing 1 percent phosphate-buffered saline for 10 minutes at 25 degrees C.

7. Blot the slides dry with bibulous paper.

8. To each slide add one drop of buffered glycerol, cover with a coverslip, and examine under a fluorescence microscope as directed by the instructor.

Observations and Results

EXPERIMENT **76**

1. Based on your microscopic observations, indicate in the chart the color of each smear and the presence (+) or absence (−) of fluorescence.

Culture	Color	Fluorescence (+) or (−)
Group A *S. pyogenes*		
Group D *S. faecalis*		
Mixed unknown with group A antibodies		
Mixed unknown with group D antibodies		

2. Based on your observations, identify your mixed unknown culture.

Culture #: _____

Organisms: _____

REVIEW QUESTIONS

1. What does the presence of fluorescence indicate?

2. Why is it necessary to wash away excess labeled antibody before viewing the preparation microscopically?

3. Briefly explain the direct fluorescent antibody procedure.

Sexually Transmitted Diseases: Rapid Immunodiagnostic Procedures

Sexually transmitted diseases (STDs) represent a diverse group of infectious syndromes that share their mode of transmission, direct sexual contact. Their etiological agents represent a broad spectrum of pathogenic microorganisms that include bacteria, viruses, yeasts, and protozoa. The bacterial STDs include **gonorrhea, syphilis, nongonococcal urethritis,** and **lymphogranuloma venereum.** The representative viral infections are **genital herpes, genital warts, hepatitis B,** and the latest member of this group, **AIDS.** The protozoan and fungal infections, namely **trichomoniasis** and **candidiasis,** are diseases of lesser magnitude in the spectrum of STDs.

The experimental procedures that follow have been chosen to demonstrate some of the rapid tests that are currently available for the diagnosis of selected STDs, specifically syphilis, genital herpes, and the chlamydial infections. In the methods described below, modified procedures will be performed in the absence of clinical specimens. Instead, commercially available positive and negative controls will be used to simulate clinical materials.

PART A: Rapid Plasma Reagin Test for Syphilis

PURPOSE

To perform a rapid screening procedure for diagnosis of syphilis.

PRINCIPLE

Treponema pallidum, the causative agent of *syphilis,* is a tightly coiled, highly motile, delicate spirochete that can be cultivated only in rabbit tissue cultures or rabbit testes. The organisms are resistant to common staining procedures and are best observed under darkfield microscopy.

Syphilis is a systemic infection that, if untreated, progresses through three clinical stages. The first stage, primary syphilis, is characterized by the formation of a painless papule, called a **chancre,** at the site of infection. Secondary syphilis represents the systemic extension of the infection that presents itself in the form of a **maculopapular rash,** malaise, and lymphadenopathy. Following this stage, the disease becomes self-limiting, and the patient appears asymptomatic until the development of the tertiary syphilis. In this final stage, life-threatening complications may develop as a result of the extensive cardiovascular and nervous tissue damage that has ensued.

The rapid plasma reagin (RPR) test, which has to a large extent replaced the VDRL (Venereal Disease Research Laboratory) agglutination test, determines the presence of **reagin,** the nonspecific antibody, present in the plasma of individuals with a syphilitic infection. The reagin appears in the plasma within two weeks of infection and will remain at high concentrations until the disease is eradicated. In the RPR test, if the reagin is present in the blood it will react with a soluble antigen bound to carbon particles to produce a macroscopically visible antigen, on carbon-antibody complex. This procedure has several advantages over the VDRL test, such as:

1. The serum does not require inactivation by heat for 30 minutes.

2. The serum may be obtained from a finger puncture, unlike the VDRL test that requires a venous blood sample.

3. The required materials, which include the antigen suspension with a dispensing bottle, diagnostic cards, and capillary pipettes, are all contained in individual kits that do not require additional equipment and are disposable.

In the qualitative form of the RPR test, the patient's blood serum and the carbon-bound antigen suspension are mixed within a circle on the diagnostic card. In the presence of a positive (reactive) serum the antigen-antibody complex will produce a macroscopically visible black agglutination reaction. The macroscopic appearance of a light gray suspension, devoid of any form of agglutination, is indicative of a negative (nonreactive) serum.

Since this is a nonspecific test, false positive results may be obtained. It is believed that the reagin is an antibody against tissue lipids in general. Therefore it may be present in uninfected individuals due to the release of lipids resulting from normally occurring wear and tear of body tissues. It has also been found that serum levels of reagin are elevated during the course of other infectious diseases such as viral pneumonia, lupus erythematosus, infectious mononucleosis, yaws, and pinta. The serum of patients with a reactive RPR result is subjected to additional serological testing, such as the FTA-ABS test, using the *Treponema* bacterium as an antigen to detect specific antibodies that are also present in the serum during syphilitic infection.

MATERIALS

Reagents

Commercially prepared syphilitic serum 4$^+$ and non-syphilitic serum.

Equipment

RPR test kit (Hynson, Westcott and Dunning Inc., Baltimore, Md.) and rotating machine (optional).

PROCEDURE

1. Label circles on the diagnostic plastic card reactive and nonreactive.

2. Using a capillary pipette with an attached rubber bulb, draw the reactive serum up to the indicated mark (0.05 ml).

3. Expel the serum directly onto the card in the circle labeled reactive serum. With a clean applicator stick spread the serum to fill the entire circle.

4. Repeat steps 2 and 3 for the nonreactive serum.

5. Shake the dispensing bottle to mix the suspension. Hold the bottle with attached 20-gauge needle in a vertical position and dispense one drop onto each circle containing the test serum.

6. If mechanical rotator is available, place the card on the rotator set at 100 rpm, or rotate the card back and forth manually for eight minutes.

PART B: Genital Herpes: Isolation and Identification of Herpes Simplex Virus

PURPOSE

To perform a tissue culture procedure for the growth and identification of the herpes simplex virus.

PRINCIPLE

The double stranded, herpes simplex virus (HSV) is the etiological agent of a variety of human infections. Included among these are **herpes labialis,** fever blisters around the lips; **keratoconjunctivitis,** infection of the eyes; **herpes genitalis,** eruptions on the genitalia; **herpes encephalitis,** a severe infection of the brain; and **neonatal herpes.** The herpes simplex virus is divided into two antigenically distinct groups, **HSV-1** and **HSV-2.** The former is most frequently implicated with infections above the waist, whereas the latter is predominantly responsible for genital infections.

Primary infection with HSV-2 manifests itself with the appearance of vesicular lesions, characterized by itching, tingling, or burning sensations on or within the male and female genitalia. These vesicles heal spontaneously within two weeks. Following this symptomatic phase, however, the virus reverts to a latent state and remains quiescent until exacerbated by some environmental factor. With no chemotherapeutic cure presently available, recurrent genital herpes with subclinical symptoms is common.

Detection of the herpes simplex virus requires the use of tissue culture techniques. The presence of the virus is then determined by the development of cytopathogenic effects in these cultures, such as the detection of intranuclear inclusion bodies. In recent years, these time-consuming, specialized procedures have been greatly facilitated by the availability of immunoenzymatic reagents for the identification of this clinically significant virus.

The **Cellmatics HSV Detection System** (Difco Laboratories, Detroit, Michigan) is a self-contained system providing for both the growth and identification of the virus from clinical specimens. In this procedure, the provided tissue culture tubes are inoculated with the clinical sample. Following a 24-hour incubation period and fixation, the presence of HSV antigens is determined by the addition of anti-HSV antibodies, which specifically bind to the HSV antigens. To obviate this antibody-antigen complex, a secondary antibody, substrate, and chromagen are added. Following this staining process, HSV-positive cultures viewed microscopically will exhibit brown-black areas of viral infection on a clear background of unstained cells.

In this exercise a modified procedure will be performed. In the absence of a clinical specimen, the actual culturing and fixation process will not be performed. Instead, the positive and negative commercially available controls will be used to simulate the clinical samples.

MATERIALS

Cultures

Cellmatics HSV Positive and Negative Control Tubes.

Reagents

Cellmatics Immunodiagnostic Reagents Kit, distilled water.

Equipment

5-ml graduated pipettes, microscope.

PROCEDURE

1. Warm immunodiagnostic reagents to room temperature.

2. Drain all fluid from the positive and negative control tubes.

3. Using a 5-ml pipette, wash the culture tubes twice with 5-ml of distilled water and drain. **Caution: when washing, exercise care to prevent disruption of the monolayer.**

4. Add 10 drops of primary antiserum (vial 1). **Caution: when adding reagents, hold the vial vertically to ensure proper delivery.**

5. Incubate the **tightly capped** tubes in a **horizontal** position for 15 minutes at 37 degrees C. To ensure complete coverage of the monolayer, occasionally rock the tubes gently during incubation.

6. Wash three times with 5 ml of distilled water and drain.

7. Add 10 drops of secondary antibody (vial 2).

8. Incubate for 15 minutes at 37 degrees C as described in step 5.

9. Wash three times with 5 ml of distilled water and drain.

10. Add 10 drops of substrate (vial 3) and 2 drops of chromagen (vial 4). Mix gently.

11. Incubate for 15 minutes at 37 degrees C as described in step 5.

12. Wash 3 times with 5 ml of distilled water and drain.

13. Examine microscopically for the presence of stained cells at $40\times$ and $100\times$ magnifications.

PART C: Detection of Sexually Transmitted Chlamydial Diseases
PURPOSE

To perform an immunofluorescent procedure for diagnosis of chlamydial infections

PRINCIPLE

Members of the genus *Chlamydia* are a group of obligate intracellular parasites. Although once believed to be viruses, their morphologic and physiologic characteristics more closely resemble bacteria, and therefore they are frequently referred to as small bacteria. Chlamydiae are Gram-negative, nonmotile, thick-walled, spherical organisms possessing both DNA and RNA that reproduce by means of binary fission. Their dependence on living tissues for cultivation and their lack of an ATP-generating system emulate the characteristics of viruses, but their bacterial nature is further affirmed by their sensitivity to antibiotic therapy. *Chlamydia trachomatis,* the human pathogen, is now recognized to be responsible for two sexually transmitted diseases, **nongonococcal urethritis** (NGU) and **lymphogranuloma venereum** (LGV). The incidence of both diseases in contemporary society is increasing dramatically.

NGU is a urethritis (inflammation of the urethra) with symptoms similar to, but less severe than, those of gonorrhea. Undiagnosed and untreated infections may lead to **epididymitis** and **proctitis** in males and **cervicitis, salpingitis** and **pelvic inflammatory disease** in females. LGV, the most severe of the genital chlamydial infections, initially develops with a painless lesion at the portal of entry, the genitalia. Systemic involvement is evidenced by swelling of the regional lymph nodes, which become tender and suppurative before disseminating the organisms to other tissues. In the absence of chemotherapeutic intervention, scarring of the lymphatic vessels can cause their obstruction, leading to **elephantiasis,** enlargement of the external genetalia in males, and narrowing of the rectum in females.

MicroTrak, a Direct Specimen Test (Syva Company, Palo Alto, California) is a rapid, immunofluorescent procedure for the detection of *C. trachomatis*. The procedure circumvents the need of culturing these organisms in susceptible tissues prior to their identification. This slide test is designed to detect elementary bodies, the infectious particles produced during the life cycle of this organism, by the use of a staining reagent, a fluorescein-labeled monoclonal antibody specific for the principle *C. trachomatis* outer-membrane protein. In this procedure a slide-smear is prepared from the clinical specimen. Following fixation, when the slide is exposed to the Direct Specimen Reagent, the antibody binds to the organisms. Their presence is then determined by the appearance of apple-green chlamydiae against a red background of counterstained cells when viewed under a fluorescent microscope.

MATERIALS

Cultures

Commercially prepared positive and negative control slides.

Reagents

MicroTrak Direct Specimen Test for *Chlamydia trachomatis*.

Equipment

Fluorescent microscope.

PROCEDURE

1. Stain the positive and negative control slides with the MicroTrak reagent for 15 minutes.

2. Incubate slides for 15 minutes.

3. Rinse the slides with distilled water.

4. Air dry the slides.

5. Examine the slides under a fluorescent microscope for the presence of apple-green chlamydiae.

Observations and Results

PART A: Rapid Plasma Reagin Test for Syphilis

1. In the presence of direct light, while tilting the card back and forth, determine the presence or absence of black clumping in each of the serum-antigen mixtures. Record your observations and the reaction as (+) or (−).

2. Draw the observed reaction.

	Reactive serum	**Nonreactive serum**
Appearance of serum-antigen mixture:	_____	_____
Reaction (+) or (−):	_____	_____

PART B: Isolation and Identification of Herpes Simplex Virus

1. Scan the entire stained monolayer of both culture tubes under 40 × and 100 × magnifications for the presence of brown to blackish-brown stained cells. HSV infection is indicated by the presence of dark colored cells when viewed against an unstained background of normal cells.

2. Indicate in the chart below the presence (+) or absence (−) of these dark-stained patches.

Control cultures			
Negative		*Positive*	
40 ×	100 ×	40 ×	100 ×

PART C: Detection of Sexually Transmitted Chlamydial Diseases

1. Observe the stained control slides for the presence of apple-green particles indicative of chlamydiae. The particles are evident against a reddish background of counterstained cells.

2. Record your results below, indicating the presence (+) or absence (−) of the apple-green chlamydiae on each of the slides.

   Positive control slide: _____

   Negative control slide: _____

REVIEW QUESTIONS

1. Explain the factors responsible for the exacerbation of HSV infections.

2. Why can't a positive reaction confirm the presence of a syphilitic infection?

3. Compare and contrast the characteristics of the chlamydiae with those of bacteria and viruses.

Microbiological Media

The formulas of the media used in the exercises in this manual are listed alphabetically in **grams per liter** of distilled water unless otherwise specified. Sterilization of all media is accomplished by autoclaving at 15 lb pressure for 15 minutes unless otherwise specified. Most of the media are available commercially in powdered form, with specific instructions for their preparation and sterilization.

Ammonium sulfate broth (pH 7.3)

Ammonium sulfate	2.0
Magnesium sulfate · 7H$_2$O	0.5
Ferric sulfate · 7H$_2$O	0.03
Sodium chloride	0.3
Magnesium carbonate	10.0
Dipotassium hydrogen phosphate	1.0

Bacteriophage broth, 10× (pH 7.6)

Peptone	100.0
Beef extract	30.0
Yeast extract	50.0
Sodium chloride	25.0
Potassium dihydrogen phosphate	80.0

Bile esculin (pH 6.6)

Beef extract	3.0
Peptone	5.0
Esculin	1.0
Oxgall	40.0
Ferric citrate	0.5
Agar	15.0

Blood agar (pH 7.3)

Infusion from beef heart	500.0
Tryptose	10.0
Sodium chloride	5.0
Agar	15.0

Note: Dissolve the above ingredients and autoclave. Cool the sterile blood agar base to 45 degrees to 50 degrees C. Aseptically add 50 ml of sterile defibrinated blood. Mix thoroughly, avoiding accumulation of air bubbles. Dispense into sterile tubes or plates while liquid.

Brain heart infusion (pH 7.4)

Infusion from calf brain	200.0
Infusion from beef heart	250.0
Peptone	10.0
Dextrose	2.0
Sodium chloride	5.0
Disodium phosphate	2.5
Agar	1.0

Campy BAP agar (pH 7.0)

Trypticase peptone	10.0
Thiotone	10.0
Dextrose	1.0
Yeast extract	2.0
Sodium chloride	5.0
Sodium bisulfite	0.1
Agar	15.0
Vancomycin	10.0 mg
Trimethoprim lactate	5.0 mg
Polymixin B sulfate	2500 IU
Amphotericin B	2.0 mg
Cephalothin	15.0 mg
Defibrinated sheep blood	10.0%

Note: Aseptically add the antibiotics and defibrinated sheep blood to the sterile, cooled, and molten agar.

Chocolate agar (pH 7.0)

Proteose peptone	20.0
Dextrose	0.5
Sodium chloride	5.0
Disodium phosphate	5.0
Agar	15.0

Note: Aseptically add 5 percent defibrinated sheep blood to the sterile and molten agar. Heat at 80 degrees C until a chocolate color develops.

Deoxyribonuclease (DNase) agar (pH 7.3)

Deoxyribonucleic acid	2.0
Phytone	5.0
Sodium chloride	5.0
Trypticase	15.0
Agar	15.0

Endo agar (pH 7.5)

Peptone	10.0
Lactose	10.0
Dipotassium phosphate	3.5
Sodium sulfite	2.5
Basic fuchsin	0.4
Agar	15.0

Eosin-methylene blue agar, Levine, (pH 7.2)

Peptone	10.0
Lactose	5.0
Dipotassium phosphate	2.0
Agar	13.5
Eosin Y	0.4
Methylene blue	0.065

Gel diffusion agar

Sodium barbital buffer	100 ml
Noble agar	0.8 gm

Glycerol yeast extract agar supplemented with aureomycin (pH 7.0)

Glycerol	5.0 ml
Yeast extract	2.0
Dipotassium phosphate	1.0
Agar	15.0

Note: Aseptically add aureomycin, 10 μg per ml, to the sterile, cooled, and molten agar.

Glucose salts broth (pH 7.2)

Dextrose	5.0
Sodium chloride	5.0
Magnesium sulfate	0.2
Ammonium dihydrogen phosphate	1.0
Dipotassium hydrogen phosphate	1.0

Grape juice broth

Commercial grape or apple juice	
Ammonium biphosphate	0.25%

Note: Sterilization not required when using a large yeast inoculum.

Inorganic synthetic broth (pH 7.2)

Sodium chloride	5.0
Magnesium sulfate	0.2
Ammonium dihydrogen phosphate	1.0
Dipotassium hydrogen phosphate	1.0

KF broth (pH 7.2)

Polypeptone	10.0
Yeast extract	10.0
Sodium chloride	5.0
Sodium glycerophosphate	10.0
Sodium carbonate	0.636
Maltose	20.0
Lactose	1.0
Sodium azide	0.4
Phenol red	0.018

Lactose fermentation broth 1× and 2×* (pH 6.9)

Beef extract	3.0
Peptone	5.0
Lactose	5.0

Note: For 2× broth use twice the concentration of the ingredients.

Litmus milk (pH 6.8)

Skim milk powder	100.0
Litmus	0.75

Note: Autoclave at 12 lb pressure for 15 minutes.

MacConkey's agar (pH 7.1)

Bacto peptone	17.0
Proteose peptone	3.0
Lactose	10.0
Bile salts mixture	1.5
Sodium chloride	5.0
Agar	13.5
Neutral red	0.03
Crystal violet	0.001

Mannitol salt agar (pH 7.4)

Beef extract	1.0
Peptone	10.0
Sodium chloride	75.0
d-Mannitol	10.0
Agar	15.0
Phenol red	0.025

M-endo broth (pH 7.5)

Yeast extract	6.0
Thiotone peptone	20.0
Lactose	25.0
Dipotassium phosphate	7.0
Sodium sulfite	2.5
Basic fuchsin	1.0

Note: Heat until boiling, do not autoclave.

M-FC broth (pH 7.4)

Biosate peptone	10.0
Polypeptone peptone	5.0
Yeast extract	3.0
Sodium chloride	5.0
Lactose	12.5
Bile salts	1.5
Aniline blue	0.1

Note: Add 10 ml of rosolic acid (1 percent in 0.2N sodium hydroxide). Heat to boiling with agitation, do not autoclave.

Milk agar (pH 7.2)

Skim milk powder	100.0
Peptone	5.0
Agar	15.0

Note: Autoclave at 12 lb pressure for 15 minutes.

Minimal agar (pH 7.0)
Minimal agar, supplemented with streptomycin and thiamine

Solution A (pH 7.0)

Potassium dihydrogen phosphate	3.0
Disodium hydrogen phosphate	6.0
Ammonium chloride	2.0
Sodium chloride	5.0
Distilled water	800.0 ml

Solution B (pH 7.0)

Glucose	8.0
Magnesium sulfate 7H$_2$O	0.1
Agar	15.0
Distilled water	200.0 ml

Note: Autoclave solutions A and B separately and combine. To solution B add 0.001 gm of thiamine prior to autoclaving. To the combined sterile and molten medium, add 50 mg (1 ml of 50 mg per ml) sterile streptomycin solution before pouring agar plates.

MR-VP broth (pH 6.9)

Peptone	7.0
Dextrose	5.0
Potassium phosphate	5.0

Mueller-Hinton agar (pH 7.4)

Beef, infusion	300.0
Casamino acids	17.5
Starch	1.5
Agar	17.0

Mueller-Hinton tellurite agar (pH 7.4)

Casamino acids	20.0
Casein	5.0
l-Tryptophane	0.05
Potassium dihydrogen phosphate	0.3
Magnesium sulfate	0.1
Agar	20.0

Note: Aseptically add 12.5 ml of tellurite serum to the sterile, 50-degree C molten agar.

Nitrate broth (pH 7.2)

Peptone	5.0
Beef extract	3.0
Potassium nitrate	5.0

Nitrite broth (pH 7.3)

Sodium nitrite	2.0
Magnesium sulfate · $7H_2O$	0.5
Ferric sulfate · $7H_2O$	0.03
Sodium chloride	0.3
Sodium carbonate	1.0
Dipotassium hydrogen sulfate	1.0

Nitrogen-free mannitol broth and agar* (pH 8.3)

Mannitol	15.0
Dipotassium hydrogen phosphate	0.5
Magnesium sulfate	0.2
Calcium sulfate	0.1
Sodium chloride	0.2
Calcium carbonate	5.0
Agar*	15.0

Nutrient agar* and broth (pH 7.0)

Peptone	5.0
Beef extract	3.0
Agar*	15.0

Nutrient gelatin (pH 6.8)

Peptone	5.0
Beef extract	3.0
Gelatin	120.0

Peptone broth (pH 7.2)

Peptone	4.0

Phenol red dextrose broth (pH 7.3)

Trypticase	10.0
Dextrose	5.0
Sodium chloride	5.0
Phenol red	0.018

Note: Autoclave at 12 lb pressure for 15 minutes.

Phenol red inulin broth (pH 7.3)

Trypticase	10.0
Inulin	5.0
Sodium chloride	5.0
Phenol red	0.018

Note: Autoclave at 12 lb pressure for 15 minutes.

Phenol red lactose broth (pH 7.3)

Trypticase	10.0
Lactose	5.0
Sodium chloride	5.0
Phenol red	0.018

Note: Autoclave at 12 lb pressure for 15 minutes.

Phenol red sucrose broth (pH 7.3)

Trypticase	10.0
Sucrose	5.0
Sodium chloride	5.0
Phenol red	0.018

Note: Autoclave at 12 lb pressure for 15 minutes.

Sabouraud agar and Sabouraud agar supplemented with aureomycin (pH 5.6)

Peptone	10.0
Dextrose	40.0
Agar	15.0

Note: Aseptically add aureomycin, 10 μg per ml, to the sterile, cooled, and molten medium.

Salt medium—Halobacterium

Sodium chloride	250.0
Magnesium sulfate · $7H_2O$	10.0
Potassium chloride	5.0
Calcium chloride · $6H_2O$	0.2
Yeast extract	10.0
Tryptone	2.5
Agar	20.0

Note: The quantities given are for preparation of one liter final volume of the medium. In preparation, make up two solutions, one involving the yeast extract and tryptone, and the other salts. Adjust the pH of the nutrient solution to 7. Sterilize separately. Mix and dispense aseptically.

SIM agar (pH 7.3)

Peptone	30.0
Beef extract	3.0
Ferrous ammonium sulfate	0.2
Sodium thiosulfate	0.025
Agar	3.0

Simmons citrate agar (pH 6.9)

Ammonium dihydrogen phosphate	1.0
Dipotassium phosphate	1.0
Sodium chloride	5.0
Sodium citrate	2.0
Magnesium sulfate	0.2
Agar	15.0
Brom thymol blue	0.08

Snyder test agar (pH 4.8)

Tryptone	20.0
Dextrose	20.0
Sodium chloride	5.0
Brom cresol green	0.02
Agar	20.0

Sodium chloride broth, 6.5%

Brain-heart infusion broth	100.0 ml
Sodium chloride	6.0

Starch agar (pH 7.0)

Peptone	5.0
Beef extract	3.0
Starch (soluble)	2.0
Agar	15.0

Thioglycollate, fluid (pH 7.1)

Peptone	15.0
Yeast extract	5.0
Dextrose	5.0
l-Cystine	0.75
Thioglycollic acid	0.3 ml
Agar	0.75
Sodium chloride	2.5
Resazurin	0.001

Top agar (for Ames test)

Sodium chloride	5.0
Agar	6.0

Tributyrin agar (pH 7.2)

Peptone	5.0
Beef extract	3.0
Agar	15.0
Tributyrin	10.0

Note: Dissolve peptone, beef extract, and agar while heating. Cool to 90 degrees C, add the tributyrin, and emulsify in a Waring blender.

Triple sugar-iron agar (pH 7.4)

Beef extract	3.0
Yeast extract	3.0
Peptone	15.0
Proteose-peptone	5.0
Lactose	10.0
Saccharose	10.0
Dextrose	1.0
Ferrous sulfate	0.2
Sodium chloride	5.0
Sodium thiosulfate	0.3
Phenol red	0.024
Agar	12.0

Trypticase nitrate broth (pH 7.2)

Trypticase	20.0
Disodium phosphate	2.0
Dextrose	1.0
Agar	1.0
Potassium nitrate	1.0

Trypticase soy agar (pH 7.3)

Trypticase	15.0
Phytone	5.0
Sodium chloride	5.0
Agar	15.0

Tryptone agar

Tryptone	10.0
Calcium chloride (reagent)	0.01–0.03 M
Sodium chloride	5.0
Agar	11.0

Tryptone broth

Tryptone	10.0
Calcium chloride (reagent)	0.01–0.03 M
Sodium chloride	5.0

Tryptone soft agar

Tryptone	10.0
Potassium chloride (reagent)	5
Agar	9.0

Urea broth

Urea broth concentrate (filter-sterilized solution)	10 ml
Sterile distilled water	90 ml

Note: Aseptically add the urea broth concentrate to the sterilized and cooled distilled water. Under aseptic conditions, dispense 3-ml amounts into sterile test tubes.

Yeast extract broth (pH 7)

Peptone	5.0
Beef extract	3.0
Sodium chloride	5.0
Yeast extract	5.0

Biochemical Test Reagents

Barrit's reagent, for detection of acetylmethylcarbinol
Solution A

Alpha-naphthol	5 gm
Ethanol, absolute	95 ml

Note: Dissolve the alpha-naphthol in the ethanol with constant stirring.

Solution B

Potassium hydroxide	40 gm
Creatine	0.3 gm
Distilled water	100 ml

Note: Dissolve the potassium hydroxide in 75 ml of distilled water. The solution will become warm. Allow to cool to room temperature. Add the creatine and stir to dissolve. Add the remaining water. Store in a refrigerator.

Biotin-histidine solution, for Ames test

l-Histidine HCl	0.5 mM
Biotin	0.5 mM
Distilled water	10.0 ml

Buffered glycerol (pH 7.2), for immunofluorescence

Glycerin	90.0 ml
Phosphate buffered saline	10.0 ml

Diphenylamine reagent, for detection of nitrates

Dissolve 0.7 gm diphenylamine in a mixture of 60 ml concentrated sulfuric acid and 28.8 ml distilled water. Cool and slowly add 11.3 ml concentrated hydrochloric acid. Allow to stand for 12 hours. Sedimentation indicates that reagent is saturated.

Ferric chloride reagent

Ferric chloride	10.0 gm
Distilled water	100.0 ml

Gram's iodine, for detection of starch

As in Gram's stain.

Hydrogen peroxide, 3%, for detection of catalase activity

Note: Refrigerate when not in use.

Kovac's reagent, for detection of indole

p-Dimethylaminobenzaldehyde	5.0 gm
Amyl alcohol	75.0 ml
Hydrochloric acid (concentrated)	25.0 ml

Note: Dissolve the p-dimethylaminobenzaldehyde in the amyl alcohol. Add the hydrochloric acid.

McFarland Barium Sulfate Standards, for Staph-Ident procedure

Using the table below add the amounts of barium chloride and sulfuric acid to clean 15×150-mm screw-capped test tubes. Label the tubes 1 through 10.

Preparation of McFarland Standards

Tube	Barium chloride 1% (ml)	Sulfuric acid 1% (ml)	Corresponding approximate density of bacteria (million/ml)
1	0.1	9.9	300
2	0.2	9.8	600
3	0.3	9.7	900
4	0.4	9.6	1,200
5	0.5	9.5	1,500
6	0.6	9.4	1,800
7	0.7	9.3	2,100
8	0.8	9.2	2,400
9	0.9	9.1	2,700
10	1.0	9.0	3,000

Methyl cellulose, for microscopic observation of protozoa

Methyl cellulose	10.0 gm
Distilled water	90.0 ml

Methyl red solution, for detection of acid

Methyl red	0.1 gm
Ethyl alcohol (95%)	300.0 ml
Distilled water	200.0 ml

Note: Dissolve the methyl red in the 95 percent ethyl alcohol. Dilute to 500 ml with distilled water.

Nessler's reagent, for detection of ammonia

Potassium iodide	50.0 gm
Distilled water (ammonia-free)	35.0 ml

Add saturated aqueous solution of mercuric chloride until a slight precipitate persists.

Potassium hydroxide (50% aqueous)	400.0 ml

Dilute to 1000 ml with ammonia-free distilled water. Let stand for one week, decant supernatant liquid, and store in tightly capped amber bottle.

Nitrate test solutions, for detection of nitrites

Solution A Sulfanilic acid

Sulfanilic acid	8.0 gm
Acetic acid, 5N: 1 part glacial acetic acid to 2.5 parts distilled water	1000.0 ml

Solution B Alpha-naphthylamine

Alpha-naphthylamine	5.0 gm
Acetic acid, 5N	1000.0 ml

Orthonitrophenyl-β-D-galactoside (ONPG), for enzyme induction

0.1 M sodium phosphate buffer (pH 7.0)	50.0 ml
ONPG (8×10^{-4} M)	12.5 mg

Phosphate-buffered saline, 1% (pH 7.2–7.4), for immunofluorescence

Solution A

Disodium phosphate	1.4 gm
Distilled water	100.0 ml

Solution B

Sodium dihydrogen phosphate	1.4 gm
Distilled water	100.0 ml

Add 84.1 ml of solution A to 15.9 ml of solution B. Add 8.5 gm of sodium chloride and q.s. to one liter.

Sodium barbitol buffer, for immunodiffusion

Sodium barbitol	6.98 gm
Sodium chloride	6.0 gm
1N hydrochloric acid	27.0 ml
Distilled water, q.s. to 1000 ml	

Tetramethyl-para-phenylenediamine dihydrochloride, for detection of oxidase activity

Tetramethyl-para-phenylenediamine dihydrochloride	0.5 gm
Distilled water	50.0 ml

Note: Prepare immediately prior to use. Darkened solution should be discarded.

Trommsdorf's reagent, for detection of nitrite

Slowly add 100 ml of 20 percent aqueous zinc chloride solution to a mixture of 4 gm of starch in water. Heat until the starch is dissolved as much as possible and the solution is almost clear. Dilute with water and add 2 gm of potassium iodide. Dilute to 1000 ml, filter, and store in an amber bottle.

Vaspar, for cultivation of viruses in chick embryo

Melt together 1 part petroleum jelly and 1 part paraffin.

Staining Reagents

ACID-FAST STAIN

Carbol fuchsin (Ziehl's)

Solution A

Basic fuchsin (90% dye content)	0.3 gm
Ethyl alcohol (95%)	10 ml

Solution B

Phenol	5 gm
Distilled water	95 ml

Mix solutions A and B.

Note: add 2 drops of Triton X per 100 ml of stain for use in heatless method.

Acid alcohol

Ethyl alcohol (95%)	97 ml
Hydrochloric acid (concentrated)	3 ml

Methylene blue

Methylene blue	0.3 gm
Distilled water	100 ml

CAPSULE STAIN

Crystal violet (1%)

Crystal violet (85% dye content)	1 gm
Distilled water	100 ml

Copper sulfate solution (20%)

Copper sulfate ($CuSO_4 \cdot 5H_2O$)	20 gm
Distilled water	80 ml

FUNGAL STAINS

Water-iodine solution

Gram's iodine (as in Gram's stain)	10 ml
Distilled water	30 ml

Lactophenol cotton blue solution

Lactic acid	20 ml
Phenol	20 gm
Glycerol	40 ml
Distilled water	20 ml
Aniline blue	0.05 gm

Note: Heat gently to dissolve in hot water (double boiler). Then add aniline blue dye.

GRAM STAIN

Crystal violet (Hucker's)

Solution A

Crystal violet (90% dye content)	2 gm
Ethyl alcohol (95%)	20 ml

Solution B

Ammonium oxalate	0.8 gm
Distilled water	80 ml

Mix solutions A and B.

Gram's iodine

Iodine	1 gm
Potassium iodine	2 gm
Distilled water	300 ml

Ethyl alcohol (95%)

Ethyl alcohol (100%)	95 ml
Distilled water	5 ml

Safranin

Safranin O (2.5% solution in 95% ethyl alcohol)	10 ml
Distilled water	100 ml

NEGATIVE STAIN

Nigrosin

Nigrosin, water-soluble	10 gm
Distilled water	100 ml

Immerse in boiling water bath for 30 minutes.

Formalin	0.5 ml

Filter twice through double filter paper.

SPORE STAIN

Malachite green

Malachite green	5 gm
Distilled water	100 ml

Safranin

Same as in Gram stain

Experimental Microorganisms

CULTURES

Bacteria

Alcaligenes faecalis
Alcaligenes viscolactis
Aquaspirillum itersonii
Bacillus cereus
Bacillus stearothermophilus
Clostridium butyricum
Corynebacterium xerosis
Enterobacter aerogenes
Escherichia coli
Escherichia coli ATCC e 23725
Escherichia coli ATCC e 23735
Escherichia coli ATCC e 23724
Escherichia coli ATCC e 23740
Escherichia coli B
Halobacterium salinarium
Klebsiella pneumoniae
Micrococcus luteus
Mycobacterium smegmatis
Peptococcus anaerobius
Proteus vulgaris
Pseudomonas aeruginosa
Pseudomonas fluorescens
Pseudomonas savastanoi
Salmonella typhimurium
Salmonella typhimurium ATCC e 29631
Serratia marcescens
Shigella dysenteriae
Staphylococcus aureus ATCC e 27660
Staphylococcus aureus ATCC e 27691
Staphylococcus aureus ATCC e 27693
Staphylococcus aureus ATCC e 27697
Staphylococcus aureus ATCC e 25923
Staphylococcus epidermidis
Staphylococcus saprophyticus ATCC 15305
Streptococcus agalactiae

Streptococcus bovis
Streptococcus faecalis
Streptococcus lactis
Streptococcus mitis
Streptococcus pneumoniae
Streptococcus pyogenes
Streptococcus var. Lancefield group E

Fungi

Alternaria sp.
Aspergillus niger
Cephalosporum sp.
Cladosporium sp.
Fusarium sp.
Mucor sp.
Penicillium notatum
Rhizopus sp.
Saccharomyces cerevisiae
Saccharomyces cerevisiae var. *ellipsoideus*
Streptomyces griseus

Viruses

Newcastle disease virus
Staphylophages types 47, 52A, 71, and 81
T_2 Coliphage

PREPARED SLIDES

Bacteria

Bacillus subtilis
Spirillum itersonii
Staphylococcus aureus

Fungi

Saccharomyces cerevisiae

Mammalian Tissues

Human red blood smear

Protozoa

Balantidium coli
Entamoeba histolytica
Giardia lamblia
Paramecium caudatum
Plasmodium vivax
Trypanasoma gambiense

Index